Methods in Molecular Biology

Series Editor
John M. Walker
School of Life and Medical Sciences
University of Hertfordshire
Hatfield, Hertfordshire, AL10 9AB, UK

For further volumes:
http://www.springer.com/series/7651

Lysosomes

Methods and Protocols

Edited by

Karin Öllinger

Experimental Pathology, Department of Clinical and Experimental Medicine, Linköping University, Linköping, Sweden

Hanna Appelqvist

Department of Physics, Chemistry and Biology, Linköping University, Linköping, Sweden

Humana Press

Editors
Karin Öllinger
Experimental Pathology
Department of Clinical and Experimental Medicine
Linköping University
Linköping, Sweden

Hanna Appelqvist
Department of Physics, Chemistry and Biology
Linköping University
Linköping, Sweden

ISSN 1064-3745 ISSN 1940-6029 (electronic)
Methods in Molecular Biology
ISBN 978-1-4939-6932-6 ISBN 978-1-4939-6934-0 (eBook)
DOI 10.1007/978-1-4939-6934-0

Library of Congress Control Number: 2017935483

Printed on acid-free paper

This Humana Press imprint is published by Springer Nature
The registered company is Springer Science+Business Media LLC
The registered company address is: 233 Spring Street, New York, NY 10013, U.S.A.

Preface

The endo-lysosomal system is central to the degradation and recycling of macromolecules delivered by endocytosis, phagocytosis, and autophagy [1–3]. As the major digestive compartment within cells, lysosomes harbor around 60 acidic hydrolases, responsible for the cellular digestion of most macromolecules. The lysosomal function goes far beyond the degradation activity and lysosomes are identified as important regulators of nutrient sensing, exocytosis, receptor recycling and regulation, cell death, and cholesterol homeostasis [4–7]. A significant finding recognized lysosomes as important signaling organelles that sense nutrient availability and generate an adaptive response to maintain cellular homeostasis, mainly through activation of the transcription factor EB (TFEB) [8]. The discovery of TFEB as a master regulator of lysosomal biogenesis, regulator of autophagic function and energy metabolism has greatly impacted our view of lysosomes as important hubs for interpretation of environmental alterations [9]. In addition, the lysosomes function as a Ca^{2+} store that participates in the signal transduction eventually leading to the nuclear translocation of TFEB [10]. The importance of lysosomes for cellular cholesterol homeostasis was identified through the inherited lysosomal storage disorder Niemann-Picks disease type C, which is caused by mutation in either of the two proteins NPC1 and NPC2 [11]. Furthermore, a lysosomal hydrolase-mediated digestion of LDL and subsequent cholesterol release from the lysosomes through the action of NPC1 and NPC2, by a not yet fully defined mechanism, has also recognized the importance of lysosomes in atherosclerosis [12].

Moreover, the lysosome is centrally involved in the regulation and control of cell death and survival. Due to their high content of hydrolytic enzymes, lysosomes are potentially harmful to cells. Christian de Duve termed the lysosomes "suicide bags" as massive lysosomal rupture may cause cytosolic acidification followed by necrosis [13]. Present knowledge has however shown that partial and selective lysosomal membrane permeabilization (LMP) could trigger several forms of controlled cell death [14]. LMP results in the release of lysosomal content to the cytosol and the main lysosomal hydrolases implicated in triggering of cell death are the cathepsins, which have been shown in several in vitro system but also in vivo [5, 15–17]. The mechanism of LMP is not clarified and most likely lysosomal permeabilization is due to alteration in both lysosomal membrane proteins and lipids causing destabilization of the membrane. Interestingly, in addition to the role of lysosomes in cell death they are also involved in the repair of the plasma membrane. In response to plasma membrane rupture, lysosomes are able to rescue the cell by rapid translocation to the damage site of the plasma membrane and donation of the membrane [18, 19]. This exocytosis process is triggered by Ca^{2+} influx from the extracellular compartment and requires the ubiquitously expressed lysosomal membrane protein synaptotagmin 7 [20]. Besides conventional lysosomes, lysosome-related organelles (LRO), including melanosomes, lytic granules, and platelet-dense granules, exist in certain cell types and have acquired special functions [21].

Over the last decade, advances in lysosome research have established a broad role for the lysosomes in the pathophysiology of disease. The most obvious are the lysosomal storage diseases (LSD), which include approximately 70 distinct disorders. Although individually rare, they collectively account for 14% of all inherited metabolic diseases. The main biochemical hallmark of LSD is the accumulation of un- or partially digested metabolites in the lysosomes. The pathologic mechanisms include malfunction of the degradation, the transport across the lysosomal membrane, or trafficking between endosomes and lysosomes [22]. Noteworthy, recent studies have observed that lysosomal alterations and malfunction are also players in some of the most common conditions nowadays including cancer and neurodegenerative diseases. The neurodegenerative hallmarks of the rare early-onset lysosomal storage diseases resemble late-onset neurodegenerative diseases such as Alzheimer's and Parkinson's diseases. It has been shown that type 1 Gaucher disease patients have a higher risk of developing Parkinson's disease [23]. Frontotemporal dementia is caused by mutation in one allele of progranulin. However if both alleles are mutated, it will lead to the neuronal ceroid lipofuscinogenesis (CLN11) [24]. Thus a theory of a general mechanism of dysfunctional clearance of cellular cargo through the secretory-endosomal-autophagic-lysosomal-exocytic (SEALE) network has been formed to explain the common underlying feature relating lysosomal dysfunction to seemingly different diseases [25].

Advanced tumor cells are highly dependent on effective lysosomal function. Thus, cancer progression and metastasis are associated with striking alterations in lysosomal compartments including changes in lysosome volume, composition, cellular distribution, and lysosomal enzyme activity. Release of cathepsins from a cancer cell into the extracellular space can promote tumor growth through their proteolytic effect on the basement membrane and activation of other pro-tumorigenic proteins [26–28]. Moreover, elevated expression of wild-type TFEB protein is sufficient for driving the oncogenic mechanism [29]. Resistance of cancer cells towards traditional therapies may be overcome by agents that trigger LMP and engage lysosomal cell death pathways [26]. On the other hand, therapeutic strategies to restrain proteolytic activity of secreted hydrolases would be a way to suppress tumor invasion. The development of techniques for control and manipulation of lysosomal function will generate future treatments of the wide variety of common and rare pathological conditions involving lysosomes.

After several groundbreaking discoveries, our knowledge has increased tremendously and the lysosome is now recognized as one of the central organelles for normal physiological function and during disease. In this volume of Methods in Molecular Biology, laboratory protocols for detailed studies of essential parts of lysosomal biology are provided. The protocols are straightforward and aim to guide researchers in their exploration of lysosomes, both under normal conditions and in pathological processes. We hope that the provided know-how and protocols will guide and inspire further research and generate new insights into the versatile tasks of this fascinating organelle.

Finally, we would like to thank all contributing authors for sharing their expertise. We would also express our sincere gratitude to Professor John M. Walker for support and guidance during the editing of this volume of MiMB series.

Linköping, Sweden *Karin Öllinger*
Linköping, Sweden *Hanna Appelqvist*

References

1. De Duve C (2005) The lysosome turns fifty. Nat Cell Biol 7:847–849
2. Saftig P, Klumperman J (2009) Lysosome biogenesis and lysosomal membrane proteins: trafficking meets function. Nat Rev Mol Cell Biol 10:623–635
3. Luzio J P, Pryor P R, Bright NA (2007) Lysosomes: fusion and function. Nat Rev Mol Cell Biol 8:622–632
4. Appelqvist H, Wäster P, Kågedal K, Öllinger K (2013) The lysosome: from waste bag to potential therapeutic target. J Mol Cell Biol 5:214–226
5. Repnik U, Stoka V, Turk V, Turk B (2012) Lysosomes and lysosomal cathepsins in cell death. Biochim Biophys Acta 1824(1):22–33
6. Luzio JP, Hackmann Y, Dieckmann NM, Griffiths GM (2014) The biogenesis of lysosomes and lysosome-related organelles. Cold Spring Harb Perspect Biol 6(9):a016840
7. Gómez-Sintes R, Ledesma MD, Boya P (2016) Lysosomal cell death mechanisms in aging. Ageing Res Rev 32:150-168. doi: 10.1016/j.arr.2016.02.009
8. Sardiello M, Palmieri M, di Ronza A, Medina DL, Valenza M, Gennarino VA, Di Malta C, Donaudy F, Embrione V, Polishchuk RS, Banfi S, Parenti G, Cattaneo E, Ballabio A (2009) A gene network regulating lysosomal biogenesis and function. Science 325(5939):473–477
9. Settembre C, Fraldi A, Medina DL, Ballabio A (2013) Signals from the lysosome: a control centre for cellular clearance and energy metabolism. Nat Rev Mol Cell Biol 14:283–296
10. Medina DL, Di Paola S, Peluso I, Armani A, De Stefani D, Venditti R, Montefusco S, Scotto-Rosato A, Prezioso C, Forrester A, Settembre C, Wang W, Gao Q, Xu H, Sandri M, Rizzuto R, De Matteis MA, Ballabio A (2015) Lysosomal calcium signalling regulates autophagy through calcineurin and TFEB. Nat Cell Biol 17(3):288–299
11. Carstea ED, Morris JA, Coleman KG, Loftus SK, Zhang D, Cummings C, Gu J, Rosenfeld MA, Pavan WJ, Krizman DB, Nagle J, Polymeropoulos MH, Sturley SL, Ioannou YA, Higgins ME et al (1997) Niemann-Pick C1 disease gene: homology to mediators of cholesterol homeostasis. Science 277:228–231
12. Chang TY, Chang CC, Ohgami N, Yamauchi Y (2006) Cholesterol sensing, trafficking, and esterification. Annu Rev Cell Dev Biol 22:129–157
13. De Duve C, Wattiaux R (1966) Functions of lysosomes. Annu Rev Physiol 28:435–492
14. Boya P, Kroemer G (2008) Lysosomal membrane permeabilization in cell death. Oncogene 27: 6434–6451
15. Roberg K, Öllinger K (1998) Oxidative stress causes relocation of the lysosomal enzyme cathepsin D with ensuing apoptosis in neonatal rat cardiomyocytes. Am J Pathol 152(5):1151–1156
16. Guicciardi ME, Gores GJ (2009) Life and death by death receptors. FASEB J. 23(6):1625–1637
17. Kreuzaler PA, Staniszewska AD, Li W, Omidvar N, Kedjouar B, Turkson J, Poli V, Flavell RA, Clarkson RW, Watson CJ (2011) Stat3 controls lysosomal-mediated cell death in vivo. Nat Cell Biol 13:303–309
18. Andrews NW, Almeida PE, Corrotte M (2014) Damage control: cellular mechanisms of plasma membrane repair. Trends Cell Biol. 24(12):734–742
19. Jaiswal JK Andrews NW, Simon SM (2002) Membrane proximal lysosomes are the major vesicles responsible for calcium-dependent exocytosis in nonsecretory cells. J Cell Biol 159(4):625–635
20. Reddy A, Caler EV, Andrews NW (2001) Plasma membrane repair is mediated by Ca2+–regulated exocytosis of lysosomes. Cell 106:157–169
21. Dell'Angelica EC, Mullins C, Caplan S, Bonifacino JS (2000) Lysosome-related organelles. FASEB J 14:1265–1278
22. Bellettato CM, Scarpa M (2010) Pathophysiology of neuropathic lysosomal storage disorders. J Inherit Metab Dis 33(4):347–362
23. Beavan MS, Schapira AH (2013) Glucocerebrosidase mutations and the pathogenesis of Parkinson disease. Ann Med 45:511–521
24. Smith KR, Damiano J, Franceschetti S, Carpenter S, Canafoglia L, Morbin M, Rossi G, Pareyson D, Mole SE, Staropoli JF, Sims KB, Lewis J, Lin WL, Dickson DW, Dahl HH, Bahlo M, Berkovic SF (2012) Strikingly different clinicopathological phenotypes determined by progranulin-mutation dosage. Am J Hum Genet 90(6):1102–1107
25. Boland B, Platt FM (2015) Bridging the age spectrum of neurodegenerative storage diseases. Best Pract Res Clin Endocrinol Metab 29(2):127–143
26. Petersen NH, Olsen OD, Groth-Pedersen L, Ellegaard AM, Bilgin M, Redmer S, Ostenfeld MS, Ulanet D, Dovmark TH, Lønborg A, Vindeløv SD, Hanahan D, Arenz C, Ejsing CS, Kirkegaard T, Rohde M,

Nylandsted J, Jäättelä M (2013) Transformation-associated changes in sphingolipid metabolism sensitize cells to lysosomal cell death induced by inhibitors of acid sphingomyelinase. Cancer Cell 24:379–393

27. Hämälistö S, Jäättelä M (2016) Lysosomes in cancer-living on the edge (of the cell). Curr Opin Cell Biol 39:69–76
28. Saftig P, Sandhoff K (2013) Cancer: Killing from the inside. Nature 502(7471):312–313
29. Palmieri M, Impey S, Kang H, di Ronza A, Pelz C, Sardiello M, Ballabio A (2011) Characterization of the CLEAR network reveals an integrated control of cellular clearance pathways. Hum Mol Gene 20: 3852–3866

Contents

Contributors

Ravi K. Amaravadi • *Department of Medicine and Abramson Cancer Center, Perelman School of Medicine, University of Pennsylvania, Philadelphia, PA, USA*
Norma W. Andrews • *Department of Cell Biology and Molecular Genetics, University of Maryland at College Park, College Park, MD, USA*
Hanna Appelqvist • *Division of Chemistry, Department of Physics, Chemistry and Biology, Linköping University, Linköping, Sweden*
Andrea Ballabio • *Telethon Institute of Genetics and Medicine (TIGEM), Pozzuoli (NA), Italy; Medical Genetics, Department of Translational Medicine, Federico II University, Naples, Italy; Department of Molecular and Human Genetics, Baylor College of Medicine, Houston, TX, USA; Jan and Dan Duncan Neurological Research Institute, Texas Children's Hospital, Houston, TX, USA*
Sandro Banfi • *Medical Genetics, Department of Biochemistry, Biophysics and General Pathology, Second University of Naples, Naples, Italy; Telethon Institute of Genetics and Medicine (TIGEM), Pozzuoli (NA), Italy*
Ptissam Bergam • *Institut Curie, PSL Research University, CNRS, Paris, France; Sorbonne Universités, UPMC Univ Paris 06, CNRS, Paris, France; Cell and Tissue Imaging Core Facility PICT-IBiSA, Institut Curie, Paris, France; Characterization Core Lab, King Abdullah University of Science and Technology (KAUST), Thuwal, Kingdom of Saudi Arabia*
Sunita Bijarnia-Mahay • *Institute of Medical Genetics and Genomics, Sir Ganga Ram Hospital, New Delhi, India*
Mesut Bilgin • *Cell Death and Metabolism Unit, Center for Autophagy, Recycling and Disease, Danish Cancer Society Research Center, Copenhagen, Denmark*
Thomas Braulke • *Department of Biochemistry, Children's Hospital, University Medical Center Hamburg-Eppendorf, Hamburg, Germany*
Beibei Deng • *Institute of Fluorescent Probes for Biological Imaging, School of Chemistry and Chemical Engineering, School of Materials Science and Engineering, University of Jinan, Jinan, Shandong, P.R. China*
Giuseppina Di Fruscio • *Medical Genetics, Department of Biochemistry, Biophysics and General Pathology, Second University of Naples, Naples, Italy*
Ida Eriksson • *Experimental Pathology, Department of Clinical and Experimental Medicine, Linköping University, Linköping, Sweden*
Anna Erlandsson • *Department of Public Health and Caring Sciences/Molecular Geriatrics, Rudbeck Laboratory, Uppsala University, Uppsala, Sweden*
Jeffrey D. Esko • *Cellular and Molecular Medicine, University of California, San Diego, La Jolla, CA, USA*
Colin Fennelly • *Department of Medicine and Abramson Cancer Center, Perelman School of Medicine, University of Pennsylvania, Philadelphia, PA, USA*
Volkmar Gieselmann • *Institute for Biochemistry and Molecular Biology, Rheinische-Friedrich-Wilhelms-University, Bonn, Germany*
Ana Maria Vilamill Giraldo • *Experimental Pathology, Department of Clinical and Experimental Medicine, Linköping University, Linköping, Sweden*

ANNA S. GUKOVSKAYA • *David Geffen School of Medicine, University of California at Los Angeles, Los Angeles, CA, USA; VA Greater Los Angeles Healthcare System, Los Angeles, CA, USA*

KRISTINA M. HAMILL • *Department of Chemistry and Biochemistry, University of California, San Diego, La Jolla, CA, USA*

YOKO HASHIMOTO • *Department of Biochemistry, School of Dentistry, Aichi-Gakuin University, Chikusa-ku, Nagoya, Japan*

XAVIER HEILIGENSTEIN • *Institut Curie, PSL Research University, CNRS, Paris, France; Sorbonne Universités, UPMC Univ Paris 06, CNRS, Paris, France; Cell and Tissue Imaging Core Facility PICT-IBiSA, Institut Curie, Paris, France*

ILSE HURBAIN • *Institut Curie, PSL Research University, CNRS, Paris, France; Sorbonne Universités, UPMC Univ Paris 06, CNRS, Paris, France; Cell and Tissue Imaging Core Facility PICT-IBiSA, Institut Curie, Paris, France*

MARJA JÄÄTTELÄ • *Cell Death and Metabolism Unit, Center for Autophagy, Recycling and Disease, Danish Cancer Society Research Center, Copenhagen, Denmark*

XIUQI KONG • *Institute of Fluorescent Probes for Biological Imaging, School of Chemistry and Chemical Engineering, School of Materials Science and Engineering, University of Jinan, Jinan, Shandong, P.R. China*

WEIYING LIN • *Institute of Fluorescent Probes for Biological Imaging, School of Chemistry and Chemical Engineering, School of Materials Science and Engineering, University of Jinan, Jinan, Shandong, P.R. China*

VESA LOITTO • *Core Facility Microscopy Unit, Medical Faculty, Linköping University, Linköping, Sweden*

CAMILLA LÖÖV • *MassGeneral Institute for Neurodegeneration, Massachusetts General Hospital, Harvard Medical School, Charlestown, MA, USA*

FREDERIK W. LUND • *Department of Biochemistry and Molecular Biology, University of Southern Denmark, Odense M, Denmark; Department of Biochemistry, Weill Medical College of Cornell University, New York, NY, USA*

KENJI MAEDA • *Cell Death and Metabolism Unit, Center for Autophagy, Recycling and Disease, Danish Cancer Society Research Center, Copenhagen, Denmark*

OLGA A. MARENINOVA • *David Geffen School of Medicine, University of California at Los Angeles, Los Angeles, CA, USA; VA Greater Los Angeles Healthcare System, Los Angeles, CA, USA*

VINCENZO NIGRO • *Medical Genetics, Department of Biochemistry, Biophysics and General Pathology, Second University of Naples, Naples, Italy; Telethon Institute of Genetics and Medicine (TIGEM), Pozzuoli (NA), Italy*

JESPER NYLANDSTED • *Cell Death and Metabolism Unit, Center for Autophagy, Recycling and Disease, Danish Cancer Society Research Center, Copenhagen, Denmark*

KARIN ÖLLINGER • *Experimental Pathology, Department of Clinical and Experimental Medicine, Linköping University, Linköping, Sweden*

TAKANOBU OTOMO • *Department of Genetics, Osaka University Graduate School of Medicine, Osaka, Japan; Laboratory of Intracellular Membrane Dynamics, Osaka University Graduate School of Frontier Biosciences, Osaka, Japan; Research Center for Autophagy, Osaka University Graduate School of Medicine, Osaka, Japan*

SOFIA RAMSTRÖM • *Department of Clinical Chemistry and Department of Clinical and Experimental Medicine, Linköping University, Linköping, Sweden; Department of Clinical Medicine, Örebro University, Örebro, Sweden*

Graça Raposo • *Institut Curie, PSL Research University, CNRS, Paris, France; Sorbonne Universités, UPMC Univ Paris 06, CNRS, Paris, France; Cell and Tissue Imaging Core Facility PICT-IBiSA, Institut Curie, Paris, France*
Neil D. Rawlings • *European Molecular Biology Laboratory, European Bioinformatics Institute (EMBL-EBI), Wellcome Genome Campus, Hinxton, Cambridge, UK*
Mingguang Ren • *Institute of Fluorescent Probes for Biological Imaging, School of Chemistry and Chemical Engineering, School of Materials Science and Engineering, University of Jinan, Jinan, Shandong, P.R. China*
Maryse Romao • *Institut Curie, PSL Research University, CNRS, Paris, France; Sorbonne Universités, UPMC Univ Paris 06, CNRS, Paris, France; Cell and Tissue Imaging Core Facility PICT-IBiSA, Institut Curie, Paris, France*
Ebru Demirel Sezer • *Department of Medical Biochemistry and Metabolism Laboratory, Ege University Faculty of Medicine, Izmir, Turkey*
Anna L. Södergren • *Clinical Chemistry, Department of Clinical and Experimental Medicine, Linköping University, Linköping, Sweden*
Eser Yildirim Sozmen • *Department of Medical Biochemistry and Metabolism Laboratory, Ege University Faculty of Medicine, Izmir, Turkey*
Yonghe Tang • *Institute of Fluorescent Probes for Biological Imaging, School of Chemistry and Chemical Engineering, School of Materials Science and Engineering, University of Jinan, Jinan, Shandong, P.R. China*
Melanie Thelen • *Institute for Biochemistry and Molecular Biology, Rheinische-Friedrich-Wilhelms-University, Bonn, Germany*
Elmira Tokhtaeva • *David Geffen School of Medicine, University of California at Los Angeles, Los Angeles, CA, USA; VA Greater Los Angeles Healthcare System, Los Angeles, CA, USA*
Wenyong Tong • *Cellular and Molecular Medicine, University of California, San Diego, La Jolla, CA, USA*
Yitzhak Tor • *Department of Chemistry and Biochemistry, University of California, San Diego, La Jolla, CA, USA*
Olga Vagin • *David Geffen School of Medicine, University of California at Los Angeles, Los Angeles, CA, USA; VA Greater Los Angeles Healthcare System, Los Angeles, CA, USA*
Ishwar C. Verma • *Institute of Medical Genetics and Genomics, Sir Ganga Ram Hospital, New Delhi, India*
Jyotsna Verma • *Institute of Medical Genetics and Genomics, Sir Ganga Ram Hospital, New Delhi, India*
Ezequiel Wexselblatt • *Department of Chemistry and Biochemistry, University of California, San Diego, La Jolla, CA, USA*
Dominic Winter • *Institute for Biochemistry and Molecular Biology, Rheinische-Friedrich-Wilhelms-University, Bonn, Germany*
Daniel Wüstner • *Department of Biochemistry and Molecular Biology, University of Southern Denmark, Odense M, Denmark*
Tamotsu Yoshimori • *Department of Genetics, Osaka University Graduate School of Medicine, Osaka, Japan; Laboratory of Intracellular Membrane Dynamics, Osaka University Graduate School of Frontier Biosciences, Osaka, Japan; Research Center for Autophagy, Osaka University Graduate School of Medicine, Osaka, Japan*

Chapter 1

SILAC-Based Comparative Proteomic Analysis of Lysosomes from Mammalian Cells Using LC-MS/MS

Melanie Thelen, Dominic Winter, Thomas Braulke, and Volkmar Gieselmann

Abstract

Mass spectrometry-based proteomics of lysosomal proteins has led to significant advances in understanding lysosomal function and pathology. The ever-increasing sensitivity and resolution of mass spectrometry in combination with labeling procedures which allow comparative quantitative proteomics can be applied to shed more light on the steadily increasing range of lysosomal functions. In addition, investigation of alterations in lysosomal protein composition in the many lysosomal storage diseases may yield further insights into the molecular pathology of these disorders. Here, we describe a protocol which allows to determine quantitative differences in the lysosomal proteome of cells which are genetically and/or biochemically different or have been exposed to certain stimuli. The method is based on stable isotope labeling of amino acids in cell culture (SILAC). Cells are exposed to superparamagnetic iron oxide particles which are endocytosed and delivered to lysosomes. After homogenization of cells, intact lysosomes are rapidly enriched by passing the cell homogenates over a magnetic column. Lysosomes are eluted after withdrawal of the magnetic field and subjected to mass spectrometry.

Key words Lysosome, Magnetic particle, Lysosomal proteome, Lysosomal storage disorder, Mass spectrometry

1 Introduction

Lysosomes are membrane-limited organelles with an acidic pH maintained by integral vacuolar H^+ ATPases. For a long time, it was assumed that the sole function of lysosomes is the degradation of a wide variety of macromolecules and the release of degradation products into the cytosol, where they can be reused for biosynthetic or energy-producing pathways. During the last years, however, it became clear that lysosomes are not merely degradative compartments but communicate with their environment and play important roles in secretion, plasma membrane repair, and antigen presentation [1]. Furthermore, protein complexes localized at the cytosolic surface of lysosomes, such as mTORC1

Karin Öllinger and Hanna Appelqvist (eds.), *Lysosomes: Methods and Protocols*, Methods in Molecular Biology, vol. 1594, DOI 10.1007/978-1-4939-6934-0_1,

(mammalian target of rapamycin complex 1) and BORC (biogenesis of lysosome-related organelles complex 1-related complex), mediate nutrient signaling and lysosomal positioning, respectively [2, 3]. This wide spectrum of lysosomal functions requires the action of many proteins.

Lysosomal degradation of macromolecules depends on about 60 soluble acid hydrolases residing in the lysosomal lumen [4]. The surrounding lysosomal membrane contains numerous integral membrane proteins encompassing transporters for delivery of degradation products, ion channels or highly glycosylated proteins securing lysosomal integrity by protecting the lysosomal membrane from self-digestion [5, 6]. Peripheral membrane proteins and soluble proteins bound to the cytoplasmic side of lysosomes by protein–protein interactions allow for communication or fusion with other cellular compartments and integrate the lysosome into the overall cellular metabolism. Proteomics of soluble lysosomal proteins revealed important aspects of lysosomal function. In these studies, lysosomal proteins were enriched using an affinity matrix that specifically binds the unique mannose 6-phosphate (M6P) residues found on N-linked oligosaccharide side chains of soluble proteins in the lysosomal lumen. These investigations provided novel insights in (1) cell type-specific lysosomal enzymes [7, 8], (2) binding preferences of the two M6P receptors for subpopulations of lysosomal enzymes [9, 10], and secreted lysosomal enzymes found in the circulation [11] or urine [12]. This approach has also been successfully applied to identify enzyme defects underlying lysosomal storage disorders such as the cholesterol binding lysosomal protein NPC2 being deficient in a form of Niemann Pick Type C disease [13]. The success of proteomics of soluble lysosomal hydrolases promises that studies on the comparatively underexplored proteome of the lysosomal membrane, and proteins associated with the cytosolic surface of lysosomes, will yield new important insights into lysosomal function. Multi-step subcellular fractionation techniques for enriching lysosomal membranes and subsequent mass spectrometric analyses led to the description of 140–300 membrane proteins of variable abundancy with known or presumed lysosomal localization [5, 14]. Most of these studies were targeted at the identification of novel bona fide lysosomal proteins, and have indeed revealed previously unknown proteins of lysosomal localization. Although the precise function of most of these proteins is still unknown, their investigation should increase our understanding of this multifaceted organelle in the future.

Irrespective of the goal of the study, lysosomal proteomics requires the enrichment of lysosomes which can be achieved by various techniques. These include subcellular fractionation methods using different density gradient materials like sucrose, percoll, nycodenz and metrizamide [15, 16], or immunoprecipitation of lysosomes from LAMP1-FLAG overexpressing cells [17].

Alternatively, lysosomes can be isolated by using an in vivo approach employing a density shift of mouse or rat liver lysosomes after injection of Triton WR-1339 [16]. One disadvantage of most density-gradient-based approaches, is the contamination with other organelles, in particular mitochondria. For mass spectrometric investigations we have adapted a technique using superparamagnetic iron oxide particles for rapid enrichment of lysosomes from cell homogenates recovering routinely up to 80% of intact lysosomes. For a detailed description of this method see Walker and Lloyd-Evans [18]. The method relies on the endocytic uptake of dextran-coated iron oxide particles with 10 nm diameter, which are subsequently enriched in the lysosomal compartment and can be isolated by passing the cell homogenate over a magnetic column. This allows for a rapid enrichment of intact lysosomes in amounts sufficient for mass spectrometric analysis from as little as two confluent 10 cm dishes of cells. The method is fast, avoids unnecessary manipulations and is thus likely to preserve the original protein composition of the sample. Due to the high sensitivity of mass spectrometers, in any proteomic dataset of subcellular fractions, a considerable number of proteins identified cannot be assigned to the enriched organelles. Some of these proteins are true contaminants but in case of lysosomes may also co-purify because they are functionally linked to the diverse functions of lysosomes. Lysosomes have contact sites with the endoplasmic reticulum and mitochondria [19–21], are tightly connected to the cytoskeleton to ensure their migration within the cytoplasm [22] and through selective and non-selective autophagy are the ultimate destination of ribosomes, mitochondria, parts of the ER and many cytosolic proteins [23–25]. Consequently, when applying a highly sensitive analytical approach, the identification of a wide range of non-lysosomal proteins cannot be avoided and it can therefore be difficult to identify the proteins which are affected by the experiment. This problem can be considerably reduced if proteomic datasets of two samples are quantitatively compared. These samples can differ in genetic background, be cultured under different conditions or be exposed to specific pharmacological or biochemical stimuli. In the comparative quantitative proteomic dataset only the amount of those proteins which are somehow functionally connected to the investigated alteration of the system will change, whereas the numerous irrelevant or contaminating proteins detected in both samples remain unchanged. Moreover, the nature of the applied stimulus or condition may allow developing a working hypothesis on how the identified proteins are functionally connected to the lysosome. To allow for quantitative proteomic comparison of lysosomal proteins, we have used stable isotope labeling of amino acids in cell culture (SILAC) [26]. Cells with unlabeled amino acids can serve as controls, those labeled with heavy amino acids can differ, e.g. genetically or biochemically, and vice versa.

We have recently used this technique successfully for the comparative investigation of the proteome of lysosomal hydrolases in mouse cells lacking M6P targeting signals. This led to the identification of alternative M6P-independent transport pathways and the receptors involved [27].

2 Materials

2.1 Isolation of Lysosomes

1. Human embryonic kidney (HEK) 293 cells (German Collection of Microorganisms and Cell Cultures, DSMZ).
2. Phosphate buffered saline (PBS).
3. Amino acids L-Lysine-2HCl ($^{13}C_6$, $^{15}N_2$ labeled and unlabeled, respectively), 17.45 g/l in PBS (200-fold stock solution) and L-Arginine-HCL ($^{13}C_6$, $^{15}N_4$ labeled and unlabeled, respectively) 36.24 g/l in PBS (200-fold stock solution) (*see* **Note 1**).
4. Fetal Bovine Serum for SILAC (dialyzed).
5. DMEM for SILAC: Dulbecco's modified Eagle Medium (DMEM) high glucose (4.5 g/l) for SILAC, without lysine (Lys) and arginine (Arg) supplemented with 10% FBS, GlutaMax™ (200 mM), Penicillin (100 U/ml)/streptomycin (0.1 mg/ml) and either conventional light Arg/Lys or heavy isotope labeled Arg ($^{13}C_6{}^{15}N_4$)/Lys ($^{13}C_6{}^{15}N_2$). The final concentrations of Arg should be 87.8 mg/l and Lys: 181.2 mg/l (*see* **Note 1**).
6. Magnetite solution: EndoMAG40, 40 kDa dextran-coated magnetite particles from Liquids Research (http://www.liquidsresearch.com/).
7. PLL solution: Poly-L-Lysine 0.5 mg/ml in PBS.
8. Isolation buffer: 250 mM sucrose, 10 mM HEPES/OH pH 7.4, 1 mM $CaCl_2$, 1 mM $MgCl_2$, 1.5 mM MgAc, 1 mM dithiothreitol (DTT), 1× protease inhibitor cocktail (PIC, Halt protease inhibitor cocktail). Add DTT and PIC always immediately before using the buffer.
9. Tight-fitting 7 ml Dounce homogenizer.
10. BSA solution: 0.5 mg/ml bovine serum albumin (BSA) in PBS.
11. DNAse 1 solution: 10 μl DNAse 1 (1 Unit/μl) in 1 ml of isolation buffer.
12. Miltenyi LS Separation columns.
13. Miltenyi MidiMACS Magnetic Separator.

2.2 β-Hexosaminidase Enzyme Assay

1. Substrate solution: 10 mM 4-Nitrophenyl N-acetyl-β-D-glucosaminide in 0.1 M sodium citrate, pH 4.6, containing 0.2% BSA.
2. 10% Triton X-100 (v/v) in water.
3. Stop solution: 400 mM glycine/OH pH 10.4.

2.3 Mass Spectrometry Sample Preparation

Prepare all solutions for mass spectrometry sample preparation with MS grade water and use MS-grade/ultrapure chemicals.

1. Maximum recovery pipet tips.
2. Maximum recovery microcentrifuge tubes.
3. Amicon Ultra Centrifugal Filters Ultracel 3K, Merck Millipore Ltd.
4. Laemmli buffer 4× concentrated (modified from [28]): 250 mM Tris, 8% (w/v) SDS, 40% (v/v) glycerol, 10% (v/v) β-mercaptoethanol, 0.004% (w/v) bromophenol blue.
5. 40% (w/v) acrylamide in water.
6. SDS-PAGE Gel (10% acrylamide).
7. PageBlue™ protein staining solution, Fermentas, or any other Coomassie G250 gel staining solution.
8. Solution A: 30% (v/v) acetonitrile (ACN) in 100 mM NH_4HCO_3, pH 7.8.
9. 100 mM NH_4HCO_3, pH 7.8.
10. Trypsin solution: 0.5 ng/μl sequencing grade trypsin in 50 mM NH_4HCO_3, pH 7.8. To prepare this solution, resuspend 20 μg of trypsin in 200 μl of water, then mix with 200 μl of 100 mM NH_4HCO_3, pH 7.8 (*see* **Note 2**).
11. Solution B: 0.1% (v/v) trifluoroacetic acid, 50% ACN.
12. Solution C: 5% (v/v) ACN, 5% formic acid (FA).
13. StageTips, prepared with Solid Phase Extraction Disk Octadecyl C18 (3 M, www.3m.com), 4 layers in a 200 μl Maximum recovery pipet tip [29].
14. 100% methanol.
15. 0.5% (v/v) acetic acid in 80% ACN.
16. 0.5% (v/v) acetic acid.
17. Vacuum centrifuge.

2.4 LC-MS-MS Measurement

1. C_{18} Analytical column: ESI spray tip produced in house with a Sutter P2000 laser puller device from 360 μm OD, 100 μm ID fused silica capillary packed with 5 μm particles [Dr. Maisch, Reprosil C-18 AQ]. Commercially available columns can alternatively be used.
2. Thermo EASY-nLC 1,000 or similar nanoflow high or ultra-high performance liquid chromatography systems.
3. Thermo Orbitrap Velos Mass Spectrometer or any other suitable mass spectrometer.
4. Running buffer A: water with 0.1% FA.
5. Running buffer B: ACN with 0.1% FA.

3 Methods

The general workflow of the described method is graphically displayed in Fig. 1. Perform all steps involving cell culture in a sterile laminar flow hood. For cell culture, pre-warm all solutions to 37 °C.

3.1 Isolation of Lysosomes

1. Cultivate the cells for at least six passages (to ensure complete SILAC labeling of the cells' proteome) in DMEM for SILAC.
2. The following procedures are described for HEK293 cells. When you want to use another cell type, please refer to **Notes 3–6**.
3. For HEK293 cells, 10 cm cell culture dishes need to be coated in advance with PLL-solution for 10 min at room temperature (RT) to ensure proper cell attachment. Thereafter, remove the liquid and wash thrice with PBS before plating the cells (*see* **Note 7**).

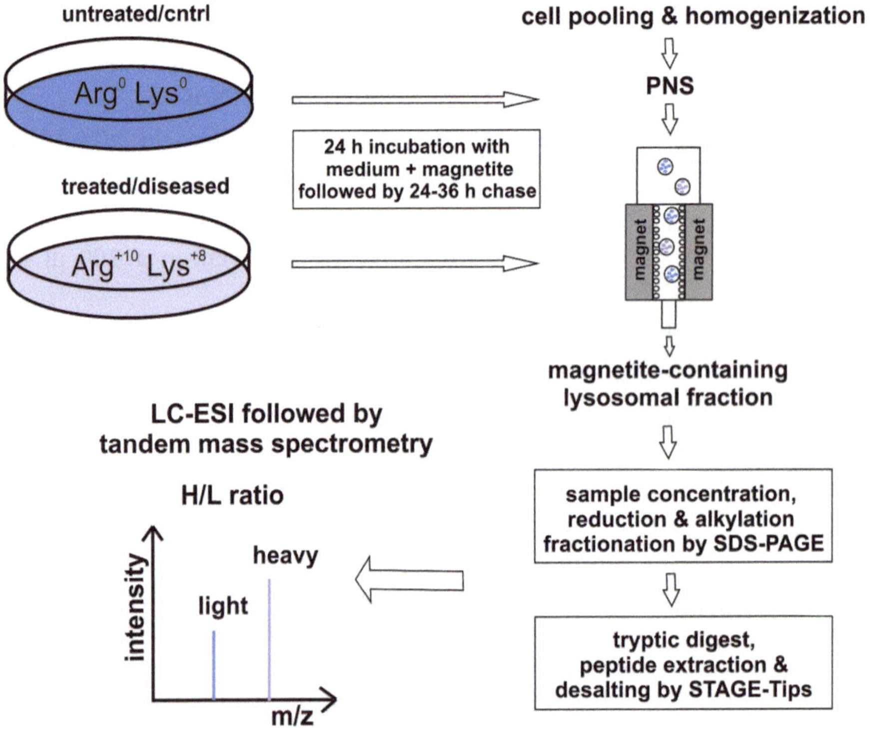

Fig. 1 Schematic representation of the experimental workflow. Two cell populations are labeled with either light or heavy arginine or lysine. They can be exposed to different conditions at any time between seeding and harvest. After 24 h of incubation with magnetite-containing medium and 24-36 h of chase time, cells are pooled immediately after harvesting, homogenized, the postnuclear supernatants (PNS) prepared and passed over a magnetic column. The eluate is then fractionated and proteolytically digested before LC-MS/MS measurement (modified and reproduced from [27] with permission from Wiley)

4. For each experiment, you will need 2 confluent 10 cm dishes of HEK293 cells, 1 dish of unlabeled (light), and one dish of cells labeled with Arg($^{13}C_6{}^{15}N_4$)/Lys($^{13}C_6{}^{15}N_2$) (heavy). Seed 5×10^6 cells in each 10 cm dish in culture medium supplemented with either light or heavy labeled amino acids containing 10% magnetite solution (*see* **Note 8**). Incubate for 24 h pulse time at 37 °C, 5% CO_2.
5. After pulse time, aspirate the medium containing magnetite and carefully wash the cells thrice with PBS (*see* **Note 9**).
6. After washing, add regular culture medium for a 24–36 h chase.
7. Treat your cells with the desired stimulus prior to isolation (*see* **Note 10**).
8. Before starting the isolation procedure, cool your cells, as well as the isolation buffer and PBS, on ice.
9. Wash cells twice with PBS to remove proteins found in the culture medium.
10. Add 2 ml of isolation buffer to each 10 cm dish and detach cells using a cell scraper. Light and heavy labeled cells should have the same cell density and can be merged at this point. If cell densities are evidently different, refer to **Note 11**.
11. Homogenize the cell suspension in a tight-fitting Dounce homogenizer for 25 strokes placing it in ice water (*see* **Note 5**).
12. Transfer homogenate to a 15 ml Falcon tube. Pellet nuclei and unbroken cells at 600 × *g* for 10 min at 4 °C.
13. Transfer the postnuclear supernatant (PNS) to a fresh tube and keep it on ice. Resuspend the cell/nuclear pellet in 4 ml isolation buffer, transfer back into the Dounce homogenizer and repeat the homogenization procedure. After centrifugation, merge both supernatants.
14. Insert the LS column into the magnetic stand and add 1 ml of BSA solution to the column. Let the column empty by gravity flow.
15. Apply combined postnuclear supernatants (input) to the column and let it pass by gravity flow. Collect non-bound material (flow-through).
16. Add 1 ml DNAse solution to the column and incubate for 10 min at 25 °C.
17. Wash the column with 5 ml of isolation buffer.
18. Remove the column from the magnetic stand. It may be that the iron beads of the column retain a certain amount of magnetic field prohibiting efficient elution of the lysosomes. In order to remove this residual magnetism hit the column against a hard surface before elution.

19. Retrieve the lysosomal fraction by adding 500 μl of isolation buffer to the column and eluting using the plunger. Repeat this procedure three times (*see* **Note 12**).
20. Remove and save around 250 μl of each fraction (input, flowthrough, wash, and eluate) to determine the β-hexosaminidase activity in all fractions after the isolation of lysosomes has been completed.

3.2 β-Hexosaminidase Enzyme Assay

To control for efficacy and quality of lysosome enrichment, the activity measurement of β-hexosaminidase is routinely performed after each experiment. Its activity serves as indicator for the quantity and integrity of lysosomes during the enrichment procedure. A representative result for the isolation of lysosomes from HEK293 cells is displayed in Fig. 2. To determine how much of the enzymatic activity is contained in intact lysosomes, the assay is performed for each sample with and without detergent. The difference between both values equals the activity that is present in intact lysosomes and is therefore available for purification.

1. Pipet 4 × 25 μl of each fraction into a microtiter plate.
2. Add 2 μl of 10% Triton X-100 to two of the samples.
3. Add 50 μl of substrate solution.
4. As blank, mix 25 μl of isolation buffer with 50 μl of substrate solution.
5. Incubate for 15 min at 37 °C. Depending on the amount of cells used incubation time can be prolonged (up to 24 h).
6. Stop the reaction by addition of 200 μl of stop solution. The β-hexosaminidase-containing samples should turn yellow during this step.
7. Measure absorbance at 405 nm.
8. Calculate enzymatic activity using the Lambert–Beer law.
9. When your isolation is not successful, please *see* **Notes 6** and **13**.

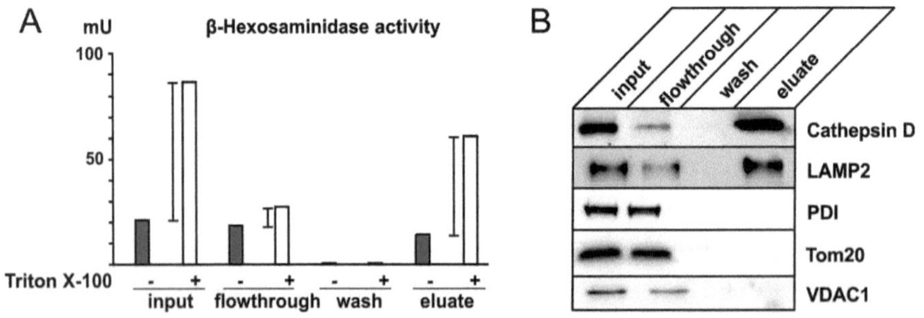

Fig. 2 Efficiency of lysosomal enrichment. (**a**) The total activity of β-hexosaminidase was determined in each fraction by an enzymatic activity assay with (*white bar*) and without (*grey bar*) Triton X-100. Lines next to the columns represent the portion of intact lysosomes. (**b**) Western blot image from 0.25% of each fraction. Membranes were probed with antibodies against cathepsin D (lysosomal lumen), LAMP2 (lysosomal membrane), protein disulfide isomerase (endoplasmic reticulum), Tom20 and VDAC1 (mitochondria)

3.3 Mass Spectrometry Sample Preparation

To avoid contamination of the sample with keratin or other contaminants, always use new plastic ware, pre-stacked tips and MS grade solvents and chemicals. During work, gloves should be always worn and frequently exchanged. To avoid the loss of sample, use maximum recovery tubes and pipet tips.

1. The protein concentration of the lysosomal fraction is frequently low (around 0.15–0.4 mg/ml). To concentrate the eluate, centrifuge in a centrifugal filter unit until you reach around 10% of the original volume (usually around 100 μl).
2. Add Laemmli buffer to 1× concentration and denature the samples at 95 °C for 5 min (*see* **Note 14**).
3. Cool samples to 25 °C and add 40% acrylamide to a final concentration of 1% (w/v). Incubate at 25 °C for 30 min for alkylation of cysteines (*see* **Note 15**).
4. Separate your sample by SDS-PAGE.
5. Wash the gel thrice for 5 min with deionized water on a shaker.
6. Stain your gel with PageBlue™ solution for several hours or overnight and destain with deionized water.
7. Cut the gel lane in ten equal slices and cut each gel slice subsequently in ~1 mm^3 pieces with a scalpel (*see* **Note 16**). Transfer pieces into separate microcentrifuge tubes, one for each slice. If you do not immediately proceed with the in-gel digestion, cover the gel pieces with water to avoid drying.
8. Add 500 μl solution A to each fraction, if the gel pieces were stored in water remove it first.
9. Incubate for 30 min at 25 °C and 1000 rpm in a thermomixer (*see* **Note 17**). Remove and discard the liquid and repeat **step 9** twice or thrice until all gel pieces are colorless (*see* **Note 18**).
10. Add 500 μl 100% ACN and incubate for 15 min at 25 °C and 1000 rpm in a thermomixer. The gel pieces should turn white during this step.
11. Remove and discard the liquid and dry the gel pieces in a vacuum centrifuge, this takes typically between 5 and 15 min.
12. Add 10 μl trypsin solution to each sample and wait until the liquid is soaked up by the dry gel pieces.
13. Add 50 μl of 100 mM NH_4HCO_3, pH 7.8 and incubate for 10 min at 25 °C.
14. If the gel pieces are not completely covered by liquid, add 100 mM NH_4HCO_3, pH 7.8 until they are covered.
15. Incubate the samples overnight at 37 °C (*see* **Note 19**).
16. Transfer the supernatant of each sample, which contains your digested peptides, to a fresh microcentrifuge tube (*see* **Notes 20** and **21**).

17. Add 50 µl of solution B to the gel pieces and incubate for 15 min at 25 °C and 1000 rpm in a thermomixer. Transfer the supernatants to the microcentrifuge tubes from **step 16**.
18. Add 50 µl of 100 mM NH_4HCO_3 pH 7.8 and incubate for 15 min at 25 °C and 1000 rpm in a thermomixer. Do not remove the liquid at the end of this step.
19. Add 100 µl 100% ACN to the samples and incubate for 15 min at 25 °C and 1000 rpm in a thermomixer. Transfer the supernatants to the microcentrifuge tubes from **step 16**.
20. Dry the samples using a vacuum centrifuge.
21. Resuspend your samples in 20 µl of solution C.
22. Add 20 µl of Methanol (MeOH) to your Stage Tips and centrifuge at 5000 × *g* for 30 s. If the liquid does not pass completely, increase centrifugation time accordingly.
23. Add 20 µl of 0.5% (v/v) acetic acid 80% ACN and centrifuge at 5000 × *g* for 30 s.
24. Add 20 µl 0.5% (v/v) acetic acid and centrifuge at 5000 × *g* for 30 s.
25. Apply each sample to one StageTip and centrifuge at 5000 × *g* for 30 s.
26. Wash Stage tips with 20 µl 0.5% (v/v) acetic acid and centrifuge at 5000 × *g* for 30 s.
27. Transfer Stage tips to fresh microcentrifuge tubes (*see* **Note 22**).
28. Elute sample with 2 × 20 µl of 0.5% (v/v) acetic acid 80% ACN by centrifugation for 30 s at 5000 × *g* and dry the eluate using a vacuum centrifuge.
29. Resuspend samples in 20 µl of solution C, sonicate in an ultrasonic water bath for 5 min and centrifuge at 20,000 × *g* for 15 min. Load 5 µl of this solution to an autosampler vial and proceed to MS analysis. Take the sample from the top in order to avoid small particles which may have accumulated at the bottom.

3.4 LC-MS-MS Measurement

1. Load 5 µl of sample on the analytical column using a nanoLC system (e.g. Thermo EASY-nLC 1000).
2. Load with 100% A at a flow rate of 1 µl/min followed by a washing step for 10 min with 100% A at a flow rate of 1 µl/min.
3. Elute with a linear gradient from 100% A to 65% A/35% B in 60 min.
4. Set positive ion mode and a capillary voltage of 1800 V for the ionization of peptides eluting from the column.
5. Acquire survey scan at a mass range m/z 400 to m/z 1200 and a resolution of 60,000 in the Orbitrap mass analyzer, followed by fragmentation of the ten most abundant ions in the ion trap part of the instrument.
6. Set the repeat count to one and the dynamic exclusion window to 60 s.

3.5 Data Analysis

The analysis of the raw files is performed with Proteome Discoverer (Thermo Scientific) using the MASCOT search engine (www.matrixscience.com) using databases from www.uniprot.org. Alternatively, other programs like MaxQuant [30] can be used.

1. Set propionamide at cysteine residues as fixed modification.
2. Set as variable modifications: protein N-acetylation, methionine oxidation, isotopic labeling of arginine ($^{13}C_6{}^{15}N_4$) and lysine ($^{13}C_6{}^{15}N_2$) and N-terminal conversion of glutamic acid and glutamine to pyroglutamic acid.
3. Set accepted missed cleavages to two. Here a mass tolerance of 10 ppm for the precursor ion and 0.6 Da for the fragment ions were applied, but these properties should be adjusted to the performance of the mass spectrometer used.
4. As a quantification method, use SILAC 2plex and normalize results to mean protein amount.
5. Process search results with a false discovery rate of 0.01 and only consider proteins with at least two unique peptides and peptide spectral matches with high confidence for quantification.

We usually recover ~80% of intact lysosomes in the lysosomal eluate fraction as determined by β-hexosaminidase assay (Fig. 2a). In the corresponding mass spectrometric analysis, however, we identify about 3000 proteins. Western blots for marker proteins covering lysosome, endoplasmic reticulum, and mitochondria show that the eluate of the magnetic column supposedly contains solely lysosomes and is free of non-lysosomal proteins (*see* Fig. 2b) as the respective marker proteins, when comparing equal volume percentages of each fraction, are only detected in the input but not in the eluate fraction. This is apparently not the case, otherwise we would only detect the lysosomal proteins (*see* Table 1) and nothing else. One must keep in mind, however, that the high sensitivity of modern mass spectrometers can exceed that of a Western blot and that the samples used for mass spectrometric analysis are concentrated using spin filters. Therefore, it is not unusual that the majority of proteins identified in a mass spectrometric dataset generated from such samples are contaminating non-lysosomal proteins from all cellular compartments.

Therefore, in order to evaluate the suitability of a lysosomal enrichment procedure for mass spectrometric analysis, one should not only consider the number of non-lysosomal proteins but rather how many lysosomal proteins can be detected in the fraction of enriched lysosomes. To assess this for the method described here, we prepared a list containing proteins which are currently verified to be of lysosomal localization (*see* Table 1). Some of these proteins may be specific to a cell type or a cellular condition. In a routine experiment we detect 136 of 186 verified lysosomal or lysosome-associated proteins corresponding to a coverage rate of ~73%. The high number of non-lysosomal proteins in our routine data sets makes it difficult to determine whether proteins which have so far

Table 1
Overview of all proteins with experimentally verified exclusive or partial lysosomal localization

Uniprot ID	Gene	Identified	Uniprot ID	Gene	Identified
Q9BZC7	ABCA2		Q9UJQ1	LAMP5	
Q9NP58	ABCB6	×	Q6IAA8	LAMTOR1	×
Q9NP78	ABCB9		Q9Y2Q5	LAMTOR2	×
O14678	ABCD4	×	Q9UHA4	LAMTOR3	×
P11117	ACP2	×	Q0VGL1	LAMTOR4	×
P13686	ACP5		O43504	LAMTOR5	×
P20933	AGA	×	Q15012	LAPTM4A	×
Q96B36	AKT1S1		Q86VI4	LAPTM4B	
P28039	AOAH		Q13571	LAPTM5	
P02743	APCS		P17931	LGALS3	
Q96BM9	ARL8A	×	O00214	LGALS8	
Q9NVJ2	ARL8B	×	Q99538	LGMN	×
P15289	ARSA	×	Q99732	LITAF	
P15848	ARSB	×	P38571	LIPA	×
Q96EG1	ARSG		Q9NUN5	LMBRD1	
Q6UWY0	ARSK	×	Q969J3	LOH12CR1	×
Q13510	ASAH1	×	O00754	MAN2B1	×
Q15904	ATP6AP1	×	Q9Y2E5	MAN2B2	×
Q93050	ATP6V0A1	×	O00462	MANBA	×
P27449	ATP6V0C	×	Q5T0T0	MARCH8	
P61421	ATP6V0D1	×	Q9GZU1	MCOLN1	×
O15342	ATP6V0E1		Q96EZ8	MCRS1	
P38606	ATP6V1A	×	Q96FH0	MEF2BNB	×
P21281	ATP6V1B2	×	Q9H3U5	MFSD1	×
P21283	ATP6V1C1	×	Q8NHS3	MFSD8	×
Q9Y5K8	ATP6V1D	×	Q9BVC4	MLST8	×
P36543	ATP6V1E1	×	P05164	MPO	
Q16864	ATP6V1F	×	P42345	MTOR	×
O75348	ATP6V1G1	×	Q02083	NAAA	×
Q9UI12	ATP6V1H	×	P17050	NAGA	×
Q6UW56	ATRAID	×	P54802	NAGLU	×
Q07812	BAX	×	Q92542	NCSTN	×

(continued)

Table 1 (continued)

Uniprot ID	Gene	Identified	Uniprot ID	Gene	Identified
O43521	BCL2L11		Q99519	NEU1	×
P78537	BLOC1S1	×	Q8WWR8	NEU4	
Q6QNY1	BLOC1S2	×	O15118	NPC1	×
Q96B45	C10ORF32	×	P61916	NPC2	×
Q96GS4	C17orf59	×	Q86WC4	OSTM1	×
P86791	CCZ1		Q86VP3	PACS2	
P08962	CD63	×	Q9UHG3	PCYOX1	×
P34810	CD68		Q9Y646	PGCP	
P51798	CLCN7	×	Q8NCC3	PLA2G15	×
Q13286	CLN3	×	Q8NHP8	PLBD2	×
O75503	CLN5	×	Q8IV08	PLD3	×
Q9H3G5	CPVL	×	P50897	PPT1	×
O75629	CREG1	×	Q9UMR5	PPT2	×
Q01459	CTBS	×	Q6ZP29	PQLC2	
O60931	CTNS		P42785	PRCP	×
P10619	CTSA	×	P07602	PSAP	×
P07858	CTSB	×	P53801	PTTG1IP	×
P53634	CTSC	×	P00797	REN1	
P07339	CTSD	×	Q15382	RHEB	×
Q9UBX1	CTSF	×	O00584	RNASET2	×
P09668	CTSH	×	Q8N122	RPTOR	×
P43235	CTSK		Q7L523	RRAGA	×
P07711	CTSL1	×	Q5VZM2	RRAGB	×
P43234	CTSO	×	Q9HB90	RRAGC	×
P25774	CTSS		Q9NQL2	RRAGD	×
Q9UBR2	CTSZ	×	Q14108	SCARB2	×
Q8TB45	DEPTOR		Q9HB40	SCPEP1	×
Q96SL1	DIRC2	×	Q96SA4	SERINC2	
O00115	DNASE2	×	P51688	SGSH	×
Q9UHL4	DPP7	×	Q9HAT2	SIAE	×
Q9Y227	ENTPD4		Q8NBJ9	SIDT2	×
Q9UM22	EPDR1	×	Q9NRA2	SLC17A5	×

(continued)

Table 1
(continued)

Uniprot ID	Gene	Identified	Uniprot ID	Gene	Identified
P04066	FUCA1	×	Q86WA9	SLC26A11	×
Q9BQS8	FYCO1	×	Q8N357	SLC35F6	×
P10253	GAA	×	Q7Z2H8	SLC36A1	×
P54803	GALC		Q8TED4	SLC37A2	
P34059	GALNS	×	Q8NCC5	SLC37A3	×
P04062	GBA	×	Q9NVC3	SLC38A7	×
Q92820	GGH	×	Q8NBW4	SLC38A9	×
P06280	GLA	×	Q7Z3Q1	SLC46A3	
P16278	GLB1	×	P17405	SMPD1	×
Q8WWB7	GLMP	×	O95295	SNAPIN	×
P17900	GM2A	×	Q8TCT8	SPPL2A	×
P15586	GNS	×	Q9BZW5	TM6SF1	
P08236	GUSB	×	O60478	TM7SF1	
P12544	GZMA		Q8NE00	TMEM104	×
P10144	GZMB		O75204	TMEM127	
P06865	HEXA	×	Q9BSA9	TMEM175	×
P07686	HEXB	×	Q8IY95	TMEM192	×
Q68CP4	HGSNAT	×	Q86T03	TMEM55B	×
Q9Y251	HPSE	×	Q96NL1	TMEM74	
Q12794	HYAL1		Q9HCN3	TMEM8A	×
Q12891	HYAL2		Q9P0T7	TMEM9	×
P22304	IDS		Q9ULQ1	TPCN1	
P35475	IDUA	×	Q8NHX9	TPCN2	×
P13284	IFI30	×	O14773	TPP1	×
Q96RQ9	IL4L1		P49815	TSC2	
Q9BQD3	KXD1	×	O60635	TSPAN1	
P11279	LAMP1	×	Q9BSA4	TTYH2	
P13473	LAMP2	×	Q9C0H2	TTYH3	×
Q9UQV4	LAMP3		Q3ZAQ7	VMA21	×

Proteins that were identified in a representative HEK293 cell lysosomal fractions dataset are indicated. For references regarding this table, please contact corresponding author not been proven to be lysosomal represent new lysosomal proteins or associated proteins with functional importance for lysosomal function. If one desires to identify new bona fide lysosomal proteins from the isolated fraction, a further comparison with a non-enriched fraction and appropriate data analysis as well as follow-up experiments are required.

4 Notes

1. If other salts of Arg/Lys are used, the concentrations have to be adapted accordingly (for recipes see www.silac.org).
2. Trypsin solution can be divided in aliquots before mixing with NH_4HCO_3 and stored at −20 °C. To avoid self-hydrolysis at room temperature, trypsin can be dissolved in 50 mM acetic acid instead of water.
3. Culture conditions, plate coating, and optimal cell seeding numbers need to be optimized for each cell type. Be careful not to seed too low cell numbers, cells should be at least 50% confluent during the incubation with magnetite-containing medium.
4. The usual cultivation time for HEK293 cells is 24 h in medium containing magnetite plus 24–36 h of chase time in regular culture medium. However, for some cell types like mouse embryonic fibroblasts, it is advised to culture them for an additional 48–72 h after trypsinization before addition of the magnetic medium. This will increase lysosomal yield.
5. Depending on your cell type used, it may be necessary to optimize the cell homogenization procedure. Using another type of homogenizer or a syringe may help to attain maximal homogenization efficiency.
6. Tetramethylrhodamine or fluorescein isothiocyanate-labeled magnetite solutions are available from Liquids Research to verify uptake of magnetite beads by your cell line of interest by fluorescence microscopy.
7. PLL-solution can be stored at −20 °C and re-used up to ten times.
8. Magnetite solution is not sterile. Sterile filter your medium before use.
9. When cells detach during washing with PBS (after incubation with magnetite-containing medium), isolation efficiency will be reduced.
10. In case you want to transfect your cells prior to isolation of lysosomes, do so after changing from magnetite to standard medium.

11. We usually mix light and heavy labeled cell suspensions before beginning the homogenization and isolation procedure. This may not be advisable in cases where the differentially labeled cell populations are different in amount or homogenization properties. This could be the case when using a control and a knock-out cell line, for example, that are different in growth properties. In this case it is advisable to either first homogenize both samples or to isolate the lysosomes separately and then combine equal amounts (as determined from, e.g., a protein assay) of each postnuclear supernatant or lysosomal fraction, respectively, before starting the mass spectrometry sample preparation.
12. Depending on how you elute the column (depending on individual handling, e.g., how hard you press the plunger) and how many lysosomes are bound you may need 2–4, 500 μl eluate fractions to completely retrieve all lysosomes. The last elution step should contain 0.5% Triton X-100 to remove all residual bound lysosomes. Efficiency of isolation should be monitored by determination of β-hexosaminidase activity in all fractions. The capacity of the magnetic columns is not reached when material from 2 confluent 10 cm plates is applied. Usually, at least ten 10 cm plates can be used per column when a higher amount of protein is needed for downstream applications. The number of elution steps should be increased when using more than 2 plates.
13. Contamination of a cell line with mycoplasma can hamper the endocytic capacity of cells and will therefore lead to insufficient lysosome enrichment efficacy. Therefore, you should routinely check your cells for mycoplasma contamination.
14. The denaturation of samples can alternatively be performed at a lower temperature, e.g. 50 °C, when lysosomal membrane proteins are of special interest, as some lysosomal membrane proteins have been shown to smear in SDS-PAGE when denatured at 95 °C.
15. Reduction and alkylation of proteins can also be performed after SDS-PAGE as a part of the digestion protocol.
16. When cutting the gel pieces, use a fresh scalpel. Although this can be reused for several samples on the same day, do not use one scalpel more frequently. A white sheet of paper or a light box can be used as underlay to see the bands more clearly. Gel pieces can also be stored in water at 4 °C to continue the protocol at a later time point.
17. The gel pieces should be floating in liquid. If this is not the case either increase the liquid volume or the rpm.
18. To remove liquid from the gel pieces, stacking a 10 μl pipet tip on top of a 1 ml pipet tip allows to aspirate large volumes while avoiding unintended suction of small gel pieces into the tip.

19. It is best to use a closed incubator to avoid precipitation of liquid on the inner lid of the reaction tube. If this is not possible, add more liquid to avoid drying of the gel pieces.
20. To extract peptides from gel pieces, there are several combinations of buffers available. Alternatives to the method described here can be found in Winter and Steen [31].
21. After tryptic digestion, gel pieces can be frozen for storage at −20 °C before peptide extraction is performed.
22. Keep the flowthrough of StageTips until your samples have been successfully measured in case peptides have not bound to the column material.

References

1. Settembre C, Fraldi A, Medina DL, Ballabio A (2013) Signals from the lysosome: a control centre for cellular clearance and energy metabolism. Nat Rev Mol Cell Biol 14(5):283–296. doi:10.1038/nrm3565
2. Zoncu R, Efeyan A, Sabatini DM (2011) mTOR: from growth signal integration to cancer, diabetes and ageing. Nat Rev Mol Cell Biol 12(1):21–35. doi:10.1038/nrm3025
3. Pu J, Schindler C, Jia R, Jarnik M, Backlund P, Bonifacino JS (2015) BORC, a multisubunit complex that regulates lysosome positioning. Dev Cell 33(2):176–188. doi:10.1016/j.devcel.2015.02.011
4. Lübke T, Lobel P, Sleat DE (2009) Proteomics of the lysosome. Biochim Biophys Acta 1793(4):625–635. doi:10.1016/j.bbamcr.2008.09.018
5. Schröder B, Wrocklage C, Pan C, Jager R, Kosters B, Schafer H, Elsasser HP, Mann M, Hasilik A (2007) Integral and associated lysosomal membrane proteins. Traffic 8(12):1676–1686
6. Schröder BA, Wrocklage C, Hasilik A, Saftig P (2010) The proteome of lysosomes. Proteomics 10(22):4053–4076. doi:10.1002/pmic.201000196
7. Czupalla C, Mansukoski H, Riedl T, Thiel D, Krause E, Hoflack B (2006) Proteomic analysis of lysosomal acid hydrolases secreted by osteoclasts: implications for lytic enzyme transport and bone metabolism. Mol Cell Proteomics 5(1):134–143. doi:10.1074/mcp.M500291-MCP200
8. Sleat DE, Della Valle MC, Zheng H, Moore DF, Lobel P (2008) The mannose 6-phosphate glycoprotein proteome. J Proteome Res 7(7):3010–3021. doi:10.1021/pr800135v
9. Sleat DE, Lobel P (1997) Ligand binding specificities of the two mannose 6-phosphate receptors. J Biol Chem 272(2):731–738
10. Qian M, Sleat DE, Zheng H, Moore D, Lobel P (2008) Proteomics analysis of serum from mutant mice reveals lysosomal proteins selectively transported by each of the two mannose 6-phosphate receptors. Mol Cell Proteomics 7(1):58–70. doi:10.1074/mcp.M700217-MCP200
11. Sleat DE, Wang Y, Sohar I, Lackland H, Li Y, Li H, Zheng H, Lobel P (2006) Identification and validation of mannose 6-phosphate glycoproteins in human plasma reveal a wide range of lysosomal and non-lysosomal proteins. Mol Cell Proteomics 5(10):1942–1956. doi:10.1074/mcp.M600030-MCP200
12. Sleat DE, Zheng H, Lobel P (2007) The human urine mannose 6-phosphate glycoproteome. Biochim Biophys Acta 1774(3):368–372. doi:10.1016/j.bbapap.2006.12.004
13. Naureckiene S, Sleat DE, Lackland H, Fensom A, Vanier MT, Wattiaux R, Jadot M, Lobel P (2000) Identification of HE1 as the second gene of Niemann-Pick C disease. Science 290(5500):2298–2301. doi:10.1126/science.290.5500.2298
14. Chapel A, Kieffer-Jaquinod S, Sagne C, Verdon Q, Ivaldi C, Mellal M, Thirion J, Jadot M, Bruley C, Garin J, Gasnier B, Journet A (2013) An extended proteome map of the lysosomal membrane reveals novel potential transporters. Mol Cell Proteomics 12(6):1572–1588. doi:10.1074/mcp.M112.021980
15. De Duve C, Pressman BC, Gianetto R, Wattiaux R, Appelmans F (1955) Tissue fractionation studies. 6. Intracellular distribution patterns of enzymes in rat-liver tissue. Biochem J 60(4):604–617
16. Wattiaux R, Wibo M, Baudhuin P (1963) Effect of the injection of triton WR 1339 on the hepatic lysosomes of the rat. Arch Int Physiol Biochim 71:140–142

17. Zoncu R, Bar-Peled L, Efeyan A, Wang S, Sancak Y, Sabatini DM (2011) mTORC1 senses lysosomal amino acids through an inside-out mechanism that requires the vacuolar H(+)-ATPase. Science 334(6056):678–683. doi:10.1126/science.1207056
18. Walker MW, Lloyd-Evans E (2015) A rapid method for the preparation of ultrapure, functional lysosomes using functionalized superparamagnetic iron oxide nanoparticles. Methods Cell Biol 126:21–43. doi:10.1016/bs.mcb.2014.10.019
19. Honscher C, Mari M, Auffarth K, Bohnert M, Griffith J, Geerts W, van der Laan M, Cabrera M, Reggiori F, Ungermann C (2014) Cellular metabolism regulates contact sites between vacuoles and mitochondria. Dev Cell 30(1):86–94. doi:10.1016/j.devcel.2014.06.006
20. Hamasaki M, Furuta N, Matsuda A, Nezu A, Yamamoto A, Fujita N, Oomori H, Noda T, Haraguchi T, Hiraoka Y, Amano A, Yoshimori T (2013) Autophagosomes form at ER-mitochondria contact sites. Nature 495(7441):389–393. doi:10.1038/nature11910
21. Eden ER, White IJ, Tsapara A, Futter CE (2010) Membrane contacts between endosomes and ER provide sites for PTP1B-epidermal growth factor receptor interaction. Nat Cell Biol 12(3):267–272. doi:10.1038/ncb2026
22. Collot M, Louvard D, Singer SJ (1984) Lysosomes are associated with microtubules and not with intermediate filaments in cultured fibroblasts. Proc Natl Acad Sci U S A 81(3):788–792
23. Wei H, Liu L, Chen Q (2015) Selective removal of mitochondria via mitophagy: distinct pathways for different mitochondrial stresses. Biochim Biophys Acta 1853(10 Pt B):2784–2790. doi:10.1016/j.bbamcr.2015.03.013
24. Kraft C, Deplazes A, Sohrmann M, Peter M (2008) Mature ribosomes are selectively degraded upon starvation by an autophagy pathway requiring the Ubp3p/Bre5p ubiquitin protease. Nat Cell Biol 10(5):602–610. doi:10.1038/ncb1723
25. Khaminets A, Heinrich T, Mari M, Grumati P, Huebner AK, Akutsu M, Liebmann L, Stolz A, Nietzsche S, Koch N, Mauthe M, Katona I, Qualmann B, Weis J, Reggiori F, Kurth I, Hubner CA, Dikic I (2015) Regulation of endoplasmic reticulum turnover by selective autophagy. Nature 522(7556):354–358. doi:10.1038/nature14498
26. Ong SE, Blagoev B, Kratchmarova I, Kristensen DB, Steen H, Pandey A, Mann M (2002) Stable isotope labeling by amino acids in cell culture, SILAC, as a simple and accurate approach to expression proteomics. Mol Cell Proteomics 1(5):376–386
27. Markmann S, Thelen M, Cornils K, Schweizer M, Brocke-Ahmadinejad N, Willnow T, Heeren J, Gieselmann V, Braulke T, Kollmann K (2015) Lrp1/LDL receptor play critical roles in mannose 6-phosphate-independent lysosomal enzyme targeting. Traffic. doi:10.1111/tra.12284
28. Laemmli UK (1970) Cleavage of structural proteins during the assembly of the head of bacteriophage T4. Nature 227(5259):680–685
29. Rappsilber J, Mann M, Ishihama Y (2007) Protocol for micro-purification, enrichment, pre-fractionation and storage of peptides for proteomics using StageTips. Nat Protoc 2(8):1896–1906. doi:10.1038/nprot.2007.261
30. Cox J, Mann M (2008) MaxQuant enables high peptide identification rates, individualized p.p.b.-range mass accuracies and proteome-wide protein quantification. Nat Biotechnol 26(12):1367–1372. doi:10.1038/nbt.1511
31. Winter D, Steen H (2011) Optimization of cell lysis and protein digestion protocols for the analysis of HeLa S3 cells by LC-MS/MS. Proteomics 11(24):4726–4730. doi:10.1002/pmic.201100162

Chapter 2

Quantitative Profiling of Lysosomal Lipidome by Shotgun Lipidomics

Mesut Bilgin, Jesper Nylandsted, Marja Jäättelä, and Kenji Maeda

Abstract

Recent studies have illuminated novel roles of lysosomes that go far beyond simple catabolism and function in the coordination of cellular metabolism and signaling. Promising therapeutic strategies emerge from knowledge in the molecular mechanisms and physiological roles of lipid metabolism in lysosomes. Global monitoring of the function and dysregulation of lysosomal lipid metabolism requires a methodology that resolves the complexity of lysosomal lipidome by quantitatively detecting hundreds of lipid species of diverse physicochemical properties. We describe here a detailed protocol that couples isolation of superparamagnetic iron dextran-loaded lysosomes from cultured mammalian cell lines with quantitative mass spectrometry-based shotgun lipidomics.

Key words Lysosomes, Metabolism, Lipids, Shotgun lipidomics, Mass spectrometry, Quantitative profiling

1 Introduction

Lipids play key roles in human physiology. They form lipid bilayers and assemble networks of specific protein–lipid and lipid–lipid interactions that define the architectures and functions of the various membranes of eukaryotic cells. The lipidome—the complete repertoire of cellular lipids—is highly complex. It includes thousands of structurally diverse species [1] classified into eight major categories: fatty acyls, glycerolipids, glycerophospholipids, sphingolipids, sterol lipids, prenol lipids, saccharolipids, and polyketides, according to LIPID MAPS [2]. Approximately 5% of protein-encoding genes are involved in lipid metabolism and transport in eukaryotes [3]. They generate cellular membranes with highly heterologous lipid landscapes, which can be dynamically altered in response to intra- and extracellular cues. Such flexibility allows spatial and kinetic control of various cellular processes, including transport and signaling [1].

Karin Öllinger and Hanna Appelqvist (eds.), *Lysosomes: Methods and Protocols*, Methods in Molecular Biology, vol. 1594,
DOI 10.1007/978-1-4939-6934-0_2, © Springer Science+Business Media LLC 2017

Lipid catabolism takes place mainly in the lysosomes and in the closely related late endosomes. The limiting membrane of lysosomes forms the boundary between the cytosol and the lumen that encloses ~60 soluble hydrolases, including many lipases. Lysosomal lipases such as phospholipase, acid sphingomyelinase, acid ceramidase, hexosaminidase, galactosidase, and arylsulfatase primarily hydrolyze lipids delivered to vesicular membranes within the lysosomal lumen by endocytosis or autophagy [4]. Their main function is to assist in the recycling of plasma membrane and intracellular organelles by hydrolyzing lipids to building blocks for de novo lipid biosynthesis. Furthermore, lysosomal lipid catabolism can provide the cell with an additional energy source. For instance, the lysosomal acid lipase hydrolyzes circulating and stored nutritional lipids such as cholesteryl esters and triacylglycerols delivered to the endolysosomal system [5] to release free fatty acids for the β-oxidation. Activities of lysosomal lipases are tightly coupled to lipid transport carried out at least in part by lysosomal lipid-transfer proteins such as saposins A-D, Niemann-Pick C1 protein, Niemann-Pick C2 protein, and ganglioside GM2 activator protein [6]. They facilitate enzyme–substrate interactions and promote the efflux of degraded lipids from the lysosomes. Dysregulations in the lysosomal lipid catabolism due to mutations in genes of enzymes and transporters are causal for numerous inherited diseases, commonly termed lysosomal storage disorders. Lipids accumulate in the lysosomes of these patients, causing imbalance in the cellular lipid profiles and impairing lysosomal activity in general [4].

The exact molecular mechanisms of lipid catabolism in the lysosomes and its physiological impacts in health and disease remain largely elusive. In addition to providing building blocks and energy, lysosomes produce signaling lipids such as fatty acids and lysophospholipids. It has also recently become evident that lysosomes act as signaling hubs that coordinate metabolism and signaling by dynamically and physically interacting with regulatory proteins such as the kinase complex mammalian target of rapamycin complex 1 (mTORC1) [7] and transcription factor EB (TFEB) [8]. Accordingly, lysosomes draw increasing attention in the field of biomedical research. For example, lipid catabolism in the lysosomes of cancer cells is often dysregulated to render their limiting membrane more susceptible to permeabilization upon an increase in the sphingomyelin content [9]. This property of cancer cell lysosomes has provided a promising therapeutic strategy, employing acid sphingomyelinase-inhibiting cationic amphiphilic drugs to trigger lysosomal membrane permeabilization and cell death specifically in cancer cells [9].

The central role of lysosomes in the metabolism of lipids and the emerging linkages between lysosomal lipid metabolism and signaling in health and disease illuminate the needs to quantitatively monitor this very complex group of metabolites in the lysosomes. Here, we provide a detailed protocol for the magnetic isolation of superparamagnetic colloidal iron dextran particles (FeDEX)-loaded lysosomes from cultured mammalian cell lines [10] coupled with shotgun lipidomics that allows for quantitatively monitoring hundreds of lipid species. The magnetic isolation procedure is fast and easy to adapt to any laboratory. Cells take up the FeDEX added to common cell culture medium via endocytic pathways eventually leading to the FeDEX accumulation in the lysosomal lumen. The FeDEX-loaded lysosomes can then be specifically captured on magnetic columns after gentle lysis of the plasma membrane. The subsequent mass spectrometry-based lipid analysis resolves the complexity of lysosomal lipidome. The shotgun lipidomics methodology applied in this protocol is a powerful and robust approach that performs full range global lipid identification and quantification directly from crude lipid extracts of any biological materials [11, 12]. The benchmark of shotgun lipidomics is that there is no upstream chromatographic separation of lipid samples. The lipids are directly ionized from crude lipid extracts by nanoelectrospray (or electrospray) ionization, and discriminated by their unique mass-to-charge ratios (m/z) on a mass spectrometer [13]. The full MS survey scanning detects the intact precursor ions, while the subsequent sequential MS/MS scanning with an ion isolation width of typically 1.0 Da and collision induced dissociation provides structure-specific fragments for lipid identification and quantification. A two-step lipid extraction procedure is applied in this protocol, where nonpolar and polar lipids are primarily recovered in the first and second step of the extraction, respectively (Table 1) [14, 15]. This procedure extracts a broad range of lipids, and sorts them according to their physicochemical properties to support resolving the high number of lipid species in mass spectrometer. Lipids in the generated extracts are infused in both positive and negative ionization modes to the mass spectrometer, because different lipids have different preferences for the conditions of ionization (Table 2). The possibility of full automatic sampling in shotgun lipidomics allows for the handling of large numbers of samples [14, 16, 17]. Overall, the present protocol couples the fast and convenient magnetic isolation of lysosomes with a robust, automated, and sensitive high-throughput platform of shotgun lipidomics [18].

Table 1
Lipids extracted in the two-step liquid-liquid extraction procedure

First step: chloroform:methanol 10:1 (*v/v*)-phase	Second step: chloroform:methanol 2:1 (*v/v*)-phase
Chol, CE, LCB, Cer, HexCer, SM, DAG, LSM, TAG, LPG, LPG O-, LPE, LPE O-, LPC, LPC O-, PG, PG O-, PE, PE O-, PC, PC O-	LCBP, CerP, diHexCer, triHexCer, SHexCer, LPA, LPA O-, LPI, LPI O-, LPS, LPS O-, PA, PA O-, PI, PI O-, PS, PS O-

A list of lipids extracted in the first and second steps of the two-step liquid–liquid extraction procedure described here. O- indicates glycerophospholipids with one alkyl group attached to the glycerol moiety

Table 2
Nanoelectrospray ionization of lipids in the positive and negative modes

Positive ionization	Negative ionization
Chol, CE, LCB, Cer, HexCer, diHexCer, triHexCer, SM, LSM, DAG, TAG, LPE, LPE O-, LPC, LPC O-, PE, PE O-, PC, PC O-	LCBP, CerP, SHexCer, LPA, LPA O-, LPG, LPG O-, LPI, LPI O-, LPS, LPS O-, LPE, LPE O-, LPC, LPC O-, PA, PA O-, PG, PG O-, PI, PI O-, PS, PS O-, PE, PE O-, PC, PC O-

A list of lipids detected in the positive and negative ionization modes of the mass spectrometry analysis procedure described here. O- indicates glycerophospholipids with one alkyl group attached to the glycerol moiety

2 Materials

2.1 Preparation of FeDEX

1. 1.2 M $FeCl_2$: Dissolve 4.56 g $FeCl_2$ in water up to 30 mL.
2. 1.8 M $FeCl_3$: Dissolve 8.76 g $FeCl_3$ in water up to 30 mL.
3. 5% (*v*/*v*) NH_4OH: Pour 100 mL of 30% (*v*/*v*) NH_4OH into a graduated cylinder and add water up to 600 mL.
4. 0.3 M HCl: Pour 300 mL of 1 M HCl into a graduated cylinder and add water up to 1 L.
5. Dextran 40.
6. Dialysis sacks 35 mm width, MWCO 12,000 Da.
7. Bottle top filter 0.22 μm.

2.2 Cell Culture and Loading of FeDEX

1. Human ductal breast carcinoma cell line MCF-7, or other cell lines of interest.
2. Dulbecco's Modified Eagle Medium (DMEM) or another medium appropriate for the cell lines.
3. Cell culture medium: DMEM with 6% fetal calf serum (FCS). Add 32.5 mL of FCS to 500 mL of the DMEM to the final concentration of 6% (*v*/*v*), and 5 mL of Pen Strep (100X solution of penicillin and streptomycin).
4. Cell culture dishes with lid, 15 cm diameter and 145 cm^2 culture area.

2.3 Magnetic Isolation of Lysosomes

1. 1× DPBS (Dulbecco's phosphate buffered saline).
2. TrypLE, cell-dissociation enzymes.
3. 1 M HEPES-KOH pH 7.5 stock: Dissolve 238.30 g HEPES in 800 mL of water. Adjust the pH to 7.5 with KOH, and add water up to 1 L.
4. 1 M KCl stock: Dissolve 74.6 g of KCl in water up to 1 L.
5. 1 M $MgCl_2$ stock: Dissolve 50.8 g of $MgCl_2 \cdot H_2O$ in water up to 250 mL.

6. 0.5 M EDTA stock: Add 800 mL of water to 186.1 g of EDTA disodium salt, and dissolve it as adjusting the pH to 7 with NaOH pellets. Then add water up to 1 L.
7. 0.5 M EGTA stock: Add 40 mL of water to 9.51 g of EGTA, and dissolve it as adjusting pH to 7 with NaOH pellets. Then add water up to 50 mL.
8. 1 M sucrose: Dissolve 171.15 g of sucrose in water up to 1 L.
9. 1 M dithiothreitol (DTT) stock: Dissolve 5.0 g of DTT in water up to 32.4 mL. Store at −20 °C.
10. 100 mM Pefabloc: Dissolve 100 mg of Pefabloc in 4.18 mL of water. Store at −20 °C.
11. SCA buffer stock: 20 mM HEPES-KOH pH 7.5, 10 mM KCl, 1.5 mM $MgCl_2$, 1 mM EDTA, 1 mM EGTA, 250 mM sucrose. Mix 10 mL of 1 M HEPES-KOH pH 7.5, 2.5 mL of 1 M KCl, 375 μL of 1 M $MgCl_2$, 500 μL of 0.5 M EDTA, 500 μL of 0.5 M EGTA, and 62.5 mL of 1 M sucrose, and add water up to 250 mL. Autoclave and store at 4 °C.
12. SCA buffer: Add 20 μL of 1 M DTT, 200 μL of 100 mM Pefabloc, and 2 tablet (or the amount recommended by the manufacturer) of Complete mini EDTA-free protease inhibitor cocktail to 20 mL of the SCA buffer stock just before use.
13. LS columns (Miltenyi Biotec, Bergisch Gladbach, Germany).
14. QuadroMACS magnetic separator (Miltenyi Biotec, Bergisch Gladbach, Germany).
15. MACS multistand (Miltenyi Biotec, Bergisch Gladbach, Germany).

2.4 Lipid extraction

Prepare all solvent mixtures freshly using HPLC grade solvents and store at 4 °C. Keep plasticware and glassware at 4 °C and always use bottles and vials with lids having inner material of Teflon/PTFE when handling organic solvents.

1. Chloroform:methanol 1:1(*v/v*): Mix 10 mL of chloroform and 10 mL of methanol in a 20 mL graduated cylinder (*see* **Note 1**).
2. 10× Internal Standard Mix (10× ISM): Weigh 0.371 mg cholesteryl ester (CE 15:0-D7), 0.193 mg ceramide (Cer 18:1;2/12:0;0), 0.232 mg ceramide phosphate (CerP 18:1;2/12:0;0), 1.563 mg cholesterol (Chol-D4), 0.091 mg diacylglycerol (DAG 12:0/12:0), 0.322 mg dihexose ceramide (diHexCer18:1;2/12:0;0), 0.322 mg hexose ceramide (HexCer 18:1;2/12:0;0), 0.115 mg long-chain base (LCB 17:0;2), 0.147 mg long-chain base phosphate (LCBP 17:0;2), 0.223 mg lysophosphatidic acid (LPA 17:0), 0.204 mg lysophosphatidylcholine (LPC 17:1), 0.206 mg lysophosphatidylethanolamine (LPE 13:0), 0.156 mg lysophosphatidylglycerol

(LPG 17:1), 0.219 mg lysophosphatidylinositol (LPI 13:0), 0.213 mg lysophosphatidylserine (LPS 17:1), 0.180 mg lysosphingomyelin (LSM 17:1;2), 0.279 mg phosphatidic acid (PA 12:0/12:0), 0.303 mg phosphatidylcholine (PC-OO 18:1/18:1), 0.290 mg phosphatidylehanolamine (PE 12:0/12:0), 0.190 mg phosphatidylglycerol (PG 12:0/12:0), 0.181 mg phosphatidylinositol (PI 8:0/8:0), 0.258 mg phosphatidylserine (PS 12:0/12:0), 0.296 mg sulfatide (SHexCer 18:1;2/12:0;0), 0.259 mg sphingomyelin (SM 18:1;2/12:0;0), 0.170 mg triacylglycerol (TAG 17:0/17:0/17:0), and 0.415 mg trihexose ceramide (triHexCer 18:1;2/17:0;0), and transfer to a 20 mL screw cap glass vial. Add 20 mL of chloroform:methanol 1:1(*v/v*) and mix gently till the lipids are completely dissolved. Store at −80 °C (*see* **Note 2**).

3. 1× ISM: Mix 100 μL of 10X ISM and 900 μL of chloroform:methanol 1:1(*v/v*). Store at −20 °C (*see* **Note 3**).
4. Chloroform:methanol 10:1(*v/v*): Mix 90 mL of chloroform and 9 mL of methanol in a 100 mL graduated cylinder (*see* **Note 1**).
5. Chloroform:methanol 2:1(*v/v*): Mix 60 mL of chloroform and 30 mL of methanol in a 100 mL graduated cylinder (*see* **Note 1**).
6. 155 mM ammonium acetate: Pour 25 mL of water into a 100 mL graduated cylinder. Weigh 0.598 g ammonium acetate and transfer to the graduated cylinder. Add 25 mL of water and mix gently till the ammonium acetate is completely dissolved (*see* **Note 1**).
7. Chloroform:methanol 1:2(*v/v*). Mix 30 mL of chloroform and 60 mL of methanol in a 100 mL graduated cylinder (*see* **Note 1**).

2.5 Shotgun Lipidomics

1. Positive ion mode infusion solvent: 13.3 mM ammonium acetate in isopropanol. Add 25 mL of isopropanol to a 100 mL Blue Cap bottle. Weigh 0.102 g ammonium acetate, transfer it to the Blue Cap bottle, and add 75 mL of isopropanol. Close the cap and heat it to 50 °C till ammonium acetate is completely dissolved (*see* **Notes 1** and **4**).
2. Negative ion mode infusion solvent: 0.2% triethyl amine. Mix 15 mL of chloroform, 75 mL of methanol, and 180 μL of triethyl amine in a 100 mL graduated cylinder (*see* **Note 1**).
3. TriVersa NanoMate, a robotic nanoelectrospray ionization source (Advion Ithaca, NY, USA).
4. A Hybrid Quadrupole-Orbitrap mass spectrometer (Thermo Fisher Scientific, Waltham, MA, USA).

3 Methods

3.1 Preparation of FeDEX

Carry out all procedures at room temperature unless otherwise specified.

1. Mix 30 mL of 1.2 M $FeCl_2$ and 30 mL of 1.8 M $FeCl_3$ in a 500 mL beaker. Slowly add 30 mL of 30% (*v/v*) NH_4OH while stirring until all precipitates are dissolved.
2. Pour 300 mL of 5% (*v/v*) NH_4OH and continue stirring.
3. Place the beaker on a magnet stirrer (without stirring) and allow iron particles to settle.
4. Decant the beaker and discard the supernatant. Wash the iron particles twice with 300 mL water.
5. Resuspend the iron particles in 240 mL of 0.3 M HCl. Stir the suspension with a magnetic stirring bar for 30 min.
6. Add 12 g dextran 40 to the suspension of iron particles and stir for 30–60 min.
7. Transfer the resultant FeDEX into dialysis tubes, seal with clamps, and dialyze against 8 L water for 1–2 days at 4 °C. Change water 1–2 times during the dialysis.
8. Transfer the dialyzed FeDEX and remove large aggregates by centrifuging at 15,000 × *g* for 10 min at 4 °C. Transfer the supernatant to a new tube.
9. Filter the FeDEX through a 0.22 μm bottle top filter to sterilize the FeDEX solution. Store the prepared FeDEX at 4 °C. FeDEX is stable for at least 7–14 days. To determine the FeDEX dry weight concentration (mg/mL), vacuum dry 100 μL of the FeDEX solution, and weigh the powder (expected to be ~50 mg/mL).

3.2 Cell Culture and Loading of FeDEX

The cultured cell line is here fed with the prepared FeDEX to load the lysosomes with iron. To ensure specific loading of lysosomes, and not of endocytic vesicles and endosomes, the cell line is cultured in the presence of FeDEX for 6–24 h, and then in the absence of FeDEX for 2 h (Fig. 1a). The present protocol has been optimized for the human ductal breast carcinoma cell line, MCF-7, but should be broadly applicable, with minor adaptation.

1. Culture the cells on a 15 cm cell culture dish to 70–80% confluence. For MCF-7 cells, this corresponds to ~1–1.5 × 10^7 cells.
2. Replace cell culture medium with 14 mL of fresh medium and add 1.0 mL of the FeDEX solution. Gently shake the dish to completely mix the FeDEX solution and the cell culture medium. Incubate the cell culture dish in a CO_2 incubator at 37 °C for 6–24 h to allow cells to uptake the FeDEX (Fig. 1a).

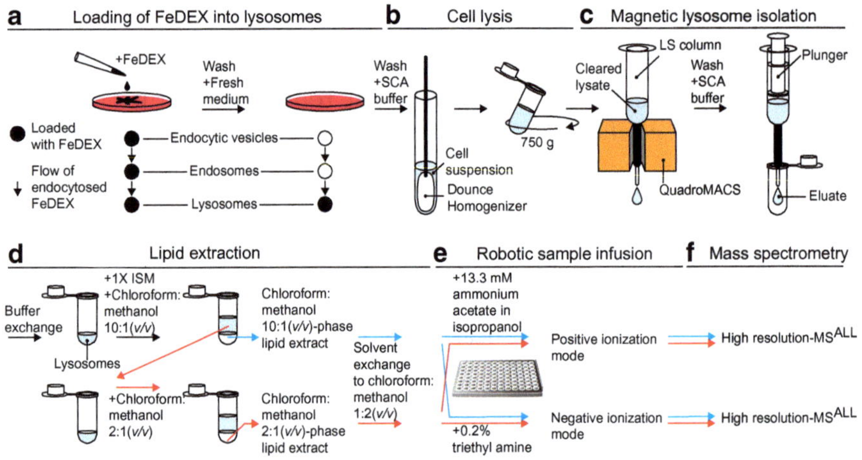

Fig. 1 Overview on the workflow of the present protocol. (**a**) FeDEX is loaded specifically into the lysosomes of cultured mammalian cells. (**b**) Cells are lysed in a buffer that maintains lysosomal membrane integrity. (**c**) The FeDEX-loaded lysosomes are magnetically captured on the LS column. (**d**) Lipids in the isolated lysosomes are extracted in a two-step lipid extraction procedure. (**e–f**) The lysosomal lipids are analyzed with shotgun lipidomics, in which the lipids in the extracts are (**e**) ionized via positive and negative ionization modes of nanoelectrospray, and (**f**) detected, identified, and quantified using mass spectrometer

3. Remove cell culture medium containing the FeDEX solution and immediately wash cells twice with 15 mL of DPBS.
4. Add 14 mL of cell culture medium to the dish and incubate it in a CO_2 incubator at 37 °C for 2 h (Fig. 1a).
5. Wash the cells twice with 15 mL of DPBS. Remove the DPBS.
6. Add 2 mL of TrypLE to the dish and keep it in a CO_2 incubator at 37 °C for 5 min to dissociate the cells from the surface. Add 10 mL of growth medium to the dish and transfer the cell suspension to a tube. Cool the tube on ice for 5 min (*see* **Note 5**).
7. Centrifuge the cell suspension at 60 × *g* for 10 min at 4 °C. Wash the pelleted cells twice with 5 mL of ice-cold DPBS. Remove the DPBS carefully after the last centrifugation.

3.3 Magnetic Isolation of Lysosomes

Carry out all procedures on ice and with ice-cold buffers. Keep the samples on ice or at 4 °C.

1. Resuspend the cell pellet with approximately 1:1 volume of the hypotonic SCA buffer. Incubate the cell suspension on ice for 5–10 min.
2. Lyse the cells with 100–150 strokes of Dounce Homogenizer (Fig. 1b) (*see* **Note 6**).
3. Centrifuge the crude cell lysate at 750 × *g* for 5 min to pellet unlyzed cells, nuclei, and large aggregates. Transfer the supernatant containing lysosomes to a new tube. Repeat the centrifugation once (Fig. 1b) (*see* **Note 7**).

4. Mount a LS column on the QuadroMACS magnet separator positioned on the MACS multistand. Pre-equilibrate the LS column by adding 1 mL of SCA buffer and allow it to flow through with gravity (Fig. 1c).
5. Load the cleared cell lysate from **step 3** on the LS column and allow it to flow through while the FeDEX-loaded lysosomes are specifically retained on the column.
6. Wash the LS column four times with 1 mL SCA buffer.
7. Remove the LS column from the QuadroMACS magnet separator, add 1 mL of the SCA buffer to the LS column, and gently push with the plunger to elute the lysosomes (Fig. 1c) (*see* **Note 8**).

3.4 Lipid Extraction

Carry out the procedure of two-step liquid–liquid lipid extraction at 4 °C and on ice with cold rack systems unless otherwise specified. For a list of which lipids are extracted in the first and second steps of the extraction, *see* Table 1.

1. To exchange the buffer, pellet the purified lysosomes at 20,000 × *g* for 20 min and resuspend the lysosomes in 1000 μL of 155 mM ammonium acetate. Repeat the procedure once (Fig. 1d) and finally resuspend in 200 μL 155 mM ammonium acetate.
2. Mix 1000 μL of chloroform:methanol 10:1(*v/v*) and 15 μL of 1× ISM with 200 μL of purified lysosomes or 200 μL of 155 mM ammonium acetate (to generate reference sample) in 2 mL Eppendorf tubes (Fig. 1d).
3. Shake the Eppendorf tubes at 2000 rpm for 60 min (*see* **Note 9**) to extract the lysosomal lipids into the organic solvents. Centrifuge the Eppendorf tubes at 2,000 × *g* for 3 min to separate the upper and lower phases.
4. Transfer the lower phase to a new 1.5 mL Eppendorf tube to recover the extracted nonpolar lipids (chloroform:methanol 10:1(*v/v*)-phase lipid extract) (Fig. 1d) (*see* **Note 10**). Save the upper phase for the subsequent extraction of polar lipids. Keep the extracted nonpolar lipids at −20 °C until **step 8**.
5. Add 1000 μL of chloroform:methanol 2:1(*v/v*) to the upper phase from **step 4** (Fig. 1d) (*see* **Note 11**).
6. Shake the Eppendorf tubes at 2000 rpm for 60 min (*see* **Note 9**). Centrifuge the Eppendorf tubes at 2000 × *g* for 3 min to separate the phases.
7. Transfer the lower phase to a new 1.5 mL Eppendorf tube to recover the extracted polar lipids (chloroform:methanol 2:1(*v/v*)-phase lipid extract) (Fig. 1d).
8. Dry the lipid extracts from **steps 4** and **7** in a vacuum centrifuge for 60 min at 37 °C (*see* **Note 12**).

9. Add 100 μL of chloroform:methanol 1:2(*v/v*) to the Eppendorf tubes containing the dried lipid extracts, and shake the tubes for 3 min at 2000 rpm to dissolve the lipids. Centrifuge the samples at 15,000 × *g* for 5 min to pellet and remove insoluble materials (*see* **Notes 13** and **14**).

3.5 Shotgun Lipidomics

Here, the extracted lipids are directly subjected to the nanoelectrospray ionization in the positive and negative modes and analyzed in the mass spectrometer. For the list of lipids analyzed in the two modes, *see* Table 2.

1. Prepare the following four mixtures in a 96-well plate (Fig. 1e). The total lipid analysis is performed on a lipidomics platform featuring a Hybrid Quadrupole-Orbitrap mass spectrometer equipped with a robotic nanoelectrospray ion source (*see* **Note 13**).
   (a) Mix 11.7 μL of the chloroform:methanol 10:1(*v/v*)-phase lipid extract and 15 μL of the positive ion mode infusion solvent (*see* **Note 15**).
   (b) Mix 11.7 μL of the chloroform:methanol 2:1(*v/v*)-phase lipid extract and 15 μL of the positive ion mode infusion solvent (*see* **Note 15**).
   (c) Mix 13 μL of the chloroform:methanol 10:1(*v/v*)-phase lipid extract and 13 μL of the negative ion mode infusion solvent (*see* **Note 16**).
   (d) Mix 13 μL of the chloroform:methanol 2:1(*v/v*)-phase lipid extract and 13 μL of the negative ion mode infusion solvent (*see* **Note 16**).
2. Prepare the robotic sample infusion set-up with the TriVersa NanoMate (a robotic nanoelectrospray ionization source) coupled to a Hybrid Quadrupole-Orbitrap mass spectrometer (*see* **Note 17**).
3. Keep the TriVersa NanoMate sample tray at 4 °C and the samples cold (*see* **Note 18**).
4. Perform high resolution-MS^{ALL} (HR-MS and HR-MS/MS) using the following approach (Fig. 1f and 2): run samples in positive or negative ion mode HR-MS using alternating (1) HR-MS analysis of the low *m/z* range, (2) HR-MS analysis of the high *m/z* range and sequential (3) HR-MS/MS analysis (*see* **Note 19**).
5. MS^{ALL} lipid detection: lipids are detected as $[M + H]^+$ and $[M + NH_4]^+$ ions in the positive ion mode and $[M–H]^-$ and $[M + CH_3OCOO]^-$ ions in the negative ion mode (for lists of precursor ions, product ions, and neutral loss detections in the two ion modes, *see* Tables 3 and 4) (*see* **Note 20**).

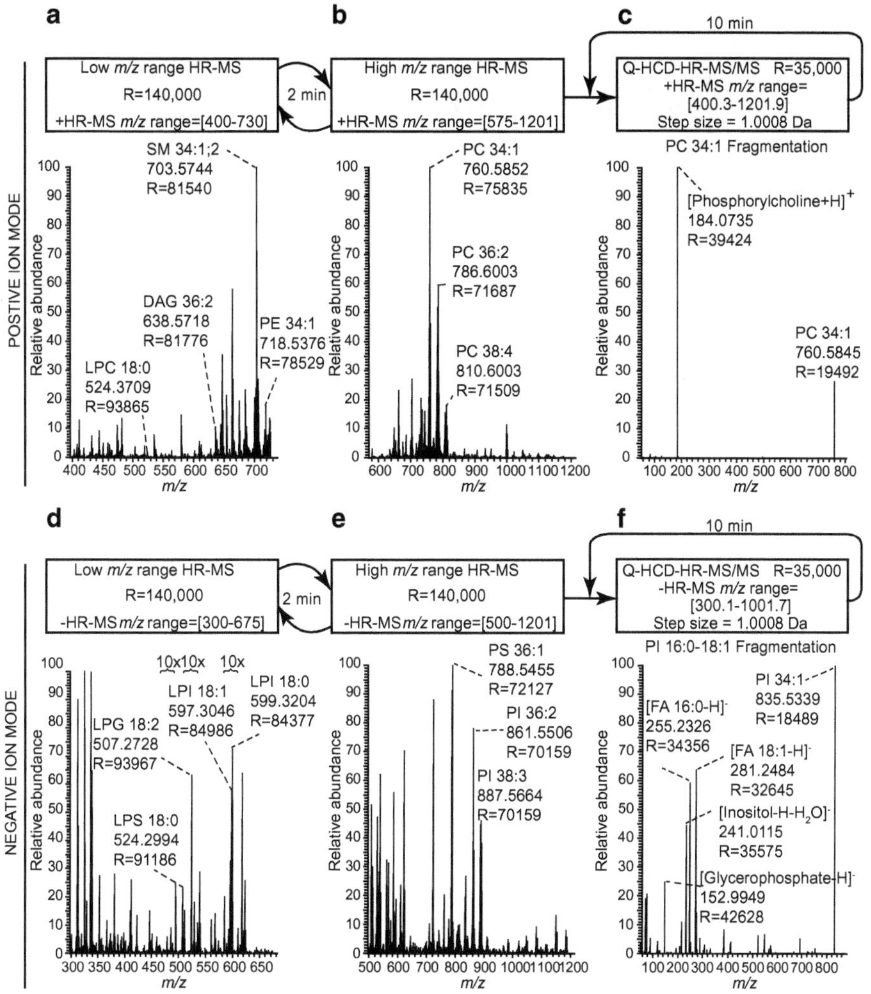

Fig. 2 Overview on the HR-MS^{ALL} methodology and examples of HR-MS and HR-MS/MS acquired data. The HR-MS^{ALL} is composed of three elements for both the positive (**a–c**) and the negative ion mode (**d–f**): (1) HR-MS analysis at the low *m/z* range, (2) HR-MS analysis at the high *m/z* range, and (3) sequential HR-MS/MS analysis using quadrupole ion isolation with higher-energy collisional dissociation (HCD). (**a** and **d**) Low *m/z* range positive and negative ion mode HR-MS spectra. Identified lipid species are annotated as sum composition. (**b** and **e**) High *m/z* range positive and negative ion mode HR-MS spectra. Identified lipid species are annotated as sum composition. (**c**) HR-MS/MS spectrum of the precursor ion of *m/z* 760.6. The lipid is annotated as PC 34:1 based on the peak of head group fragment ion of phosphorylcholine indicated in the figure. (**f**) HR-MS/MS spectrum of the precursor ion of *m/z* 835.6. The lipid is identified as PI 16:0–18:1 based on the peaks of fragment ions of glycerophosphate, head group inositol, and fatty acid moieties as indicated in the figure. *R* indicates the resolution at given m/z value

6. Requirements of lipid data analysis software. (1) Read out MS and MS/MS spectra, (2) lipid identification, (3) isotopic correction for quantification, and (4) process large size data-batch (*see* **Note 21**).

Table 3
Precursor ions, product ions, and neutral loss detections in the positive ion mode

Specific detection of precursor ion	Fragment ion	Neutral loss	Mass (fragment or neutral loss)
LPE/LPE O–/PE/PE O- $[M+H]^+$	–	[Phosphoethanolamine]	141.0191
LPC/LPC O–/PC/PC O–/SM/ LSM $[M+H]^+$	$[Phosphorylcholine+H]^+$	–	184.0733
Chol/CE $[M+NH_4]^+$	$[Chol–NH_3–H_2O]^+$	–	369.3516
LCB/Cer/HexCer/diHexCer/ triHexCer $[M+H]^+$	$[LCB+H–H_2O]^+$	–	Species-specific
LCB/Cer/HexCer/diHexCer/ triHexCer $[M+H]^+$	$[LCB+H–2H_2O]^+$	–	Species-specific
LCB/Cer/HexCer/diHexCer/ triHexCer $[M+H]^+$	$[LCB+H–CH_4O_2]^+$	–	Species-specific
DAG/TAG $[M+NH_4]^+$		[Fatty acid–H+NH_4]	Species-specific

O- indicates glycerophospholipids with one alkyl group attached to the glycerol moiety

Table 4
Precursor ions, product ions, and neutral loss detections in the negative ion mode

Specific detection of precursor ion	Fragment ion	Neutral loss	Mass (fragment or neutral loss)
CerP/LCBP $[M–H]^-$	$[Phosphoric acid–H–H_2O]^-$	–	78.9590
LPS/LPS O-/PS/PS O- $[M–H]^-$	–	$[C_3H_5NO_2]$	87.0320
SHexCer $[M–H]^-$	$[Sulfuric acid–H]^-$	–	96.9601
LPA/LPG/LPS/LPI/PA/PG/PS/PI/ LPA O–/LPG O-/LPS O-/LPI O-/ PA O-/PG O-/PS O-/PI O- $[M–H]^-$	$[Glycerophosphate–H–H_2O]^-$	–	152.9958
LPC/LPCO-/PC/PCO-$[M+CH_3OCOO]^-$	$[Cholinephosphate–H–CH_3]^-$	–	168.0431
LPE/LPE O-/PE/PE O- $[M–H]^-$	$[Ethanolaminephosphate–H–H_2O]^-$	–	196.0380
PI $[M–H]^-$	$[Inositolphosphate-H–2H_2O]^-$	–	223.0013
PI $[M–H]^-$	$[Inositolphosphate-H–H_2O]^-$	–	241.0119
PI $[M–H]^-$	$[Inositolphosphate–H]^-$	–	259.0224
Glycerophospholipid species $[M–H]^-$	$[Fatty acid–H]^-$	–	Species-specific
Glycerophospholipid species O- $[M–H]^-$	$[Fatty acid O- –H]^-$	–	Species-specific
Polyunsaturated glycerophospholipid species $[M–H]^-$	$[Fatty acid–H–CO_2]^-$	–	Species-specific

O- indicates glycerophospholipids with one alkyl group attached to the glycerol moiety

7. Quantification: The detected lipid species are quantified by comparing their peak intensities with the peak intensities of an internal standard lipid with known quantity that represents a certain lipid class (internal standard PE 12:0/12:0 represents PE and PE O- etc.) (*see* **Note 22**).

4 Notes

1. Use glassware cleaned with chloroform:methanol 1:1(*v*/*v*) for the preparation of all solvent mixtures. Avoid plasticware since organic solvents can dissolve polymers that affect the mass spectrometry data quality.
2. Store 10× ISM at −80 °C in as small aliquots as possible and in glass ampoules. The lipid concentrations in the 10× ISM are: 30 μM CE 15:0-D7, 20 μM Cer 18:1;2/12:0;0, 20 μM CerP 18:1;2/12:0;0, 200 μM Chol-D4, 10 μM DAG 12:0/12:0, 20 μM diHexCer 18:1;2/12:0;0, 25 μM HexCer 18:1;2/12:0;0, 20 μM LCB 17:0;2, 20 μM LCBP 17:0;2, 25 μM LPA 17:0, 20 μM LPC 17:1, 25 μM LPE 13:0, 15 μM LPG 17:1, 20 μM LPI 13:0, 20 μM LPS 17:1, 20 μM LSM 17:1;2, 25 μM PA 12:0/12:0, 20 μM PC-OO 18:1/18:1, 25 μM PE 12:0/12:0, 15 μM PG 12:0/12:0, 15 μM PI 8:0/8:0, 20 μM PS 12:0/12:0, 20 μM SHexCer 18:1;2/12:0;0, 20 μM SM 18:1;2/12:0;0, 10 μM TAG 17:0/17:0/17:0, and 20 μM triHexCer 18:1;2/17:0;0.
3. 1× ISM can be stored at −20 °C for up to 6 months or for longer period as aliquots in glass ampoules at −80 °C.
4. To dissolve the ammonium acetate fast, gently shake the Blue Cap bottle a few times.
5. Alternative to enzymatic cell dissociation, the cells can be harvested by scraping them off the dishes.
6. The present protocol has been optimized for the MCF-7 cell line. Use any other gentle lysis methods appropriate for your cell line of interest if necessary.
7. Carry over of unlyzed cells and large aggregates can result in low purity of the isolated lysosomes. Be careful not to disturb the pellet when transferring the supernatant containing cleared lysate and if necessary repeat the centrifugation steps.
8. The recovery of lysosomes can be estimated via assaying the lysosomal enzyme β-N-acetyl-glucosaminidase (NAG) activity as previously described [19].
9. Lower shaking speed of down to 1400 rpm can still extract efficiently in 60 min.
10. Avoid transferring any aqueous (upper) phase, since it will affect the data quality and cause problems for the stability of nanoelectrospray ionization.

11. Drying of the chloroform:methanol 10:1(*v/v*)-phase lipid extract (as described in **step 8** of Subheading 3.4.) can be started in parallel with this step.
12. When many samples are handled at the same time, this step can take longer than 60 min.
13. It is best to dissolve all samples at the same time and immediately before loading the samples for mass spectrometry analysis. This will minimize the time where the organic solvents are in contact with the Eppendorf tubes with possible release of polymers from the plastic. Dried samples can be stored at −20 °C for short time.
14. Some low abundant lipids such as LCB and LCBP are not always detectable. Furthermore, the $[LCB + H]^+$ species are isobaric with the $[\text{fatty acid} + NH_4]^+$ species, which makes these lipids harder to detect and quantify. Chemical derivatization with methane iodide or TMS-diazomethane/methanol can be used to overcome this problem [20–22].
15. The resultant mixture contains 7.5 mM ammonium acetate in chloroform:methanol:isopropanol 1:2:4(*v/v/v*). With the TriVersa NanoMate ion source, the best positive ion mode ionization can be achieved for solvents with this composition using settings of +0.96 kV and 1.25 psi. This is only valid for the TriVersa NanoMate ion source, and different ion sources need different settings.
16. The resultant mixture contains 0.1% triethyl amine in chloroform:methanol 1:3(*v/v*). With the TriVersa NanoMate ion source, the best negative ion mode ionization can be achieved for solvents with this composition using settings of −1.06 kV and 0.7 psi. This is only valid for the TriVersa NanoMate ion source, and different ion sources need different settings.
17. Another ion source such as nanoelectrospray needle infusion, syringe injection electrospray or flow-injection electrospray can be used instead of the TriVersa NanoMate ion source. A different mass spectrometer can also be used for lipid analysis.
18. Sample tray needs to be kept cold to minimize evaporation of organic solvents and contamination with polymers from plastic.
19. The analysis is performed with alternating low and high mass ranges. The low mass range should cover the lipids with low *m/z* values and their corresponding internal standards, while the high mass range the lipids with high *m/z* values and their corresponding internal standards. The HR-MS/MS should cover the range from the lowest lipid *m/z* value to the highest *m/z* value. The HR-MS^{ALL} ranges change between the positive ion mode and the negative ion mode (Fig. 2).

20. The specific settings for our Hybrid Quadrupole-Orbitrap mass spectrometer are optimized using a crude total lipid extract with robotic sample infusion. The optimal settings depend on types of mass spectrometer, and it is therefore important to optimize them prior to sample analysis.
21. Overview on available lipidomics software: http://omictools.com/lipid-identification-category.
22. Calculation of lipid concentration: [LP] = (Analyte Intensity/Internal Standard Intensity × pmol Internal Standard spiked × ρAnalyte/ρInternal Standard, where ρ is the intensity of the monoisotopic peak relative to the total intensity of all peaks in the isotopic cluster [12, 23]. ρAnalyte/ρInternal Standard can also be referred as Isotope Correction Factor, which in most lipidomics software is accounted for, when output intensities are given. Then the part ρAnalyte/ρInternal Standard of the equation is neglected.

Acknowledgement

This work was supported by the European Research Council Advance grant (M.J.) (#340751, LYSOSOME), the Danish National Research Foundation Center of Excellence grant (M.J.) (CARD), the Scientific Committee of the Danish Cancer Society (KBVU) (K.M., J.N., and M.J.) (R124-A7929-15-S2, R90-A5847-14-S2, and R90-A5783), and the Novo Nordisk Foundation (M.J.) (NNF15OC0016914). We are grateful to Inger Ødum Nielsen for crucial comments on this manuscript.

References

1. van Meer G, Voelker DR, Feigenson GW (2008) Membrane lipids: where they are and how they behave. Nat Rev Mol Cell Biol 9(2):112–124. doi:10.1038/Nrm2330
2. Fahy E, Subramaniam S, Murphy RC, Nishijima M, Raetz CR, Shimizu T et al (2009) Update of the LIPID MAPS comprehensive classification system for lipids. J Lipid Res 50(Suppl):S9–14. doi:10.1194/jlr.R800095-JLR200
3. van Meer G (2005) Cellular lipidomics. EMBO J 24(18):3159–3165. doi:10.1038/sj.emboj.7600798
4. Schulze H, Sandhoff K (2011) Lysosomal lipid storage diseases. Cold Spring Harb Perspect Biol 3(6):a004804. doi:10.1101/cshperspect.a004804
5. Anderson RA, Sando GN (1991) Cloning and expression of cDNA encoding human lysosomal acid lipase/cholesteryl ester hydrolase. Similarities to gastric and lingual lipases. J Biol Chem 266(33):22479–22484
6. Chiapparino A, Maeda K, Turei D, Saez-Rodriguez J, Gavin AC (2016) The orchestra of lipid-transfer proteins at the crossroads between metabolism and signaling. Prog Lipid Res 61:30–39. doi:10.1016/j.plipres.2015.10.004
7. Laplante M, Sabatini DM (2012) mTOR signaling in growth control and disease. Cell 149(2):274–293. doi:10.1016/j.cell.2012.03.017
8. Sardiello M, Palmieri M, di Ronza A, Medina DL, Valenza M, Gennarino VA et al (2009) A gene network regulating lysosomal biogenesis and function. Science 325(5939):473–477. doi:10.1126/science.1174447
9. Petersen NH, Olsen OD, Groth-Pedersen L, Ellegaard AM, Bilgin M, Redmer S et al (2013) Transformation-associated changes in sphingolipid metabolism sensitize cells to lysosomal cell death induced by inhibitors of acid sphingomyelinase. Cancer Cell 24(3):379–393. doi:10.1016/j.ccr.2013.08.003

10. Diettrich O, Mills K, Johnson AW, Hasilik A, Winchester BG (1998) Application of magnetic chromatography to the isolation of lysosomes from fibroblasts of patients with lysosomal storage disorders. FEBS Lett 441(3):369–372
11. Han X, Gross RW (2003) Global analyses of cellular lipidomes directly from crude extracts of biological samples by ESI mass spectrometry: a bridge to lipidomics. J Lipid Res 44(6):1071–1079. doi:10.1194/jlr.R300004-JLR200
12. Ejsing CS, Duchoslav E, Sampaio J, Simons K, Bonner R, Thiele C, Ekroos K, Shevchenko A (2006) Automated identification and quantification of glycerophospholipid molecular species by multiple precursor ion scanning. Anal Chem 78(17):6202–6214. doi:10.1021/ac060545x
13. Han X, Yang K, Gross RW (2012) Multi-dimensional mass spectrometry-based shotgun lipidomics and novel strategies for lipidomic analyses. Mass Spectrom Rev 31(1):134–178. doi:10.1002/mas.20342
14. Ejsing CS, Sampaio JL, Surendranath V, Duchoslav E, Ekroos K, Klemm RW et al (2009) Global analysis of the yeast lipidome by quantitative shotgun mass spectrometry. Proc Natl Acad Sci U S A 106(7):2136–2141. doi:10.1073/pnas.0811700106
15. Sampaio JL, Gerl MJ, Klose C, Ejsing CS, Beug H, Simons K, Shevchenko A (2011) Membrane lipidome of an epithelial cell line. Proc Natl Acad Sci U S A 108(5):1903–1907. doi:10.1073/pnas.1019267108
16. Graessler J, Schwudke D, Schwarz PE, Herzog R, Shevchenko A, Bornstein SR (2009) Top-down lipidomics reveals ether lipid deficiency in blood plasma of hypertensive patients. PLoS One 4(7):e6261. doi:10.1371/journal.pone.0006261
17. Surma MA, Herzog R, Vasilj A, Klose C, Christinat N, Morin-Rivron D et al (2015) An automated shotgun lipidomics platform for high throughput, comprehensive, and quantitative analysis of blood plasma intact lipids. Eur J Lipid Sci Technol 117(10):1540–1549. doi:10.1002/ejlt.201500145
18. Brugger B, Erben G, Sandhoff R, Wieland FT, Lehmann WD (1997) Quantitative analysis of biological membrane lipids at the low picomole level by nano-electrospray ionization tandem mass spectrometry. Proc Natl Acad Sci U S A 94(6):2339–2344
19. Nylandsted J, Gyrd-Hansen M, Danielewicz A, Fehrenbacher N, Lademann U, Hoyer-Hansen M et al (2004) Heat shock protein 70 promotes cell survival by inhibiting lysosomal membrane permeabilization. J Exp Med 200(4):425–435. doi:10.1084/jem.20040531
20. Bilgin M, Markgraf DF, Duchoslav E, Knudsen J, Jensen ON, de Kroon AI, Ejsing CS (2011) Quantitative profiling of PE, MMPE, DMPE, and PC lipid species by multiple precursor ion scanning: a tool for monitoring PE metabolism. Biochim Biophys Acta 1811(12):1081–1089. doi:10.1016/j.bbalip.2011.09.018
21. Ejsing CS, Bilgin M, Fabregat A (2015) Quantitative profiling of long-chain bases by mass tagging and parallel reaction monitoring. PLoS One 10(12):e0144817. doi:10.1371/journal.pone.0144817
22. Narayanaswamy P, Shinde S, Sulc R, Kraut R, Staples G, Thiam CH, Grimm R, Sellergren B, Torta F, Wenk MR (2014) Lipidomic "deep profiling": an enhanced workflow to reveal new molecular species of signaling lipids. Anal Chem 86(6):3043–3047. doi:10.1021/ac4039652
23. Han X, Gross RW (2005) Shotgun lipidomics: electrospray ionization mass spectrometric analysis and quantitation of cellular lipidomes directly from crude extracts of biological samples. Mass Spectrom Rev 24(3):367–412. doi:10.1002/mas.20023

Chapter 3

Analysis of *N*- and *O*-Glycosylation of Lysosomal Glycoproteins

Elmira Tokhtaeva, Olga A. Mareninova, Anna S. Gukovskaya, and Olga Vagin

Abstract

The vast majority of lysosomal proteins are heavily glycosylated. The present protocol describes the method of analyzing *N*- and *O*-linked glycans in lysosomal proteins of interest. The method is based on using deglycosylating enzymes, endoglycosidases, and exoglycosidases. Endoglycosidases catalyze the cleavage of an internal bond in an oligosaccharide, while exoglycosidases remove terminal carbohydrates from glycans. Different types of carbohydrate residues or chains can be removed by specific glycosidases. Removing oligosaccharides with glycosidases increases the electrophoretic mobility of a protein. This increase in mobility depends on the size and number of removed carbohydrate chains. Therefore, the treatment of lysosomal proteins with specific glycosidases followed by a western blot analysis of a protein of interest provides a way to determine which types of glycans are present in the protein by comparing the gel mobility before and after treatment.

Key words Glycosidases, Glycosylation, Lysosomes, *N*-glycans, *O*-glycans

1 Introduction

Lysosomes are composed of soluble and transmembrane proteins that are targeted to lysosomes in a signal-dependent manner [1]. More than 60 soluble and more than 100 transmembrane lysosomal proteins have been identified [2, 3], but recent proteomic studies suggest that this list should be largely expanded [4, 5]. The most abundant transmembrane proteins of the lysosomal membrane are the lysosomal-associated membrane proteins LAMP-1 and LAMP-2. LAMPs play critical role in the function of lysosomes, and deficiency of LAMP proteins is associated with disease development [3]. Lysosomal proteins have to withstand intralysosomal acidic pH < 5 and proteolytic activity of multiple acidic hydrolases and hence are usually heavily glycosylated [2, 4, 5].

Both soluble and transmembrane lysosomal proteins are synthesized in the rough ER, where they are modified by *N*-linked

Karin Öllinger and Hanna Appelqvist (eds.), *Lysosomes: Methods and Protocols*, Methods in Molecular Biology, vol. 1594, DOI 10.1007/978-1-4939-6934-0_3, © Springer Science+Business Media LLC 2017

glycans that are added to the asparagine of NXS or NXT (where X is any amino acid except proline) motifs in the nascent protein co-translationally. Immediately after coupling, *N*-glycans are modified by the ER-resident enzymes. The modification of *N*-glycans continues during the transit of glycoproteins through the Golgi complex until they reach the trans-Golgi network (TGN) (Fig. 1a). Many lysosomal proteins, including LAMP-1 and LAMP-2, are additionally modified by *O*-linked glycans that are added to selected serine or threonine residues [7]. Sorting of lysosomal transmembrane proteins depends on the presence of sorting signals in their cytosolic domains. These signals interact with clathrin adaptors that facilitate transport of these proteins to lysosomes either directly or via surface delivery followed by clathrin-dependent endocytosis [1, 8] (Fig. 1a). The majority of soluble acid hydrolases are modified with mannose 6-phosphate residues in the Golgi, allowing their recognition by mannose 6-phosphate receptors that facilitate transport of these proteins from the TGN to the endosomal/lysosomal compartments [9].

This protocol describes methods of analysis of *N*- and *O*-glycans attached to lysosomal proteins using deglycosylating enzymes. Treatment with glycosidases increases the electrophoretic mobility of a protein. This increase is more or less profound depending on the size and number of removed carbohydrate residues or chains. Therefore, following the gel migration before and after treatment

A

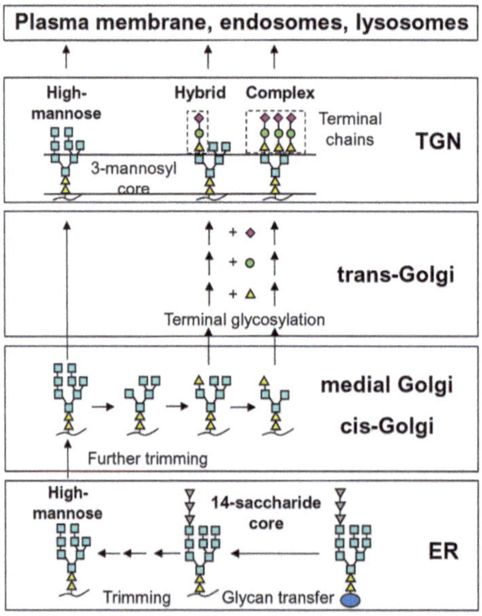

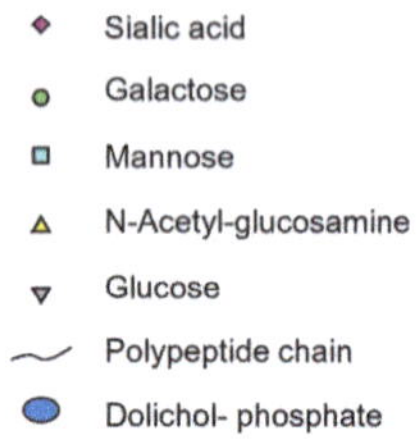

B

	Cleaved or modified by		
Type of chains	α-mannosidase	Endo H	PNGase F
High-mannose	Yes	Yes	Yes
Hybrid	No	Yes/No	Yes
Complex	No	No	Yes

Fig. 1 Analysis of *N*-glycosylated proteins using glycosidases. (**a**) Cartoon showing *N*-glycosylation and trafficking pathways of lysosomal membrane proteins. (**b**) Specificity of glycosidases to different type of oligosaccharide residues/chains in *N*-glycans. Modified from [6]

with glycosidases provides a convenient way to understand which types of glycans are present in a protein of interest.

Deglycosylating enzymes are classified as endoglycosidases and exoglycosidases. Endoglycosidases are enzymes that catalyze the cleavage of an internal glycoside bond in an oligosaccharide. Exoglycosidases are enzymes that remove terminal carbohydrates from the non-reducing end of a glycan, but do not cleave internal bonds between carbohydrates. Endoglycosidase H (Endo H) cleaves high-mannose and some hybrid oligosaccharide chains from *N*-linked glycoproteins [10]. The most common endoglycosidase, Peptide-*N*-Glycosidase F (PNGase F), catalyzes the deglycosylation of most *N*-linked glycoproteins and cleaves high-mannose, hybrid and complex oligosaccharide chains [10] (Fig. 1b). Exoglycosidase α-Mannosidase cleaves terminal mannose residues [11] and hence can be used as a tool to determine the presence of high-mannose oligosaccharide chains in *N*-linked glycans (Fig. 1b). Neuraminidase is an exoglycosidase that cleaves terminal *N*-acetyl neuraminic acid (sialic acid) residues from glycoproteins. *O*-glycosidase is an endoglycosidase that cleaves Core 1 and Core 3 *O*-linked glycans from glycoproteins, but this cleavage is effective only after removing terminal sialic residues by a neuraminidase. The treatment with these two enzymes removes some but not all *O*-glycans from glycoproteins [12]. Unfortunately, the enzymes that remove all *O*-linked glycans from glycoproteins are not available. Therefore, the analysis of glycoproteins using *O*-glycosidase and neuraminidase cannot detect all *O*-glycans linked to a protein of interest. In other words, the increase in the electrophoretic mobility of a protein after treatment with these two glycosidases is indicative of the presence of *O*-glycans. However, a lack of change in the electrophoretic mobility cannot be interpreted as the complete absence of *O*-glycans.

2 Materials

1. **Enzymes:** Peptide-*N*-Glycosidase F (PNGase F) from *Flavobacterium meningosepticum* (New England BioLabs), endoglycosidase H (Endo H) from *Streptomyces plicatus* (Glyco-Prozyme Inc.), Jack bean α-Mannosidase (Sigma-Aldrich) and *O*-glycosidase & Neuraminidase Bundle (New England BioLabs).
2. Denaturing Solution (10×): 5% SDS (sodium dodecyl sulfate), 400 mM DTT (dithiothreitol) (*see* **Note 1**).
3. NP40 (10×): 10% NP40 (*see* **Note 2**).
4. Reaction Buffer for PNGase F and for *O*-glycosidase & Neuraminidase Bundle (10×): 500 mM Sodium Phosphate buffer pH 7.5 (*see* **Note 1**).

5. Reaction Buffer for Endo H (10×): 500 mM sodium acetate buffer pH 6.0 (*see* **Note 1**).
6. Reaction Buffer for α-Mannosidase (10×): 250 mM sodium acetate buffer pH 5.0, 10% Triton X-100.
7. SDS-PAGE sample buffer (2×): 0.1 M Tris–HCl buffer pH 6.8, 4% SDS, 0.05% bromophenol blue, 20% glycerol, 1% β-mercaptoethanol.
8. Primary antibodies for western blot analysis: rabbit polyclonal antibody against the C-terminal region of rat and mouse LAMP-1 (ab24170; Abcam, Cambridge, MA), rat monoclonal antibody against full length mouse LAMP-1 (1D4B; Developmental Studies Hybridoma Bank), University of Iowa, Iowa City, IA), and rat monoclonal antibody against full length mouse LAMP-2 (GL2A7; Developmental Studies Hybridoma Bank).
9. Secondary antibodies for western blot analysis: horseradish peroxidase linked goat anti-rat, horseradish peroxidase linked goat anti-rabbit.

3 Methods

3.1 Denaturing Proteins (See Note 3)

1. Combine 1–9 μl of microsome membranes containing 2–30 μg of protein (*see* **Note 4**), 1 μl 10× of Denaturing Solution and H_2O (if necessary) to make a 10 μl total reaction volume.
2. Denature proteins by heating the reaction mixture at 100 °C for 10 min.
3. Chill denatured proteins on ice and spin for 10 s.

3.2 Deglycosylation Using PNGase F

1. Combine 10 μl of chilled denatured proteins from Subheading 3.1, **step 3** (*see* **Note 5**) with 2 μl of 10× Reaction Buffer for PNGase F, 2 μl 10× NP40 (*see* **Note 2**) and 5 μl H_2O and 1 μl PNGase F (*see* **Note 6**) to make a total reaction volume of 20 μl.
2. Incubate reaction at 37 °C for 1 h.

3.3 Deglycosylation Using Endo H

1. Combine 10 μl of chilled denatured proteins from Subheading 3.1, **step 3** with 2 μl of 10× Reaction Buffer for Endo H, 1–5 μl Endo H and H_2O to make a total reaction volume of 20 μl (*see* **Note 7**).
2. Incubate reaction at 37 °C for 3 h.

3.4 Deglycosylation Using α-Mannosidase

1. Combine 10 μl of chilled denatured proteins from Subheading 3.1, **step 3** with 2 μl of 10× Reaction Buffer for Jack been α-mannosidase, 1–3 μl Jack been α-mannosidase and H_2O to make a total reaction volume of 20 μl (*see* **Note 8**).
2. Incubate reaction at 37 °C for 3 h.

3.5 Deglycosylation Using O-Glycosidase & Neuraminidase Bundle

1. Combine 10 μl of chilled denatured proteins from Subheading 3.1, **step 3** with 2 μl of 10× Reaction Buffer for PNGase F, 2 μl 10× NP40 (*see* **Note 2**), H_2O, 2 μl Neuraminidase, and 1-5 μl *O*-Glycosidase (*see* **Note 9**) to make a total reaction volume of 20 μl.
2. Incubate reaction at 37 °C for 1 h.

3.6 Western Blot Analysis of Deglycosylated Proteins

1. After completing the incubation with enzymes, place the reaction mixture on ice and combine it with equal volume of 2× SDS-PAGE sample buffer.
2. Separate proteins by SDS-PAGE.
3. Analyze the lysosomal glycoproteins by western blot using appropriate antibodies (*see* **Note 10**).

4 Notes

1. Most vendors of deglycosylating enzymes provide 10× denaturing and reaction buffers.
2. PNGase F is inhibited by SDS, thus enough NP40 needs to be added after denaturation to prevent PNGase F inactivation.
3. Deglycosylation is usually more effective under denaturing conditions (using SDS and heat). However, some glycoproteins aggregate as a result of boiling. Optimal denaturing conditions for each glycoprotein should be determined experimentally. For several glycoproteins, including LAMP-1, we found the best results after incubating in the denaturing buffer at 80 °C for 5 min. If necessary, deglycosylation can be performed under native conditions without the addition of detergents. However, longer incubation time and/or more enzyme may be required to deglycosylate native glycoproteins. Also, when deglycosylation of membrane proteins is performed, the presence of non-ionic detergents is required to keep these proteins in solution. The control reaction in denaturing conditions is recommended to confirm that deglycosylation in native conditions is efficient.
4. Instead of membrane preparations, purified glycoproteins, immunoprecipitated protein complexes, or proteins of total cell lysates can be used. When total lysates are used, precautions should be taken to inhibit cytoplasmic and lysosomal proteases during the deglycosylation reaction. The following protease inhibitors can be added to the reaction mixture: aprotinin (10 μg/ml), benzamidine (1 mM), pepstatin (10 μg/ml), leupeptin (1 μM), EGTA (1 mM), EDTA (1 mM), PMSF (phenylmethylsulfonyl fluoride; 1 mM), or Pefabloc SC (1 mM). A 1000× concentrated stock of each inhibitor can be made in

water except for pepstatin and PMSF, which should be dissolved in methanol or DMSO (dimethyl sulfoxide). Note that PMSF has the ability to modify basic residues on glycoprotein substrates.

5. Some lysis buffers contain high NaCl concentrations that can inhibit PNFase F. It is recommended to keep the total molarity of the reaction below 200 mM, and the total SDS concentration not higher than 1%.
6. When purchasing PNGase F, check if it contains glycerol. Since glycerol can inhibit the enzyme activity, limit the amount of the enzyme in the total reaction mixture so that the final glycerol concentration is equal to (or less than) 5%.
7. Endo H enzymatic activity is not affected by SDS, thus there is no need to add NP40.
8. To prevent inactivation of the enzyme by SDS, 1% Triton X-100 is present in the 1× Reaction Buffer for Jack bean α-Mannosidase.
9. *O*-glycosidase removes *O*-linked disaccharides from serine and threonine residues at the attachment site, while Neuraminidase is an exoglycosidase that cleaves *N*-acetyl neuraminic acid (sialic acid) residues from glycoproteins. The former needs trimming of glycoproteins by the latter for its enzymatic action. The treatment with the two enzymes removes some but not all *O*-glycans from glycoproteins [12]. Unfortunately, there is no endoglycosidase that can completely remove all *O*-linked glycans. A more efficient removal of *O*-glycans can be achieved by adding several different exoglycosidases to *O*-glycosidase. A complete removal of *O*-glycans can be achieved by using chemical methods, but they often result in the degradation of the protein.
10. The selection of the right antibody to determine the efficiency of deglycosylation by western blot analysis is crucial. Many glycoprotein-reacting antibodies have different affinities to glycosylated and deglycosylated forms of a protein. An example is provided in Fig. 2. Microsome membranes isolated from a mouse pancreas were analyzed by western blot analysis using two different anti-LAMP-1 antibodies. Both antibodies detect a single wide band at 100 kDa corresponding to the predicted molecular mass of the fully glycosylated LAMP-1 (Fig. 2a, b, left lane). The treatment with Endo H or Jack bean α-mannosidase hardly affected the electrophoretic mobility of the band, while the treatment with PNGase F resulted in a significant increase in mobility, producing a band at 50 kDa as detected with the antibody reacting with the cytoplasmic C-terminal domain of LAMP-1. These data indicate that LAMP-1 has predominantly complex-type *N*-glycans

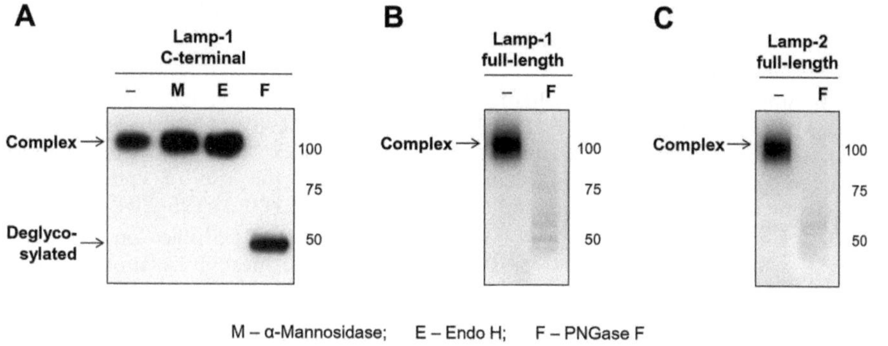

Fig. 2 Deglycosylation of LAMP-1 and LAMP-2 as analyzed by different antibodies. Microsomal membrane proteins were treated with glycosidases as indicated and analyzed using antibodies against the (**a**) C-terminus of LAMP-1, (**b**) full-length LAMP-1 or (**c**) full-length LAMP-2. Modified from [13]

that are resistant to cleavage by Endo H or α-mannosidase (Fig. 1b). Surprisingly, the band corresponding to the deglycosylated LAMP-1 was not detected by the antibody raised against the full-length LAMP-1 (Fig. 2b), suggesting that this antibody reacts exclusively with the glycosylated form of LAMP-1. Similarly, the antibody raised against the full-length LAMP-2 reacts with the fully glycosylated protein (Fig. 2c, left lane), but not with the deglycosylated product (Fig. 2c, right lane). By contrast, many other antibodies react better with the deglycosylated or less glycosylated forms of proteins than with their fully glycosylated forms. Therefore, it is important to confirm that the antibody of your choice equally reacts with glycosylated and deglycosylated forms of the glycoprotein of interest. Usually, it is safer to use the antibodies that react with cytoplasmic epitopes that are not affected by *N*- or mucin-type *O*-glycosylation.

Acknowledgement

This study was funded, fully or in part, by National Institute of Health grants R01HL113350 (O.V.), P01DK098108 (to A.S.G), and R01AA19730 (to O.A.M.)

References

1. Braulke T, Bonifacino JS (2009) Sorting of lysosomal proteins. Biochim Biophys Acta 1793(4):605–614. doi:10.1016/j.bbamcr.2008.10.016
2. Schröder BA, Wrocklage C, Hasilik A, Saftig P (2010) The proteome of lysosomes. Proteomics 10(22):4053–4076. doi:10.1002/pmic.201000196
3. Schwake M, Schroder B, Saftig P (2013) Lysosomal membrane proteins and their central role in physiology. Traffic 14(7):739–748. doi:10.1111/tra.12056
4. Chapel A, Kieffer-Jaquinod S, Sagne C, Verdon Q, Ivaldi C, Mellal M, Thirion J, Jadot M, Bruley C, Garin J, Gasnier B, Journet A (2013) An extended proteome map of the lysosomal

membrane reveals novel potential transporters. Mol Cell Proteomics 12(6):1572–1588. doi:10.1074/mcp.M112.021980

5. Sleat DE, Sun P, Wiseman JA, Huang L, El-Banna M, Zheng H, Moore DF, Lobel P (2013) Extending the mannose 6-phosphate glycoproteome by high resolution/accuracy mass spectrometry analysis of control and acid phosphatase 5-deficient mice. Mol Cell Proteomics 12(7):1806–1817. doi:10.1074/mcp.M112.026179
6. Vagin O, Turdikulova S, Sachs G (2004) The H,K-ATPase beta subunit as a model to study the role of N-glycosylation in membrane trafficking and apical sorting. J Biol Chem 279(37):39026–39034. doi:10.1074/jbc.M405453200
7. Fukuda M (1991) Lysosomal membrane glycoproteins. Structure, biosynthesis, and intracellular trafficking. J Biol Chem 266(32):21327–21330
8. Coutinho MF, Prata MJ, Alves S (2012) A shortcut to the lysosome: the mannose-6-phosphate-independent pathway. Mol Genet Metab 107(3):257–266. doi:10.1016/j.ymgme.2012.07.012
9. Coutinho MF, Prata MJ, Alves S (2012) Mannose-6-phosphate pathway: a review on its role in lysosomal function and dysfunction. Mol Genet Metab 105(4):542–550. doi:10.1016/j.ymgme.2011.12.012
10. Maley F, Trimble RB, Tarentino AL, Plummer TH Jr (1989) Characterization of glycoproteins and their associated oligosaccharides through the use of endoglycosidases. Anal Biochem 180(2):195–204
11. Li YT (1967) Studies on the glycosidases in jack bean meal. I Isolation and properties of alpha-mannosidase. J Biol Chem 242(23):5474–5480
12. Koutsioulis D, Landry D, Guthrie EP (2008) Novel endo-alpha-N-acetylgalactosaminidases with broader substrate specificity. Glycobiology 18(10):799–805. doi:10.1093/glycob/cwn069
13. Mareninova OA, Sendler M, Malla SR, Yakubov I, French SW, Tokhtaeva E, Vagin O, Oorschot V, Lullmann-Rauch R, Blanz J, Dawson D, Klumperman J, Lerch MM, Mayerle J, Gukovsky I, Gukovskaya AS (2015) Lysosome associated membrane proteins maintain pancreatic acinar cell homeostasis: LAMP-2 deficient mice develop pancreatitis. Cell Mol Gastroenterol Hepatol 1(6):678–694. doi:10.1016/j.jcmgh.2015.07.006

Chapter 4

Analyzing Lysosome-Related Organelles by Electron Microscopy

Ilse Hurbain, Maryse Romao, Ptissam Bergam, Xavier Heiligenstein, and Graça Raposo

Abstract

Intracellular organelles have a particular morphological signature that can only be appreciated by ultrastructural analysis at the electron microscopy level. Optical imaging and associated methodologies allow to explore organelle localization and their dynamics at the cellular level. Deciphering the biogenesis and functions of lysosomes and lysosome-related organelles (LROs) and their dysfunctions requires their visualization and detailed characterization at high resolution by electron microscopy. Here, we provide detailed protocols for studying LROs by transmission electron microscopy. While conventional electron microscopy and its recent improvements is the method of choice to investigate organelle morphology, immunoelectron microscopy allows to localize organelle components and description of their molecular make up qualitatively and quantitatively.

Key words Lysosome-related organelles, Transmission electron microscopy, Chemical fixation, High pressure freezing, Freeze substitution, Immunolabeling, Tokuyasu

1 Introduction

Selected cell types host organelles sharing features with lysosomes, so called lysosome-related organelles (LROs) [1, 2]. Several genetic disorders and cell transformation are accompanied by dysfunctional LRO formation and/or motility [2]. For over 60 years, transmission electron microscopy (TEM) has been a unique method to investigate LRO morphology and essential to understand how their host cells generate them [3–5]. "Conventional" EM that involves chemical fixation of cells and tissues, dehydration, embedding in resins, and ultrathin sectioning is the method of choice to appreciate ultrastructural details. The protocols have remained basically unchanged for over five decades but progresses have been made through the development

Karin Öllinger and Hanna Appelqvist (eds.), *Lysosomes: Methods and Protocols*, Methods in Molecular Biology, vol. 1594, DOI 10.1007/978-1-4939-6934-0_4, © Springer Science+Business Media LLC 2017

of high pressure freezing (HPF). HPF followed by freeze substitution optimally preserves the ultrastructure of cells and tissues in a near to native state [6, 7]. While cytochemistry has been used often to detect enzymatic activities present in LROs (e.g., acid phosphatase, tyrosinase) [8, 9], immunoelectron microscopy (IEM) is required to visualize and analyze the distribution of organelle components and allows their identification with specific antibodies that are detected using "secondary" antibodies or protein A from *Staphylococcus aureus* both conjugated to gold particles [10]. A statistical analysis of gold particle distribution is an asset to obtain information on the presence and relative enrichment of molecules [11–13]. Immunocytochemistry can be performed on cells and tissues embedded in hydrophobic resins (e.g., Lowicryl or LR White), though such methods rely on the abundance of antigens often masked by the resins [12]. Therefore, the method of choice is the Tokuyasu method based on ultrathin cryosectioning of cells and tissues that combines an optimal visualization of membranes and retention of antigenicity [13, 14].

In this chapter, we describe in detail classical TEM and IEM (immunogold labeling on ultrathin cryosections) that are routinely used in our laboratory to investigate LRO morphology, alterations in disease, biogenetic pathways, and molecular mechanisms involved [15–22].

It should be mentioned that other methods are of interest to obtain additional and complementary information to investigate organelle morphology, composition, and close interactions with other subcellular structures. Electron tomography (ET) adds the third dimension (reviewed [23]) and allows to investigate inter-organelle contacts, continuities between compartments [20, 24] and subcellular features that can only be appreciated after 3D modeling [25, 26]. As TEM gives ultrastructural details, organelle distribution at the cellular level and organelle dynamics can only be investigated by optical imaging using fluorescent probes (confocal microscopy, 3D deconvolution after immunofluorescence, several novel super-resolution methods such as SIM, PALM, live cell imaging, TIRF among others). Therefore, optimally, one should investigate the organelles of interest by both fluorescence and EM. This can be accomplished by correlative light electron microscopy (CLEM), which has been developed already for several years [19, 27–33]. Novel improvements on CLEM methods made in our group [34, 35] and other laboratories [36–38] combining several optical and electron microscopy modalities are described elsewhere.

2 Materials

2.1 Chemical Fixation of Tissue and Cells Containing Lysosome-Related Organelles

Prepare all solutions using analytical grade reagents. All preparations are done under a chemical hood (unless indicated). Follow all security and waste disposal rules displayed in the products information sheet and institutional procedures.

2.1.1 Material for Fixation

1. Single edge razor blades.
2. Fixation buffer 2×: 200 mM sodium cacodylate buffer pH 7.2. Weigh 21.4 g cacodylic acid ($C_2H_6AsNaO_2{\cdot}3H_2O$) and add 450 mL of water in a glass beaker. Mix and adjust pH with HCl. Make up to 500 mL with water. Store at 4 °C.
3. Fixation and rinsing buffer 1×: 100 mM sodium cacodylate buffer pH 7.2. Add 50 mL water to 50 mL 200 mM sodium cacodylate buffer pH 7.2. Store at 4 °C.
4. Fixative: 2.5% glutaraldehyde and 2% formaldehyde in 100 mM cacodylate buffer pH 7.2. For 10 mL fixative mix 5 mL 200 mM sodium cacodylate buffer pH 7.2, 1 mL 25% glutaraldehyde, 1.25 mL 16% formaldehyde and additional 2.75 mL water (*see* **Note 1**).
5. Post-fixative: 1% OsO_4, 1.5% potassium ferrocyanide in 100 mM sodium cacodylate buffer pH 7.2 (*see* **Note 2**). For 10 mL prepare two tubes.
   (a) Mix 2.5 mL 4% osmic acid anhydride, and 2.5 mL sodium cacodylate buffer 200 mM, pH 7.2. Protect the solution against light with aluminum foil around the tube.
   (b) Weigh 150 mg potassium ferrocyanide ($C_6FeK_4N_6{\cdot}3H_2O$) and dissolve it in 5 mL sodium cacodylate buffer 100 mM pH 7.2. Protect the solution against light with an aluminum foil around the tube (*see* **Note 3**).

2.1.2 Material for Dehydration and Embedding

1. Ethanol in water (40, 70, 90, 95, and 100%).
2. 100% Acetone.
3. EPON.
   (a) Solution A: Mix 8.25 g EPON 812 resin and 12.5 g DDSA (dodecyl succinic anhydride) for 15 min at room temperature. Centrifuge the solution 1 min at 600 × *g* to remove air bubbles. Store solution A in aliquots of 4 mL at −20 °C.
   (b) Solution B: Mix 25 g EPON 812 resin and 21 g NMA (methyl-5-norbornene-2-3-dicarboxylic anhydride) for 15 min at room temperature. Centrifuge the solution 1 min at 600 × *g*. Store solution B in aliquots of 7 mL in centrifuge tubes of 15 mL at −20 °C (*see* **Note 4**).

   (c) For use, warm a tube A and a tube B to room temperature. Add 3 mL of solution A to 7 mL solution B, mix for 10 min. Add 170 μL DMAE (2-dimethylaminoethanol) and mix for 10 min. Use immediately and discard unused resin.
4. EPON 30, 50, and 70% in acetone.
5. Flat embedding silicon mold.

2.1.3 Material for Sectionning and Contrast

1. Uranyl acetate: Prepare 4% aqueous solution by weighing 800 mg uranyl acetate and dissolve in 20 mL of water. Store solution in a syringe closed with a 0.22 μm filter at 4 °C. Protect uranyl acetate solution against light by wrapping the syringe in aluminum foil.
2. Sodium hydroxide (NaOH) 1 N: dissolve 400 mg NaOH in 10 mL water.
3. Lead citrate: Dissolve 332 mg lead nitrate ($Pb(NO_3)_2$) in 7.5 mL boiled or autoclaved cooled water in a 15 mL centrifuge tube (*see* **Note 5**). Add 444 mg sodium citrate ($Na_3C_6H_5O_7{\cdot}2H_2O$) and mix well. The solution will become white, continue the stirring intermittently during 30 min so that lead nitrate is transformed into lead citrate. Add 2 mL 1 N NaOH and complete with water up to 12.5 mL. Store the solution in a syringe closed with a 0.22 μm filter at 4 °C. All air bubbles in the syringe should be removed.
4. Ultramicrotome.
5. Electron microscopy copper grids 100 Mesh or 200 Mesh.

2.2 High-Pressure Freezing and Freeze Substitution of Cells Containing Lysosome-Related Organelles

Prepare all solutions using analytical-grade reagents. Follow all security and waste disposal regulations. High-pressure freezer and accessories are from Leica.

2.2.1 Material for Cell Culture

1. Sapphire disks 1.4 mm diameter (Leica).
2. Finder grids.
3. Carbon coater.
4. Ethanol 100%.

2.2.2 Material for HPF

1. HPM 100 or any other high pressure freezing machine (*see* **Note 6**).
2. Specimen carrier 3.0 mm diameter, Type A.
3. Specimen carrier 3.0 mm diameter, Type B.
4. Sample holder half cylinder 3.0 mm diameter.
5. Sample holder middle plate 3.0 mm diameter.
6. Long forceps with insulation coating.

7. Sample Carrier Release Device.
8. 60 L of liquid nitrogen.
9. Appropriate cell culture medium.
10. Fetal calf serum.
11. Cryovials 2 mL, perforated several times in the upper part.

2.2.3 Material for Freeze Substitution

1. Freeze substitution apparatus (*see* **Note 7**).
2. Plastic capsules D 5 × H 15 mm.
3. Cryotubes 2 mL.
4. 100% ethanol.
5. Crystalized osmic acid anhydride (OsO_4).
6. 100% acetone.
7. Solution of 1% OsO_4/2% H_2O in acetone for cryosubstitution: complete 10 mL water-free acetone with 200 μL of distilled water and 100 mg of osmic acid.
8. EPON:
   (a) Solution A: Mix 8.25 g EPON 812 resin and 12.5 g DDSA for 15 min at room temperature. Centrifuge the solution 1 min at 600 × *g* to remove air bubbles. Store solution A in aliquots of 4 mL at −20 °C.
   (b) Solution B: Mix 25 g EPON 812 resin and 21 g NMA for 15 min at room temperature. Centrifuge the solution 1 min at 600 × *g*. Store solution B at aliquots of 7 mL in a centrifuge tube of 15 mL at −20 °C.
   (c) For use, warm a tube A and a tube B to room temperature. Add 3 mL of solution A to 7 mL solution B, mix 10 min. Add 170 μL DMAE and mix for 10 min. Use immediately and discard unused resin.
9. EPON 30, and 75% in acetone.
10. Pyramid shaped flat mold or flat embedding silicon mold.
11. Hot plate.
12. Liquid nitrogen.
13. Razor blade.
14. Magnification glass 4×, long working distance.
15. Oven 60 °C.

2.2.4 Material for Contrast

1. Uranyl acetate: Prepare 4% aqueous solution by weighing 800 mg uranyl acetate and dissolve in 20 mL of water. Store the solution in a syringe closed with a 0.22 μm filter at 4 °C. Protect uranyl acetate solution against light by wrapping the syringe in aluminum foil.

2. Sodium hydroxide (NaOH) 1 N: dissolve 400 mg NaOH in 10 mL water.
3. Lead citrate: Dissolve in a 15 mL centrifuge tube 332 mg lead nitrate ($Pb(NO_3)_2$) in 7.5 mL boiled or autoclaved cooled water (*see* **Note 5**). Add 444 mg sodium citrate ($Na_3C_6H_5O_7{\cdot}2H_2O$) and mix well. The solution will become white. Continue the stirring intermittently during 30 min so that lead nitrate is transformed into lead citrate. Add 2 mL 1 N NaOH and complete with water up to 12.5 mL. Store the solution in a syringe closed with a 0.22 μm filter at 4 °C. All air bubbles in the syringe should be removed.
4. Ultramicrotome.
5. Electron microscopy copper/palladium grids 100 Mesh or 200 Mesh.

2.3 Immunogold Labeling on Ultrathin Cryosections, the Tokuyasu Method

Prepare all solutions with distilled water and analytical-grade reagents. Follow all safety and waste disposal regulations using appropriate laboratory equipment (wear lab coat, gloves, protection glasses, etc.).

2.3.1 Materials

1. Razor blades.
2. Toothpick.
3. Thin tweezers.
4. Eutectic plate.
5. Pins holder.
6. Pins (Leica).
7. Isolated coated forceps for handling liquid nitrogen.
8. Magnifying glass.
9. Ultracryomicrotome (Leica).
10. Dalmatian dog hair (or eyelash) on a bamboo stick (glued with nail polish).
11. Pickup loop.
12. Diamond trimmer and cryo immuno 35° diamond knife.
13. 200 Mesh formvar/carbon coated grids (*see* **Note 8**).
14. Several drying loops.
15. Filter paper (Whatman n°1).

2.3.2 Solutions and Buffers

All solutions are prepared in advance and stored unless indicated

1. Phosphate buffer (PB) 0.2 M, pH 7.4:
   (a) 0.2 M NaH_2PO_4: add 27.59 g of $NaH_2PO_4{\cdot}H_2O$ in 1 L distilled water.

(b) 0.2 M Na_2HPO_4: add 35.60 g of $Na_2HPO_4 \cdot 2H_2O$ in 1 L distilled water.

(c) Mix 1 volume of 0.2 M NaH_2PO_4 and 4 volumes of 0.2 M Na_2HPO_4. Use it filtered extemporaneously.

2. Phosphate buffered saline (PBS): 10 mM phosphate buffer, 150 mM NaCl.
We use PBS powder without Ca^{2+} and Mg^{2+} ready to dissolve in sterile water.
3. Gelatin 2% in 30 mm diameter petri dishes:
Weigh 2 g of gelatin powder (from porcine skin) and add 80 mL 0.1 M PB. Incubate at 60 °C until the gelatin is dissolved. Add sodium azide 0.02% and complete the volume to 100 mL with 0.1 M phosphate buffer and store in small petri dishes at 4 °C sealed with a parafilm to avoid drying.
4. Gelatin 10% for cell embedding:
Weigh 10 g of gelatin powder (from porcine skin) and add 80 mL 0.1 M PB. Incubate at 60 °C until dissolved. Add sodium azide 0.02% and complete the volume to 100 mL with 0.1 M PB. Store at 4 °C in 2 mL tubes.
5. 50 mM PBS/glycine: Mix 0.18 g of glycine in 50 mL PBS. Filtrate with Stericup 0.22 μm from Millipore. Prepare fresh solution each time.
6. Blocking buffer for immunolabeling: 0.1% CWFS (cold water fish skin) gelatin in PBS. Prepare fresh solution each time.
7. BSA (bovine serum albumin): 10% in PBS can be used as blocking buffer in case of weak labeling. 1% BSA dilution for antibodies and 0.1% BSA in PBS in the washing steps.
8. 2.3 M sucrose: Add 78.73 g in 100 mL distilled water. Stir until dissolved. Store at 4 °C.
9. 2% methylcellulose: Add 4 g methylcellulose (25 centipoises) in 196 mL distilled water at 90 °C. While stirring cool the solution on ice rapidly then seal it and leave it overnight at 4 °C still mixing. Stop stirring and leave it for 2 days at 4 °C for polymer maturation. Add distilled water to final volume of 200 mL and centrifuge at 100,000 × *g* during 1.5 h. Store the supernatant as aliquots in tubes at 4 °C.
10. 4% uranyl acetate: Add 4 g in 100 mL distilled water. Let it mix on a rotor (rotation speed 120 rpm) until it dissolves. Store it covered with aluminum foil at 4 °C.
11. Embedding and contrast: Mix gently 900 μL of 2% methylcellulose and 100 μL of 4% uranyl acetate. Keep it in the dark at 4 °C. Prepare it fresh.

3 Methods

3.1 Chemical Fixation of Tissue and Cells Containing Lysosome-Related Organelles

3.1.1 Chemical Fixation

1. Preparation of tissue, cells in suspension or cell monolayers before fixation:
   (a) Tissue: place it in culture medium or 100 mM phosphate buffer, pH 7.2 and cut it in 1 mm^3 cubes with single-edged razor blade.
   (b) Isolated cells: centrifuge 1–2 × 10^6 cells in a 1.5 mL Eppendorf tube at 1200 rpm (0.1 × *g*) for 5 min at room temperature and discard the supernatant.
   (c) Cell monolayer cultured on coverslips: discard culture medium (*see* **Note 9**).
2. Fix four tissue cubes per condition or two cell pellets in Eppendorf tubes with fixative overnight at 4 °C. Cells cultured as monolayers are fixed in their culture dish.
3. Rinse 3× 5 min in 100 mM cacodylate buffer pH 7.2 at room temperature (*see* **Note 10**).
4. Perform post fixation in 1% OsO_4, 1.5% potassium ferrocyanide in 100 mM cacodylate buffer pH 7.2, 45 min at 4 °C (*see* **Note 11**).
5. Rinse 3× 5 min in water at room temperature.

3.1.2 Dehydration and Embedding of Tissue Blocs and Cell Pellets in EPON Resin

1. Dehydrate the samples at room temperature successively for 15 min in 40% ethanol, 15 min in 70% ethanol, 15 min in 90% ethanol, 15 min in 95% ethanol, 3× 15 min in 100% ethanol, and finally 10 min in acetone (*see* **Note 12**).
2. Infiltrate with Epon resin successively starting with 30% EPON for 30 min (*see* **Notes 13** and **14**), 50% EPON for 30 min, 70% EPON for 1 h and pure EPON overnight (*see* **Note 15**).
3. Deposit tissue samples in a flat embedding mold filled with EPON resin (*see* **Note 16**). For cell pellets in Eppendorf tubes EPON resin is added to a height of approximately 1.2 cm.
4. Polymerize EPON resin for 24 h in an oven at 60 °C (*see* **Note 17**).

3.1.3 Dehydration and Embedding of Cell Monolayers in EPON Resin

1. Dehydrate the samples in the culture dish at room temperature successively for 5 min in 40% ethanol, 5 min in 70% ethanol, 5 min in 90% ethanol, 2× 10 min in 95% ethanol, 3× 6 min in 100% ethanol.
2. Infiltrate with EPON resin: Deposit one droplet of EPON resin on a microscope slide (**star-mark in** Fig. 1a). Deposit one droplet of EPON resin on the coverslip containing the monolayer of cells (on the side where the cells are) (**arrow in** Fig. 1a). Catch the coverslip with a tweezer, deposit it on the

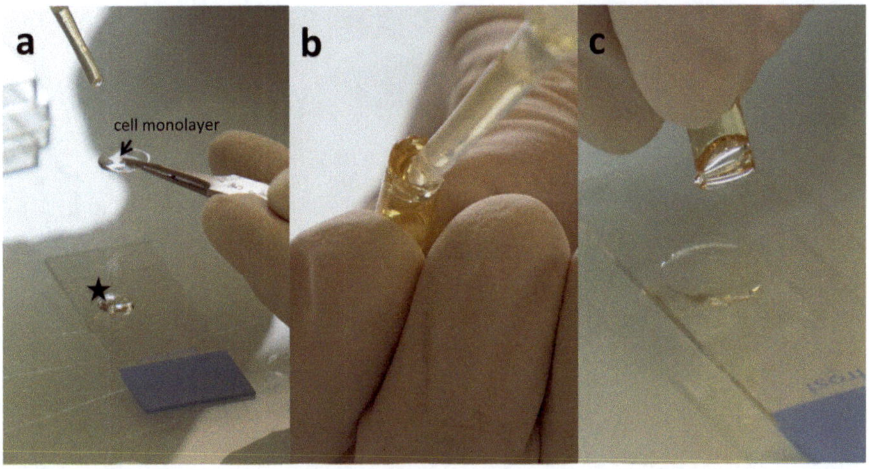

Fig. 1 Preparation of cell monolayers for polymerization in EPON resin. A droplet of EPON is put on a microscope slide (*star-mark* in **a**). One or two droplets of EPON are deposited on the coverslip, on the side containing the cell monolayer. The coverslip is then put on the resin droplet on the microscope slide, cells upward. A gelatin capsule is filled with resin (**b**) and deposited on top of the coverslip (**c**)

resin droplet on the microscopy slide, cells upward. Let the residual ethanol evaporate under the fume hood for at least 4 h.

3. Overlay an inverted gelatine cap filled with EPON (Fig. 1b) on the region of interest (Fig. 1c). Leave the sample under the fume hood at room temperature for 6–8 h (or overnight).
4. Polymerize EPON resin for 24 h in an oven at 60 °C (*see* **Note 17**).

3.1.4 Cutting and Contrast

1. Pull out the tissue samples and cell pellets in the polymerized resin from the mold or the Eppendorf tube. For cell monolayers the gelatine capsule is separated from the coverslip by heating the base of the microscopy slide during 20 s directly under the gelatine capsule (Fig. 2a), with a flame (matches). Perform a turning movement with the flame. Apply pressure at one side of the gelatine capsule (Fig. 2b) that then should detach easily with cells on the capsule side (star-mark Fig. 2c). If it is not the case, start again with heating very briefly.
2. Cut 60 nm thick sections using an ultramicrotome according to the manufacturer's instructions. Collect the sections on 100 Mesh or 200 Mesh copper grids.
3. Contrast sections with uranyl acetate: prepare one droplet of filtered uranyl acetate and five droplets of autoclaved and filtered water per grid on a clean parafilm. Deposit the grid for 10 min on the uranyl acetate droplet, facing the side that contains the section. Protect incubating grids from light. Take the grid with a tweezers and rinse it for 30 s in each of

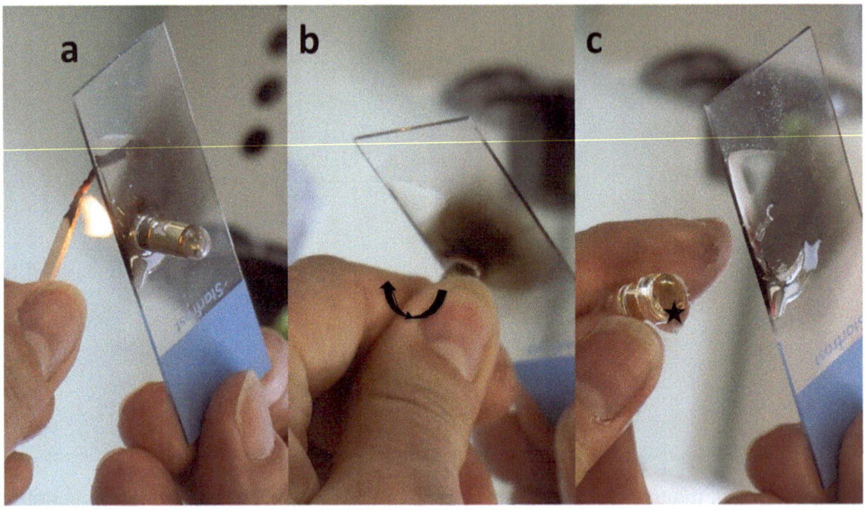

Fig. 2 Separation of the cell monolayer from the glass slide. The base of the microscope slide is warmed for 20 s with a flame (**a**), then pressure is applied on the resin filled gelatin capsule (**b**) and the capsule will detach (**c**). The surface on the bottom of the capsule (*star-mark*) contains the resin embedded cell monolayer

the five water droplets. Soak water away from the grid with a filter paper. Touch the filter paper with the edge of the grid inclined at 45°.

4. Contrast sections with lead citrate: prepare five droplets of autoclaved, filtered water on parafilm. Put a culture dish of 30 mm diameter containing some NaOH pellets near the droplets. Deposit one droplet of filtered lead citrate near the NaOH pellets (*see* **Note 18**). Put the grid for 1 min on the lead citrate droplet, facing the side that contains the section. Rinse and dry grids as described in **step 3** above.
5. Observe in a transmission electron microscope at 80 kV according to the manufacturer's instructions.

3.2 High-Pressure Freezing and Freeze Substitution of Cells Containing Lysosome-Related Organelles

3.2.1 Cell Culture

1. Put six finder grids on six sapphire disks (*see* **Note 19**). Deposit them in a carbon coater and make a 20 nm thick carbon layer (Fig. 3a, b). Fix the carbon layer by curing the sapphire disk at 60 °C during 24 h.
2. Sterilize the sapphire disks by dipping them in 100% ethanol (*see* **Note 20**) under sterile conditions (microbiological safety cabinet). Put three sapphire disks per condition in an empty 30 mm diameter culture dish carbon matrix upward, and let the ethanol evaporate under the microbiological safety cabinet. Seed approximately 80×10^4 cells per dish and culture them for 3–5 days in a CO_2 incubator to reach 70–100% confluency.

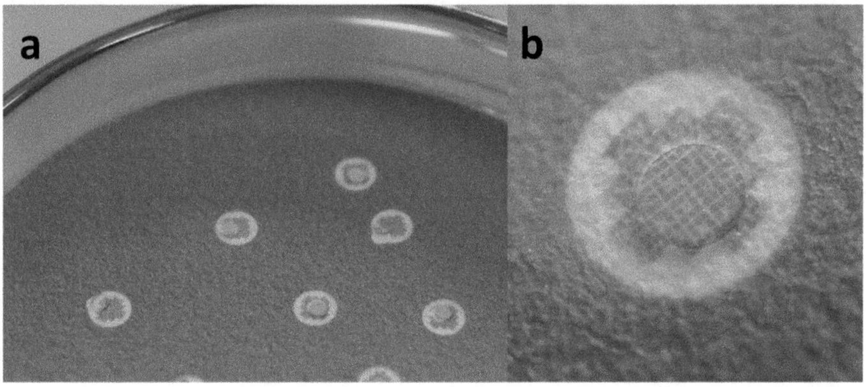

Fig. 3 Carbon coating of sapphire disks. Finder grids are deposited on sapphire disks in a 60 mm glass Petri dish containing a filter paper (**a**). After the sapphire disks are coated with a thin, 20 nm layer of carbon, the finder grids are taken off (**b**)

3.2.2 High-Pressure Freezing

1. Fill the high-pressure freezer with liquid nitrogen according to the manufacturer's indications and cool the machine down.
2. Insert a sample carrier A, with one side with a cavity of 100 and the other side of 200 μm, in the hole of the middle plate, the 100 μm cavity upside. Take one sapphire disk using tweezers No 2 and tip it into 100% fetal bovine serum. Transfer it to the sample carrier (Fig. 4a) and put it in the 0.1 mm deep indentation, cell layer on sapphire must be upside (*see* **Note 21**).
3. Take a 2-μL pipette and fill the cavity of the sample carrier A with 100% fetal bovine serum (*see* **Note 22**). Close the carrier A with the flat side of a carrier B (Fig. 4b, c), forming a 100 μm deep cavity containing the 50 μm thick sapphire. Close the half cylinders together and freeze (consult the EM HPM100 manual for details). The half cylinders and middle plate containing carriers and sapphire disk fall in the LN_2 Dewar.
4. Pick up the middle plate using two isolated straight long forceps (*see* **Note 23**) and separate the carriers from the middle plate in liquid nitrogen using the Sample Carrier Release Device (Fig. 5a, b). Store the carrier sandwich in a cryotube containing several holes (**star-mark in** Fig. 5b) in liquid nitrogen.
5. Meanwhile, store the culture dish with the remaining sapphire disks in the CO_2 incubator.

3.2.3 Freeze Substitution

1. Fill the freeze substitution apparatus with liquid nitrogen. Regulate the substitution program: temperature 1: –90 °C for 45 h, temperature change 2 °C/h, temperature 2: –60 °C for 8 h, temperature change 2 °C/h, temperature 3: –30 °C for 8 h (Fig. 6).

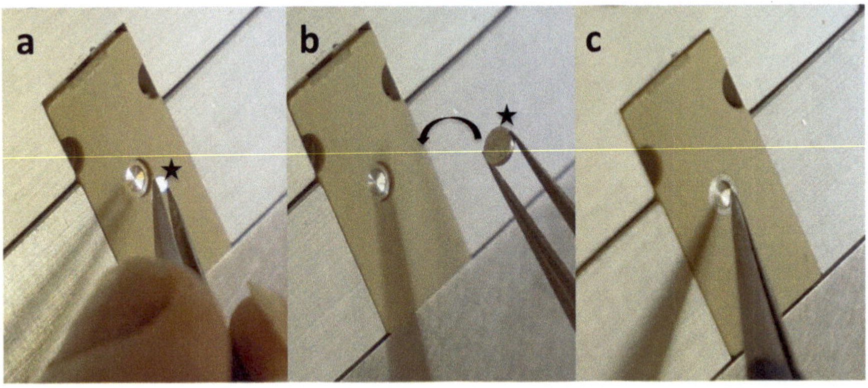

Fig. 4 Insertion of the sample in the middle plate. The sapphire disk (*star-mark*) containing the cell monolayer is deposited in a specimen carrier type A (**a**). The chamber is closed with a specimen carrier type B, the flat surface in contact with the chamber (*star-mark* in **b**) without introducing any air bubbles. Push the carrier B in place with the tweezers (**c**)

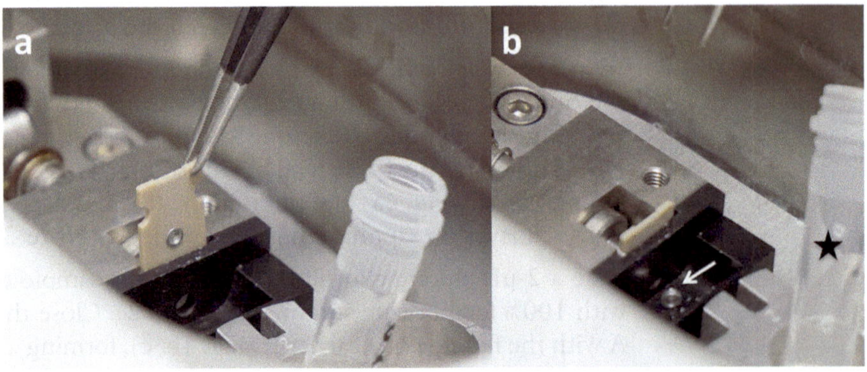

Fig. 5 Separation of carriers A and B containing the sapphire disk from the middle plate using the Sample Carrier Release Device. The middle plate is introduced in the slot (**a**). The orientation of the middle plate is not of importance. When the piston is activated the sample is falling out (*arrow* in **b**). Deposit it with pre-cooled insulated tweezers in a perforated cryovial (*star-mark* in **b**)

2. Insert the holder for cryotubes in the substitution chamber and fill the chamber with 150 mL of 100% ethanol.
3. Cool the machine down to −90 °C.
4. Prepare 10 mL substitution solution (1% OsO_4/2% H_2O/acetone), which is enough for 6 samples (*see* **Note 24**). Fill 800 μL of substitution solution in each cryotube at room temperature, close the tubes and store them immediately at −90 °C in the substitution chamber. Cool down the substitution solution to −90 °C.
5. Fill up a styrene box of maximal height of 6 cm with liquid nitrogen, put a white plastic pot of PTFE (polytetrafluoroethylene) in the box (which should be covered by

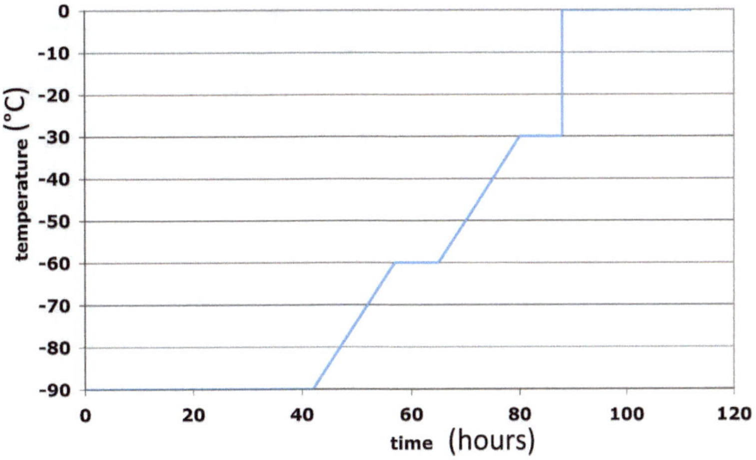

Fig. 6 Temperature curve for freeze substitution

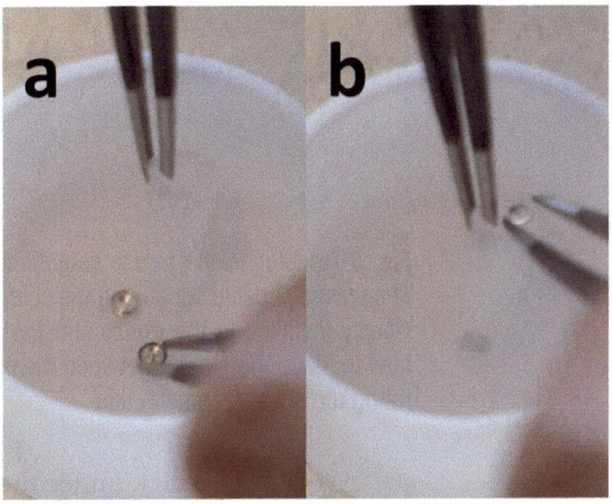

Fig. 7 Transfer of the sample. The A carriers, B carriers and the sapphire disk are poured in a PTFE pot (**a**) and are transferred to a plastic capsule (D5 × H15 mm) with pre-cooled insulated coated tweezers (**b**)

the liquid nitrogen) and pour the sapphire containing carriers into it (Fig. 7a).

6. For each sample transfer the carriers A, B and the sapphire disk in a plastic capsule D5 × H15 mm, which is perforated on its bottom, using isolated straight forceps (Fig. 7b). The plastic capsule is transported in liquid nitrogen to the cryo-substitution machine, plastic capsules are put inside the cryo vials (Fig. 8a). Start the substitution program.
7. At the end of the program continue with 1% OsO_4/2% H_2O/ acetone for 1 h at 4 °C (put the cryovials in an ice bucket) followed by 1% OsO_4/2% H_2O/acetone 1 h at RT (put the

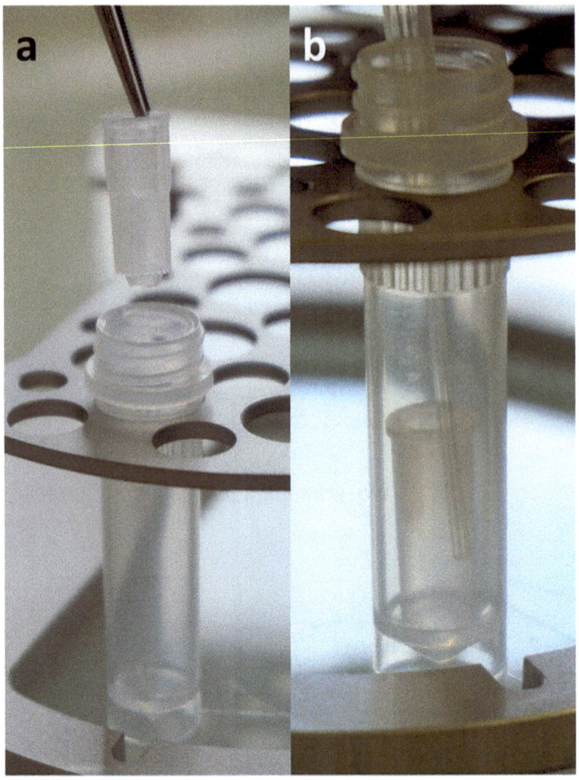

Fig. 8 Transfer to the freeze substitution apparatus and exchanging solutions. The plastic capsule containing the sample is inserted in cryovials which are previously installed in the freeze substitution apparatus at −90 °C (**a**). (**b**) Shows how to place the pipette between the plastic capsule and the cryovials for aspirating the substitution solution

cryovials in a rack under the hood), acetone 5× 5 min at RT (*see* **Note 25**; Fig. 8b), EPON 30% 3 h at RT, EPON 75% 3 h at RT and finally pure EPON resin overnight at RT.

8. Retrieve the plastic capsule of the cryovials (Fig. 9a, b), deposit it with the aperture downward (Fig. 9c) in a culture dish, perforate the bottom of the capsule (**arrowhead in** Fig. 9c) and pour out EPON resin, carriers and sapphire disk (Fig. 9d). Using a magnifying glass, pick up the sapphire disk and put it, cells upward (*see* **Note 26**), in a pyramid-shaped flat mold (Fig. 10b, c) or a flat embedding mold depending on in which orientation you decide to cut your sample (*see* **Note 27,** Figs. 10a–c). Fill up the cavity of the mold with EPON resin (Fig. 10a). For samples in a pyramid shaped flat embedding mold, deposit a gelatine capsule containing polymerized EPON resin over it (Fig. 10d, e) and polymerize for 24 h at 60 °C.
9. Mount the specimen in a specimen holder of a microtome (Fig. 11a). To separate the sapphire from the resin, remove the

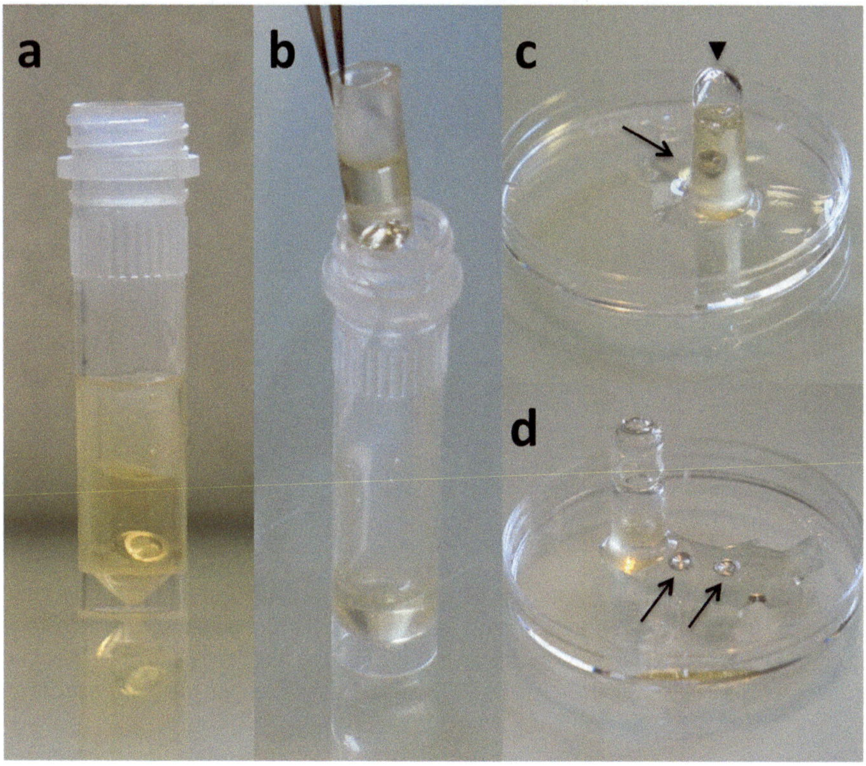

Fig. 9 Recovery of the sapphire disk for polymerization in EPON resin. The plastic capsule containing the sample is extracted from the cryovials (**a** and **b**). The capsule is perforated at the top (*arrowhead* in **c**). Resin, carriers, and sapphire disk are floating into the culture dish (*arrows* in **d**)

resin above (**star-mark** Fig. 11b) and around the sapphire disk (**star-mark** Fig. 11c). Plunge the top of the sample in liquid nitrogen for a few seconds (Fig. 11d), press the surface of the sapphire disk against a hot plate at 80 °C for a few seconds until the surface is not frozen anymore (Fig. 11e) and remove the sapphire (Fig. 11f) by inserting the edge of a razor blade (as indicated by an arrowhead in Fig. 11c) between the resin and the disk (*see* **Note 28**).

3.2.4 Cutting and Contrast

1. Cut 60 nm thick sections using a ultramicrotome according to the manufacturer's instructions. Collect sections on 100 Mesh or 200 Mesh copper grids.
2. Contrast the sections with uranyl acetate: prepare one droplet of filtered uranyl acetate and five droplets of autoclaved and filtered water per grid on a parafilm. Put the grid for 10 min on the uranyl acetate droplet with the section facing the droplet. Protect grids from light during incubation.
3. Take the grid with a forceps and rinse it for 30 s in each of the five water droplets. Soak water away from grid with a filter

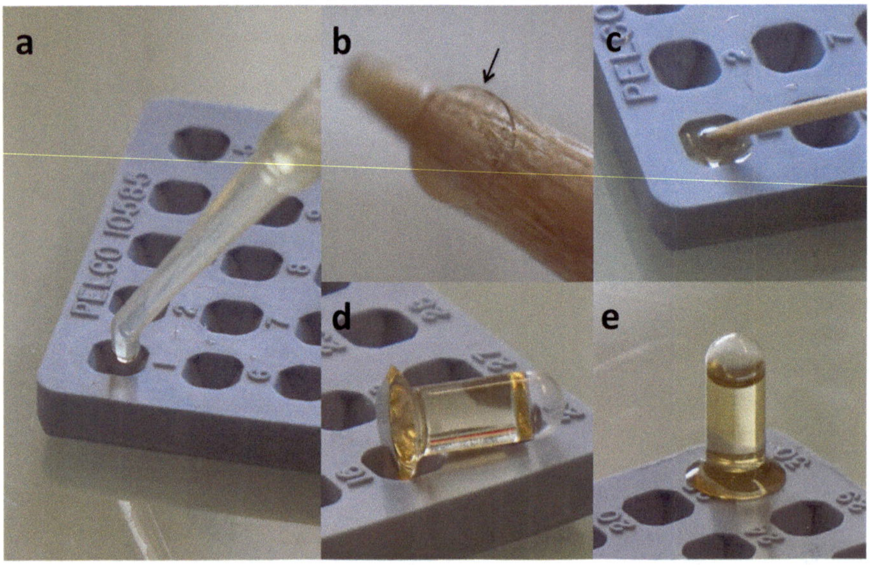

Fig. 10 Retrieving the sapphire disk. The mold is filled with EPON resin (**a**). The sapphire disk (*allow*) is recovered (**b**) and placed in the mold (**c**). Through the layer of carbon, the sapphire disk can be oriented under a magnifying glass so that the cell monolayer is located at the top. The mold is filled to the top with resin and a gelatin capsule previously filled with polymerized EPON (**d**) is deposited on the top of the sample (**e**)

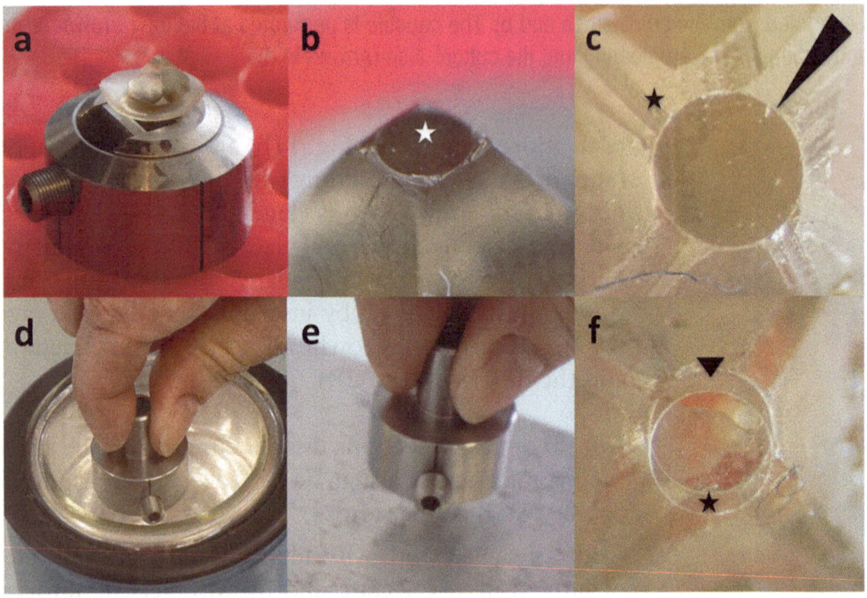

Fig. 11 Separation of the sapphire disk. The sample is fixed in a specimen holder of the ultramicrotome (**a**). The sapphire disk is released on its top and around it from the resin (*star-marks* in **b** and **c**). The sample is plunged in liquid nitrogen (**d**), heated at 80 °C on a hot plate (**e**) and the sapphire disk is taken off by introducing the edge of a razor blade between the sapphire disk and the underlying resin (*arrowhead* in **c**). The detached sapphire disk (*star-mark*) and the resin surface containing the cell monolayer (*arrowhead*) is shown in (**f**)

paper by touching the filter paper with the edge of the grid inclined at 45°.

4. Contrast the sections with lead citrate: prepare five droplets of autoclaved, filtered water on parafilm. Place a culture dish of 30 mm diameter containing some NaOH pellets near the droplets. Deposit one droplet of filtered lead citrate near the NaOH pellets (*see* **Note 18**). Put the grid on the lead citrate droplet with the section facing the liquid for 1 min. Rinse and dry the grids as described in **step 3** above.
5. Observe under a transmission electron microscope at 80 kV according to the manufacturer's instructions (Fig. 12).

3.3 Immunogold Labeling on Ultrathin Cryosections, the Tokuyasu Method

3.3.1 Sample Fixation (Cells and Tissues)

1. Prepare the fixative: either use 4% paraformaldehyde in 0.1 M PB, or a mixture of 4% paraformaldehyde (PFA) and 0.4% glutaraldehyde (GA) in 0.1 M PB.
2. Dilute the fixative two times with the cell culture medium (v/v) and add to the adherent cells for 5 min at room temperature.
3. Replace this solution with the same fixative diluted twice: 2% PFA in 0.1 M PB, or a mixture of 2% PFA and 0.2% GA in 0.1 M PB (*see* **Note 29**). Fix overnight at 4 °C.

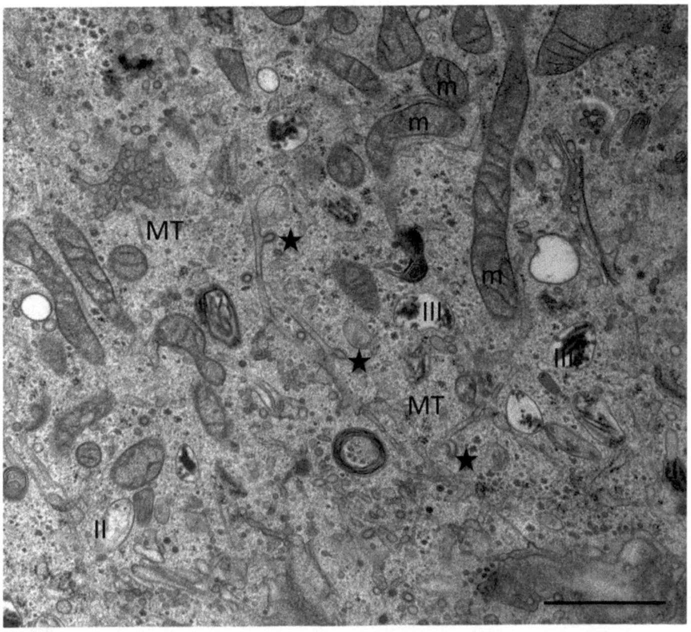

Fig. 12 High pressure frozen MNT1 cells. Different organelles are clearly visible and well preserved: mitochondria (m), microtubules (MT), melanosomes stage II (II) and stage III (III), and the endosomal network depicting clathrin-coated buds (*star-mark*). Scale bar: 1 μm

3.3.2 Embedding, Cryoprotection, and Freezing

1. Remove the fixative and wash the samples three times in PB.
2. Wash in PB with 50 mM glycine to quench aldehydes.
3. Gently scrape adherent cells in 1% gelatin, and gently centrifuge to obtain a pellet.
4. Mix the cells in 10% gelatin at 37 °C for 10 min. For tissues, you need to cut 1 mm^3 pieces in a small petri dish before immersion in 10% gelatin.
5. Centrifuge for 5 min at 600 × *g* the cell suspension in a 90° angle centrifuge.
6. Incubate the samples (cells and tissues) on ice until the gelatin has solidified (minimum 30 min).
7. Cut off the bottom of the tube with a razor blade. Take the bottom part of the tube containing the pellet and cut it lengthwise in two equal parts in cold PBS. Thus the pellet is cut in two equal parts. Finally, using a tip or a toothpick, push gently the samples out of each half bottom tubes.
8. Prepare small cubic blocks (1 mm^3) directly from the solidified pellet or tissue sample under a magnifying glass or dissection microscope and infuse in 2.3 M sucrose, on a rotator, overnight at 4 °C for cell pellets or 48 h for tissue.
9. Each block is then transferred from the tube to the specimen holders of the cryo-ultramicrotome, so called "pins", using a dissection microscope, and working at 4 °C (in a cold room or on a eutectic plate) (*see* **Note 30**; Fig. 13a).

Fig. 13 (**a**) Cubic blocks of fixed cells attached to the pin with sucrose and ready to be frozen in the liquid nitrogen. (**b**) Ultracryomicrotome prepared for cutting with the diamond knife is in front of the sample. (**c**) Ribbon of ultrathin cryosections sections on the knife with an approximate thickness of 70 nm

10. Remove the excess of sucrose with a piece of filter paper, leaving a thin film underneath the block, sufficient to "glue" it to the metal surface (*see* **Note 31**).
11. Immediately freeze the blocks mounted on the holders in a small cryogenic Dewar flask filled with liquid nitrogen (LN_2). They will then be ready for transfer into the cryochamber of the cryo-ultramicrotome (at −90 °C) or storage in LN_2 (*see* **Note 32**).

3.3.3 Cryosectioning

Cryo-ultramicrotomy requires expertise. Training and practice are needed to obtain satisfactory results. With the commercially available equipment it can become a routine method in several laboratories and all students should be able to perform cryosectioning.

1. Cool down the cryo-ultramicrotome, equipped with a cryochamber, to −90 °C (Fig. 13b).
2. Insert the knife holder with diamond knifes: trimming knife (Trim 45°, Diatome) and the cryoimmuno diamond knife 35° (Diatome).
3. When the temperature of −90 °C is reached in the cryochamber, the frozen blocks mounted on the pins are quickly transferred into it.
4. Insert the sample in the cryo specimen holder, using insulated forceps.
5. With the trimming knife, trim the sample surface until obtaining a smooth shiny surface (*see* **Note 33**). The following settings are recommended: speed, 100 mm/s and feed, 250 nm. During the trimming use the antistatic device at full voltage (*see* **Note 34**).
6. Using the edges of the trimming knife, trim the sample into a perfect rectangle of 90° angles and 200–300 μm in size to facilitate ultrathin cryosectioning (60–70 μm).
7. Decrease the temperature of the cryochamber to −115 °C, and rotate the knife holder to use the cryoimmuno knife.
8. Perform ultrathin sections with the cryoimmuno knife, using the following settings: Speed, 1.2 nm/s, and feed, 80 nm (*see* **Note 35**). An "optimal" ribbon should be as flat and shiny as possible (silver/yellowish color is indicative of the appropriate thickness) (Fig. 13c).
9. When you have a ribbon of sections, switch off the antistatic device before collecting the sections, which prevents sticking of the sections to the knife.
10. Remove the ribbon from the knife edge with an eyelash (especially if your ribbon is short, in order to prevent damage to the diamond during picking up step).

11. Introduce a loop, filled with a 1:1 fresh mixture of 2.3 M sucrose and 2% methylcellulose, into the cryochamber and pick up the sections (sections should be in the middle of the drop). The perfect timing to pick up your sample is when ice crystals appear on the edges of the loop (*see* **Note 36**).
12. After picking up the sections, the sucrose/methylcellulose drop is allowed to thaw briefly at room temperature. Then, the drop bearing the section is deposited on a 200 mesh formvar/carbon-coated grid.
13. The sections, laying on the grid, covered with sucrose/methylcellulose, can be stored at 4 °C, pending immunogold labeling (*see* **Note 37**).

3.3.4 Immunogold Labeling

1. Deposit grids with the sections in petri dishes containing 2% gelatin in 0.1 M PB with the sections facing the gelatin.
2. Place the petri dish with the grids at 37 °C for 20 min (*see* **Note 38**).
3. Unroll parafilm on the bench (*see* **Note 39**).
4. Wash in glycine 50 mM in PBS, 4× 2 min (quenching of free aldehyde groups) (Fig. 14a).
5. Incubate in 0.1% cold water fish skin gelatin in PBS (CWFS/PBS) for 5 min (saturation of non-specific binding sites, avoiding non-specific binding of antibodies and therefore background).

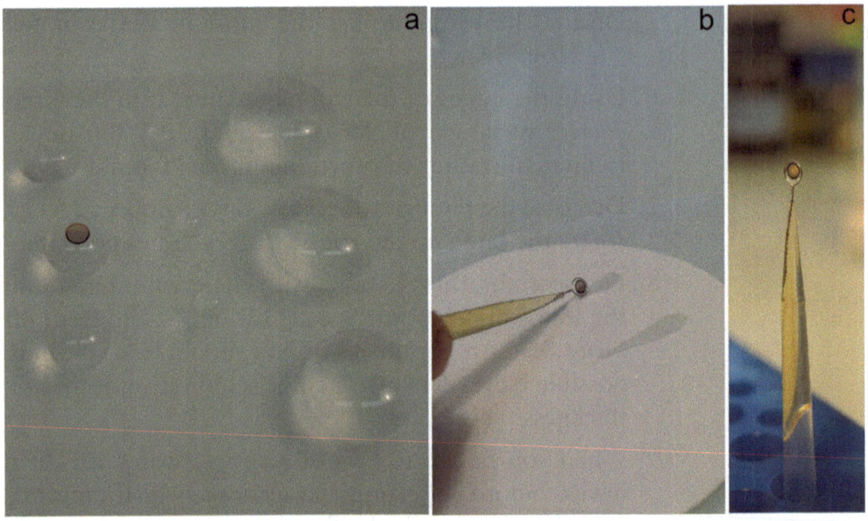

Fig. 14 (**a**) Immunolabeling steps showing a section on the grid floating on the top of a droplet. The first drop is for the blotting step, the second is for the antibody incubation and finally the washing step in big drops to allow a complete removal of excess of antibody. (**b**) Blotting step to remove the uranyl acetate and methylcellulose with an inclination of 45°. (**c**) Air-dried sample on the grid will be removed from the loop to be observed by electron microscopy

6. Incubate with the primary antibody diluted in 0.1% CWFS/PBS for 30–120 min.
7. Wash in 0.1% CWFS/PBS 5× 2 min.
8. Bridging antibody in 0.1% CWFS/PBS for 20 min (*see* Subheading 3.3.5).
9. Wash in 0.1% CWFS/PBS 5× 2 min.
10. Incubate with Protein-A-Gold (PAG) in 0.1% CWFS/PBS for 20 min (for multiple labeling, use different sizes of gold, 5–15 nm). For a single labeling we use generally 10 nm gold particles.
11. Wash with PBS 5× 2 min.
12. Wash in 1% GA in PBS 5 min. This stabilizes the immunoreaction, and denatures the IgGs, which is required for double labeling to prevent co-labeling of the secondary antibody). For a single protein labeling go directly to the **step 14**. For multiple labeling proceed to **step 13** (*see* **Note 40**).
13. Wash in PBS 2× 1 min, then go back to the **step 4**.
14. Wash with distilled water 6× 1 min.
15. Embedding and contrast step: incubate the grids on a mixture of 9:1 methylcellulose/uranyl acetate, on ice, in the dark for 10 min.
16. Remove the grids from the drop with loops that have a slightly larger diameter than the grid.
17. Soak the excess of fluid by gently pushing the loop sideways (with an inclination of 45°) on a Whatman filter paper No 1 placed on a parafilm (to avoid blotting uranyl acetate on the bench) so that a thin film is left covering the section side of the grid (*see* **Note 41;** Fig. 14b).
18. Air-dry the grid. After drying, the film should show a dark gold to blue interference color (Fig. 14c).
19. Remove the grid from the loop with a thin tweezers and store in an appropriate gridbox (plastic, numbered gridboxes are commercially available).

3.3.5 Antibody Reactivity

A bridging antibody is necessary when the primary antibody does not react with PAG or to amplify the signal of an antibody. For instance, PAG has a weak affinity for mouse IgG1 antibodies, goat and rat IgG antibodies. In those cases, we use a rabbit anti-mouse IgG, rabbit anti-goat IgG, or rabbit anti-rat IgG, respectively (Table 1).

In order to visualize tagged recombinant proteins expressed in cells, there are commercially available anti-tag antibodies. As for example anti-GFP, anti-tag HA, rat anti HA or anti-myc (clone 9E10) and anti-flag polyclonal (Fig. 15).

4 Notes

1. Glutaraldehyde 25% aqueous solution and formaldehyde 16% aqueous solution are purchased at electron microscopy grade in glass ampoules sealed under dry nitrogen. Formaldehyde, once opened should be used freshly. Glutaraldehyde 25% can be stored at 4 °C and used during 2 months.

Table 1
Affinity of protein A for different classes of antibodies

Species	Immunoglobulin	Binding to protein A
Human	IgG	++++
	IgG1	++++
	IgG2	++++
	IgG3	–
	IgG4	++++
	IgM	–
	IgA	–
	IgE	–
Mouse	IgG1	+
	IgG2a	++++
	IgG2b	+++
	IgG3	++
Rat	IgG1	–
	IgG2a	–
	IgG2b	–
	IgG2c	+
Goat	IgG	+/–
Rabbit	IgG	++++
Sheep	IgG	+/–

Fig. 15 (continued) (GA) as well as mitochondria (Mit.). Scale bar: 200 nm. (**b**) Ultrathin cryosections of highly pigmented melanoma cells (MNT-1 cells). Ultrathin cryosections were single immunogold labeled with anti-TYRP1 antibodies and protein A conjugated to 10 nm gold particles (PAG10). Note pigmented melanosomes mostly labeled at their limiting membrane (*arrowheads*). The membranes appear as negatively stained as compared to conventional microscopy where membranes appear electron dense. Melanosomes are clearly visible at different stages, stage I(I), stage II(II), and stage III(III). Scale bar: 200 nm. (**c**) Ultrathin cryosections of retinal pigment epithelium, double immunogold labeled for ApoE3 and protein A conjugated to 15 nm gold particles (PAG15) and the melanosomal protein PMEL (HMB45 monoclonal antibody) and protein A conjugated to 10 nm particles (PAG10). Scale bar: 200 nm

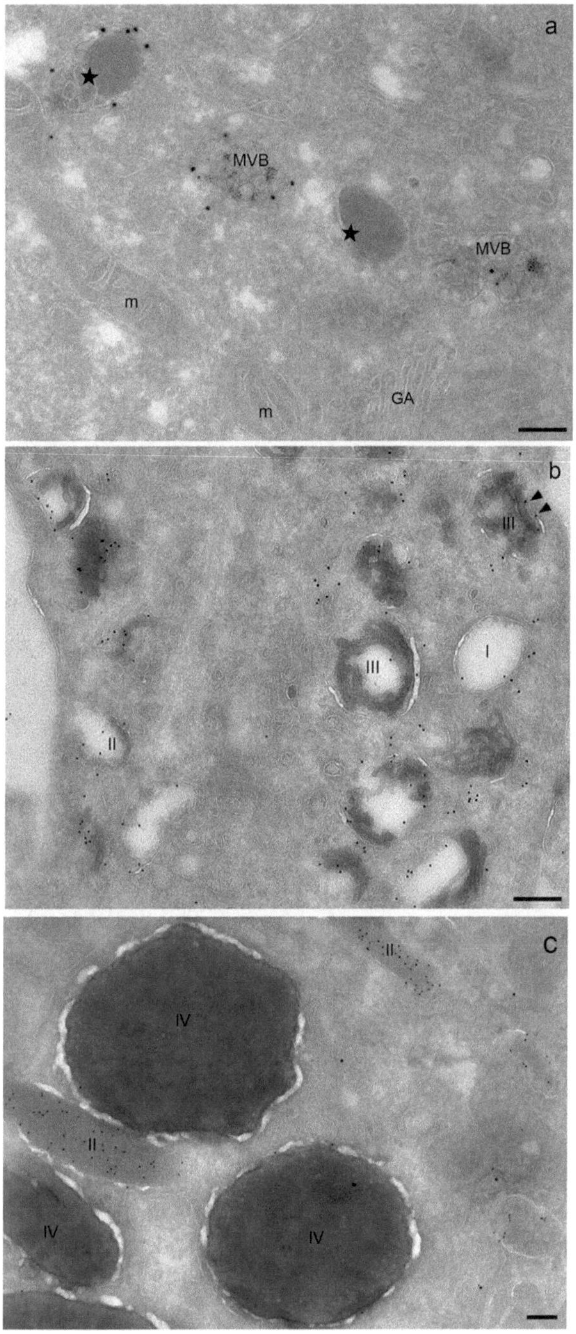

Fig. 15 (**a**) Ultrathin cryosections of cytotoxic T cells that were allowed to internalize BSA conjugated to 5 nm gold particles (BSAG) for 30 min before fixation and processing for ultrathin cryosectioning. Cytolytic granules appear as organelles with an electron dense core (*star*) surrounded by small intraluminal vesicles. Sections were labeled with Lamp-1 antibodies and Protein A conjugated to 10 nm gold particles (PAG10). Cytolytic granules depict LAMP-1 labeling on their limiting membrane and occasionally on the intraluminal vesicles. BSAG appears in endocytic multivesicular bodies (MVBs). The Golgi apparatus is also observed

2. OsO_4 is purchased as 4% aqueous solution in glass ampoules sealed under dry nitrogen. OsO_4 once opened should be stored at 4 °C in a consumable centrifuge tube which is packed in a tightly closed vial containing glycine. Glycine will capture leaking OsO_4 vapors and avoid black precipitates on the inside of the refrigerator. Aqueous OsO_4 solution can be used for several months. When the original lightly yellowish solution becomes brownish, solution should be discarded. OsO_4 should only be handled under a fume hood with protection gloves using consumable transfer pipettes and consumable 15 mL centrifuge tubes.
3. Just before use, pour the content of the ferrocyanide tube in tube with osmic acid anhydride. The solution will be brownish.
4. EPON 812 resin, DDSA, NMA, and DMAE can be bought as a kit.
5. Lead citrate precipitates in the presence of CO_2, so water used for preparation of solution should be CO_2 free. Use boiled or autoclaved water. The solution should not be in contact with air during storage. Discard air bubbles carefully from the syringe.
6. Several High-Pressure Freezers exist on the market from various manufacturers: the historical HPM010 or the newer "HPM α" and "HPM Light μ" from ABRA Fluid and CryoCapCell are distributed by RMC-Boekeler. TechnoTradeInc distributes the "HPF Compact02" from Wohlwend. The "HPM100" or the "EM ICE" are distributed by Leica microsystems.
7. Several freeze substitution automates exist on the market from various manufacturers: the "AFS" or the "AFS-2" from Leica microsystems, the "FS-8500 Freeze substitution system" from RMC-Boekeler.
8. Formvar-coated carbon-coated grids can be homemade if a carbon evaporator is available [13, 14] or purchased commercially (from Cell Microscopy Center Utrecht University or other sources, for example Aurion).
9. Attention: at the following steps the samples should always be maintained wet to preserve their structural integrity.
10. Isolated cells must be centrifuged 5 min at 600 × *g* after each exchange of solution.
11. Place the Eppendorf tubes in an ice basket and cover them to protect against light.
12. Acetone should be water-free as EPON resin do not infiltrate and polymerize well in samples containing traces of water. Use a recently opened bottle.

13. Close the tube with the cap to avoid evaporation of acetone.
14. Do not dissociate cell pellets any more. EPON resin is very viscous and the cells cannot be centrifuged to a pellet.
15. Open the cap to evaporate the residual acetone.
16. Push the tissue bloc to the border of the mold and put a paper containing the archive number of the sample written with a pencil. EPON behaves like a solvent and erases most of scripts.
17. The oven should be used only for polymerization of resins. Traces of moister can indeed prevent good polymerization.
18. To protect lead citrate from CO_2, cover the entire installation with a lid. NaOH captures CO_2. The first droplet of lead citrate coming out from the syringe should always be discarded. Never breathe on the grids during incubation in lead citrate. Use the lid as a protection screen.
19. Cells are cultured on the carbon pattern, which give indication about which side the cells are situated. For carbon coating all finder grids should be flipped into a readable orientation.
20. Ethanol should be in a flat petri dish. If a sapphire disk is lost, it can easily be found in the culture dish by discarding ethanol. The disk will stay at the bottom and can be picked up.
21. To release the sapphire from the forceps during transfer, use second forceps to push the sapphire into the carrier.
22. Build a little dome of liquid at the top of the cavity. Do not generate air bubbles! If the sapphire disk sticks to the pipette tip, push it back with forceps.
23. Forceps must be cooled down to −196 °C before touching the middle plate or the samples to avoid raise of the sample temperature and formation of ice crystals in the vitrified tissue.
24. The cryosubstitution solution should be prepared rapidly and stored quickly at −90 °C in the AFS apparatus to avoid reduction of osmium tetroxide.
25. To change the solutions, glide a fine plastic pipette between the plastic capsule containing the carriers and the wall of the cryotube. Aspire approximately ¾ of the liquid. The sapphire disk in the carrier should never dry out. Turbulences of liquids should be avoided to prevent the sapphire disk from falling out of the carrier.
26. The carbon matrix on the sapphire disk helps for orientation. The letters should be in the readable orientation. Cells will then be facing upward.
27. Protect surfaces against non-polymerized EPON resin with aluminum foil. Use a toothpick to pick up the sapphire disk, to deposit it in the mold and to orientate it.

28. Air enters between the surface of resin containing the cells and the sapphire. The sapphire disk will be entirely separated. If this is not the case, do not use force but start again by tipping the sample in liquid nitrogen.
29. The GA fixative solution could be used from 0.1% up to 0.5%. The use of GA results in a good ultrastructure preservation. However, some antibodies do not recognize antigens cross-linked by GA (even at very low concentration). For those, the fixative must be GA free. To optimize the results, the right fixative solution for each antibody has to be tested. When testing an unknown antibody, we generally fix samples in duplicates with PFA alone or a mixture of PFA and GA.
30. Mount the samples on the pins under the dissection microscope or a magnifying glass. To put down the sample block on the pins one can use a flat toothpick. Be sure that the pins are clean (you can clean them with acetone, and then let them dry). The surface of the pins should be rough (or designed with concentric rings) for a good adhesion of the sample. Also, to optimize the adhesion of the sample, the largest side of the sample should be based on the pin for ensuring stability.
31. Proceed as quickly as possible to avoid drying, during the mounting on pins. Prepare samples one by one.
32. The frozen specimens can be stored in liquid nitrogen for several years and can be used repeatedly for sectioning if they are not exposed to substantial changes of temperature or unfreezing.
33. If the sample surface appears white and not translucent and shiny, shows irregularities and/or rough surface, it is most likely the consequence of ice crystals in this area. In an ice crystal area, the cell ultrastructure is damaged, and it will be impossible to cut ultrathin sections. A possible solution is to trim further until you get a shiny and smooth surface. If not successful, try to change sample, as some samples may not be correctly infused in sucrose 2.3 M.
34. Using the antistatic device during trimming avoids sticking sections on the knife edge.
35. Use the antistatic device at full voltage, and when you obtain the first sections decrease the antistatic power, and find a balance in order to stabilize the ribbon. If the antistatic device voltage is too high, the ribbon will fly away. On the other hand, if it is too low, sections will stick on the edge of the diamond. An alternative is to manually move forward or backward the antistatic device from the knife. Finally, with an eyelash, mounted on a wooden stick, handle and maintain the ribbon during cutting.

36. Picking up ultrathin cryosections from the knife is a critical step. The ultrathin sections are collected from the knife with a stainless steel wire (0.2–0.25 mm) loop (diameter 2.5 mm), which has been mounted at the end of a 15 cm wooden stick. To get an optimal preservation of ultrastructure, the droplet of sucrose/methylcellulose should be neither frozen nor too warm before collecting. The mixture sucrose/methylcellulose is kept at 4 °C on ice, like the sections deposited on the grids after picking up.
37. If the samples are mildly fixed, such as when using only PFA, it is recommended to start the immunolabeling as soon as possible in order to minimize loss and redistribution of antigens.
38. This step allows the removal of the sucrose/methylcellulose that covers the sections. This step is extremely important, as residues of sucrose/methylcellulose will directly influence the quality of the labeling.
39. The following immunolabeling procedure is performed by floating the grids (sections facing the buffer) on successive drops of washing, blocking, and labeling solutions, disposed on the parafilm sheet placed on the bench. Each rinsing step is performed on drops of 200–500 μL (disposable plastic pipettes are convenient) while small droplets of 5–15 μL of immunoreagents are sufficient (this prevents the use of large amounts of antibodies and probes that are generally precious). It is important to keep the grids wet on the section side during all steps, but dry at the opposite side. Drying out of the sections during the incubation step can result in high background level.
40. Before performing any double labeling, perform first a single labeling with each antibody separately to get acquainted with the labeling pattern. It is possible to perform labeling with 2 polyclonal antibodies in combination with protein A-Gold or 1 polyclonal and 1 monoclonal antibody (different species). Labeling with 2 monoclonal antibodies is definitely not recommended even if they are of different isotypes, because the bridging steps will most likely cause false positives (even after the GA step).
41. During blotting, a quick removal of the uranyl acetate/methylcellulose droplet will result in a thicker film of methylcellulose, and a slower removal in a thinner film. This step is critical: if the methylcellulose film is too thin, sections can dry out and will not be stable enough under the electron beam. If the methylcellulose film is too thick, structural details on the sections will hardly be visible and will not show any contrast.

Acknowledgements

We are grateful to Institut Curie and the BioImaging platform (PICT IBiSA), member of the France-BioImaging national research infrastructure (ANR-10-INSB-04) and CNRS. Research in our group is supported by the Fondation pour la Recherche Médical (Equipes FRM), Association de Recherche pour le Cancer (ARC), Indian French cooperation (CEFIPRA), Clarins and L'Oréal.

References

1. Raposo G, Marks MS, Cutler DF (2007) Lysosome-related organelles: driving post-Golgi compartments into specialisation. Curr Opin Cell Biol 19(4):394–401
2. Marks MS, Heijnen HF, Raposo G (2013) Lysosome-related organelles: unusual compartments become mainstream. Curr Opin Cell Biol. 25(4):495–505. doi:10.1016/j.ceb.2013.04.008
3. Novikoff AB, Albala A, Biempica L (1968) Ultrastructural and cytochemical observations on B-16 and Harding-Passey mouse melanomas. The origin of premelanosomes and compound melanosomes. J Histochem Cytochem 16(5):299–319
4. Palade GE (1983) Membrane biogenesis: an overview. Methods Enzymol 96:xxix–xxlv
5. Walther P, Ziegler A (2002) Freeze substitution of high-pressure frozen samples: the visibility of biological membranes is improved when the substitution medium contains water. J Microsc 208(Pt 1):3–10
6. Studer D, Graber W, Al-Amoudi A, Eggli P (2001) A new approach for cryofixation by high pressure freezing. J Microsc 203:285–294
7. Hawes P, Netherton CL, Mueller M, Wileman T, Monaghan P (2007) Rapid freeze-substitution preserves membranes in high-pressure frozen tissue culture cells. J Microsc 226(Pt 2):182–189. doi:10.1111/j.1365-2818.2007.01767.x
8. Shio H, Farquhar MG, de Duve C (1974) Lysosomes of the arterial wall. IV. Cytochemical localization of acid phosphatase and catalase in smooth muscle cells and foam cells from rabbit atheromatous aorta. Am J Pathol 76(1):1–16
9. Novikoff AB, Novikoff PM (1977) Cytochemical contributions to differentiating GERL from the Golgi apparatus. Histochem J 9(5):525–551
10. Slot JW, Geuze HJ (1981) Sizing of protein A-colloidal gold probes for immunoelectron microscopy. J Cell Biol 90:533–536
11. Geuze HJ, Slot JW, Strous GJ, Lodish HF, Schwartz AL (1983) Intracellular site of asialoglycoprotein receptor-ligand uncoupling: double label immunoelectron microscopy during receptor-mediated endocytosis. Cell 32:277–287
12. Slot JW, Posthuma G, Chang L, Crapo JD, Geuze HJ (1988b) Quantitative assessment of immuno-gold labeling in cryosections. In: Verleij AJ, Leunissen JLM (eds) Immuno-gold labeling in cell biology. CRC Press Inc., Boca Raton, FL, pp 135–156
13. Slot JW, Geuze HJ (2007) Cryosectioning and immunolabeling. Nat Protoc 2(10):2480–2491
14. Raposo G, Kleijmeer MJ, Posthuma G, Slot JW, Geuze HJ (1997) Immunogold labeling of ultrathin cryosections: application in immunology. In: Herzenberg LA, Weir D, Herzenberg LA, Blackwell C (eds) Handbook of experimental immunology, 5th edn, vol 4. Blackwell Science Inc., Cambridge, MA, pp 1–11, Chapter 208
15. Klumperman J, Raposo G (2014) The complex ultrastructure of the endolysosomal system. Cold Spring Harb Perspect Biol 6(10):a016857. doi:10.1101/cshperspect.a016857
16. Setty SR, Tenza D, Sviderskaya EV, Bennett DC, Raposo G, Marks MS (2008) Cell-specific ATP7A transport sustains copper-dependent tyrosinase activity in melanosomes. Nature 454(7208):1142–1146
17. Theos AC, Tenza D, Martina JA, Hurbain I, Peden AA, Sviderskaya EV, Stewart A, Robinson MS, Bennett DC, Cutler DF, Bonifacino JS, Marks MS, Raposo G (2005) Functions of AP-3 and AP-1 in tyrosinase sorting from endosomes to melanosomes. Mol Biol Cell 16:5356–5372
18. Raposo G, Tenza D, Murphy DM, Berson JF, Marks MS (2001) Distinct protein sorting and localization to premelanosomes, melanosomes, and lysosomes in pigmented melanocytic cells. J Cell Biol 152(4):809–824

19. Delevoye C, Heiligenstein X, Ripoll L, Gilles-Marsens F, Dennis MK, Linares RA, Derman L, Gokhale A, Morel E, Faundez V, Marks MS, Raposo G (2016) BLOC-1 brings together the actin and microtubule cytoskeletons to generate recycling endosomes. Curr Biol 26(1):1–13. doi:10.1016/j.cub.2015.11.020
20. Delevoye C, Hurbain I, Tenza D, Sibarita JB, Uzan-Gafsou S, Ohno H, Geerts WJ, Verkleij AJ, Salamero J, Marks MS, Raposo G (2009) AP-1 and KIF13A coordinate endosomal sorting and positioning during melanosome biogenesis. J Cell Biol 187(2):247–264. doi:10.1083/jcb.200907122.
21. Wasmeier C, Romao M, Plowright L, Bennett DC, Raposo G, Seabra MC (2006) Rab38 and Rab32 control post-Golgi trafficking of melanogenic enzymes. J Cell Biol 175(2):271–281
22. van Niel G, Charrin S, Simoes S, Romao M, Rochin L, Saftig P, Marks MS, Rubinstein E, Raposo G (2011) The tetraspanin CD63 regulates ESCRT-independent and -dependent endosomal sorting during melanogenesis. Dev Cell 21(4):708–721. doi:10.1016/j.devcel.2011.08.019
23. Koster AJ, Klumperman J (2003) Electron microscopy in cell biology: integrating structure and function. Nat Rev Mol Cell Biol Suppl:SS6–SS10
24. Daniele T, Hurbain I, Vago R, Casari G, Raposo G, Tacchetti C, Schiaffino MV (2014) Mitochondria and melanosomes establish physical contacts modulated by Mfn2 and involved in organelle biogenesis. Curr Biol 24(4):393–403. doi:10.1016/j.cub.2014.01.007
25. Hurbain I, Geerts WJ, Boudier T, Marco S, Verkleij AJ, Marks MS, Raposo G (2008) Electron tomography of early melanosomes: Implications for melanogenesis and the generation of fibrillar amyloid sheets. Proc Natl Acad Sci U S A 105(50):19726–19731
26. Rochin L, Hurbain I, Serneels L, Fort C, Watt B, Leblanc P, Marks MS, De Strooper B, Raposo G, van Niel G (2013) BACE2 processes PMEL to form the melanosome amyloid matrix in pigment cells. Proc Natl Acad Sci U S A 110(26):10658–10663. doi:10.1073/pnas.1220748110
27. Kukulski W, Schorb M, Kaksonen M, Briggs JA (2012) Plasma membrane reshaping during endocytosis is revealed by time-resolved electron tomography. Cell 150(3):508–520. doi:10.1016/j.cell.2012.05.046
28. Kukulski W, Schorb M, Welsch S, Picco A, Kaksonen M, Briggs JA (2012) Precise, correlated fluorescence microscopy and electron tomography of lowicryl sections using fluorescent fiducial markers. Methods Cell Biol 111:235–257. doi:10.1016/B978-0-12-416026-2.00013-3
29. Mironov AA, Beznoussenko GV (2012) Correlative light-electron microscopy a potent tool for the imaging of rare or unique cellular and tissue events and structures. Methods Enzymol 504:201–219. doi:10.1016/B978-0-12-391857-4.00010-0
30. Muller-Reichert T, Verkade P (2012) Introduction to correlative light and electron microscopy. Methods Cell Biol 111:xvii–xxix. doi:10.1016/B978-0-12-416026-2.03001-6
31. van Rijnsoever C, Oorschot V, Klumperman J (2008) Correlative light-electron microscopy (CLEM) combining live-cell imaging and immunolabeling of ultrathin cryosections. Nat Methods 5(11):973–980. doi:10.1038/nmeth.1263. doi:nmeth.1263 [pii]
32. Polishchuk EV, Polishchuk RS, Luini A (2013) Correlative light-electron microscopy as a tool to study in vivo dynamics and ultrastructure of intracellular structures. Methods Mol Biol 931:413–422. doi:10.1007/978-1-62703-056-4_20
33. Sepulveda FE, Burgess A, Heiligenstein X, Goudin N, Menager MM, Romao M, Cote M, Mahlaoui N, Fischer A, Raposo G, Menasche G, de Saint BG (2015) LYST controls the biogenesis of the endosomal compartment required for secretory lysosome function. Traffic 16(2):191–203. doi:10.1111/tra.12244
34. Heiligenstein X, Heiligenstein J, Delevoye C, Hurbain I, Bardin S, Paul-Gilloteaux P, Sengmanivong L, Regnier G, Salamero J, Antony C, Raposo G (2014) The CryoCapsule: simplifying correlative light to electron microscopy. Traffic 15(6):700–716. doi:10.1111/tra.12164
35. Heiligenstein X, Hurbain I, Delevoye C, Salamero J, Antony C, Raposo G (2014) Step by step manipulation of the CryoCapsule with HPM high pressure freezers. Methods Cell Biol 124:259–274. doi:10.1016/B978-0-12-801075-4.00012-4
36. Markert SM, Britz S, Proppert S, Lang M, Witvliet D, Mulcahy B, Sauer M, Zhen M, Bessereau JL, Stigloher C (2016) Filling the gap: adding super-resolution to array tomography for correlated ultrastructural and molecular identification of electrical synapses at the *C. elegans* connectome. Neurophotonics 3(4):041802. doi:10.1117/1.NPh.3.4.041802
37. Wolff G, Hagen C, Grunewald K, Kaufmann R (2016) Towards correlative super-resolution fluorescence and electron cryo-microscopy. Biol Cell 108(9):245–258. doi:10.1111/boc.201600008
38. Liu B, Xue Y, Zhao W, Chen Y, Fan C, Gu L, Zhang Y, Zhang X, Sun L, Huang X, Ding W, Sun F, Ji W, Xu T (2015) Three-dimensional super-resolution protein localization correlated with vitrified cellular context. Sci Rep 5:13017. doi:10.1038/srep13017

Chapter 5

Microscopic Analysis of Lysosomal Membrane Permeabilization

Ana Maria Vilamill Giraldo, Karin Öllinger, and Vesa Loitto

Abstract

Lysosomes and lysosomal proteases have been found to participate during several forms of cell death pathways including apoptosis. A critical step in the mediation of apoptotic signaling is the release of cathepsins to the cytosol, a process known as lysosomal membrane permeabilization (LMP). In this chapter, we describe immunofluorescence detection of LMP in cell cultures stained for cathepsin B and LAMP-2 using three confocal techniques namely laser scanning, spinning disk, and aperture correlation spinning disk confocal to obtain images. Image analysis is performed using *Huygens* software for deconvolution. LMP results in a decrease in the fraction of cathepsin B colocalizing with LAMP-2, which is quantified through Manders' colocalization coefficient. Analysis of the images obtained by the three techniques show the same trend but the magnitude of the decrease differs due to the axial resolution. The observations emphasize the use of highest possible resolution when determining colocalization.

Key words Lysosomal membrane permeabilization, Confocal microscopy, Image analysis, Immunocytochemistry, Cathepsin B, LAMP-2

1 Introduction

Due to their high hydrolase content, lysosomes were referred to as "suicide bags" upon their characterization by Christian de Duve [1]. Indeed, lysosomes are potentially harmful to the cell and damage to the lysosomal membrane might result in leakage of lysosomal content to the cytosol. Depending on the duration and concentration of the lethal stimuli, the lysosome can undergo partial permeabilization of the membrane, which triggers apoptosis, or massive lysosomal rupture, which might induce necrosis [2]. The executors of lysosome-mediated apoptosis are their hydrolases, more specifically, the cathepsins [3, 4]. A critical step in the mediation of apoptotic signaling by cathepsins is their release to the cytosol, a process known as lysosomal membrane permeabilization (LMP). The importance of the cytosolic location of cathepsins for their pro-apoptotic function is highlighted by studies in which

Karin Öllinger and Hanna Appelqvist (eds.), *Lysosomes: Methods and Protocols*, Methods in Molecular Biology, vol. 1594, DOI 10.1007/978-1-4939-6934-0_5, © Springer Science+Business Media LLC 2017

microinjection of cathepsins into the cytosol was sufficient to induce apoptosis [5–7].

Immunofluorescence is a useful tool for visualizing cellular structures and intracellular processes. For successful immunofluorescence experiment, fixation and permeabilization must be optimized to preserve morphology and at the same time permeabilize the cell sufficiently to let the antibodies penetrate [8–10]. Standard protocols use fixation by crosslinkers like paraformaldehyde (PFA), or precipitating fixatives such as methanol or acetone. While methanol and acetone cause permeabilization during fixation, crosslinking fixative should be complemented with permeabilization performed by nonionic detergents such as TritonX-100, which dissolve lipids, or saponin, which is an amphipathic glycoside that removes cholesterol. Saponin provides a reversible permeabilization that preserves the membrane structures better than nonionic detergents do [11]. Next an optimal combination of specific antibodies needs to be found, which form an immune complex to mark the desired molecules, and fluorochromes, which are coupled to the immune complexes to visualize the target structures during microscopy. To perform analyzes on colocalization, the different primary antibodies must be derived from different species in order to distinguish the immune complexes. Furthermore, the fluorochromes must differ in their wavelength spectrum to discriminate their fluorescence signal in microscopic analysis [12].

To reliably analyze colocalization between two or more fluorescent-labeled proteins, high lateral and axial resolution is a prerequisite. Preferably superresolution microscopy (reviewed in [13, 14]) should be used but these are all still expensive techniques and not readily available for most users. However, most research groups have access to various forms of confocal microscopy, and although these techniques are limited by diffraction these microscopes comprise a reasonable compromise. The three confocal techniques described below have been addressed in depth in many resources [15, 16] and thus only a brief description is provided below.

In confocal laser scanning microscopy (CLSM), the sample is excited in a raster pattern with a small illumination spot focused at the focal plane of the objective. Emission from out-of-focus planes is subsequently blocked using a small, variable aperture (pinhole) in front of the detector, which is a photomultiplier tube (PMT), meaning that the detected volume primarily emanate from the in-focus plane [15]. The resulting image thereby becomes sharper with higher contrast and resolution than achieved with conventional wide-field microscopy where the whole sample is illuminated at once. The main benefits of CLSM are that it gives improved contrast and allows optical sectioning, meaning that fluorescent objects can be localized axially. It is a sensitive method with good blocking of out-of-focus light. With motorized sliders and polarizers it is also easy to combine with, e.g., differential interference contrast (DIC) microscopy. On the downside, the scanning process and the setup is moderately slow. Setup of image acquisition is very much speeded up by reusing a previously acquired setting.

In spinning disk confocal microscopy (reviewed in [17]), the illumination is passed through a rapidly rotating disk with a large number of small apertures (microlens array). As the disk rotates the sample is illuminated with a very large number of spots simultaneously, which allows visualization of the sample at video rate (~30 fps). Confocal sectioning is achieved by passing the emission through another, pinhole disk. The pinholes in spinning disk are not variable, meaning that the thickness of the optical section is fixed. On the other hand, as the disks are motorized and can be removed from the beam path, a widefield image is easily acquired in the system. Illumination in spinning disk can be obtained from high pressure mercury or xenon arc lamps, lasers or light emitting diodes (LEDs). Detection is done using a charge-coupled device (CCD), or preferably a sensitive scientific complementary metal–oxide–semiconductor (sCMOS) or an electron multiplying charge-coupled device (EMCCD) camera. The main benefits with spinning disk confocal microscopy are the acquisition speed, the reduced illumination power resulting in less phototoxicity and photobleaching, as well as real-time visualization of the confocal image on the monitor. The technique is very easy to combine with brightfield, such as DIC, which requires a motorized analyzer. On the downside, the spinning disk is moderately sensitive, does allow more out-of-focus light to reach the detector, pinhole size is not variable, and it poses a somewhat increased requirement considering labeling density.

An alternative confocal technique is aperture correlation spinning disk confocal. Although it is not a new technique, the specific methodology has not been published in detail. A very nice overview for the Zeiss VivaTome can be found at the interactive Zeiss educational website (http://zeiss-campus.magnet.fsu.edu/), and scientific results using the technique can be found in Karlsson et al. [18, 19]. Using the spinning grid-based confocal module, it is possible to acquire a live widefield and a confocal image simultaneously, side-by-side on the split camera chip. The spinning grid allows excitation illumination (metal halide source) to pass through to the sample, but then selectively filters the emission. All of the in-focus emission passes through the spinning grid, which it does together with 50% of the out-of-focus light, detected with a digital CCD or sCMOS camera. The remaining 50% of the out-of-focus light is reflected but also recorded. A widefield image is thus composed by summing up the transmitted light and the reflected part, while a confocal image is obtained by subtracting the contribution of reflected light from the transmitted part. Aperture correlation spinning disk is moderately fast, and another benefit of the technique is that both the widefield and the confocal images are observed in real time on the monitor. It is, thus, a live cell confocal as the spinning disk described above. On the downside, the setup is, depending on the actual software, moderately slow (fast when reusing a previous setting), it is not possible to simultaneously acquire bright-field images, and it has a smaller field of view as the camera chip is split between the wide-field and the confocal image.

Huygens Professional (Scientific Volume Imaging, Hilversum, The Netherlands) is an advanced software mainly focused on deconvolution and image restoration. However, it is also a very versatile image processing software and contains a large selection of available options, such as colocalization analysis, object analysis, object tracker, and track analysis. The principles of image restoration have been reviewed elsewhere [20], and an excellent guide to evaluate colocalization in biological microscopy is provided by Dunn et al. [21].

In this chapter, we utilize all three techniques to acquire images of the lysosomal protein cathepsin B labeled with Alexa Fluor 488 and the lysosomal membrane protein LAMP-2 labeled with Alexa Fluor 546 and analyze their colocalization in control samples as well as in samples where lysosomal membrane permeabilization was induced. We stress the importance of analyzing single labeled samples to determine the amount of bleed through between the green and red channel, as well as checking the amount of unspecific labeling and autofluorescence using samples incubated with the respective fluorescently tagged secondary antibody only. Due to the sensitivity to co-occurrence, i.e., the fraction of pixels with positive values for both analyzed channels, regardless of signal level, we focused our analysis on Manders' colocalization coefficient (MCC).

2 Materials

2.1 Cell Culture and Exposures

1. Normal human foreskin fibroblasts (*see* **Note 1**).
2. Cell culture medium: Eagle's minimum essential medium with GlutaMAX, 10% fetal calf serum, 50 IU/ml penicillin-G and 50 mg/ml streptomycin.
3. MSDH (O-methyl-serine dodecylamide hydrochloride); stock solution 5 mM in water, can be stored at 4 °C for several months.

2.2 Immunocytochemistry

1. 4% Paraformaldehyde (PFA) in PBS (*see* **Note 2**).
2. Incubation buffer: PBS without Ca^{2+} and Mg^{2+} supplemented with 0.1% saponine (*see* **Note 3**), 5% fetal bovine serum (*see* **Note 4**). Mix on a magnetic stirrer until the saponine is solved. Can be stored in the refrigerator for 1 week.
3. Mounting medium without DAPI (*see* **Note 5**).

2.3 Microscope

2.3.1 Confocal Laser Scanning Microscopy (CLSM)

CLSM, single point scanning confocal microscopy, was performed using a fully commercial, upright Zeiss Axio Imager.Z2 equipped with a LSM700 confocal module equipped with solid state diode lasers (405, 488, 555, and 639 nm) and Zeiss Zen 2012 software (*see* **Note 6**).

2.3.2 Spinning Disk Confocal Microscopy

Spinning disk confocal microscopy was performed using an inverted Nikon Ti-E (Tokyo, Japan) equipped with a CREST Optics X-Light V2 (Rome, Italy) spinning disk module with 60 μm pinholes illuminated by a Lumencor Spectra X (Beaverton, OR) LED-based light source. Images were captured using a 16-bit Hamamatsu Photonics ORCA-Flash4 LT sCMOS (Shizuoka, Japan). The software used in the setup is Nikon NIS Elements AR 4.5 (*see* **Note 7**).

2.3.3 Aperture Correlation Spinning Disk Confocal (VivaTome)

Aperture correlation confocal microscopy was performed using a VivaTome (Zeiss) module mounted on an inverted Zeiss Axiovert 200. Illumination for the VivaTome was done using a metal halide (HXP 120C) illumination source. Two 12-bit charged coupled devices (CCD) were used to acquire images of widefield fluorescence and VivaTome widefield/confocal, respectively. The VivaTome-module and the peripherals associated with the microscope were controlled via Zeiss AxioVision Rel. 4.8 (*see* **Note 8**).

3 Methods

3.1 Culturing of Cells and Exposure to LMP-Inducing Drugs

1. Culture human fibroblasts in cell culture medium in humidified air with 5% CO_2 at 37 °C and subculture once a week (*see* **Note 1**).
2. Trypsinize, count and seed cells at a density of 20,000 cells/cm^2 (*see* **Note 9**) in ∅ 35 mm cell culture dishes with glass bottom or on coverslips (*see* **Note 10**). Allow the cells to attach and stretch out for 24 h.
3. Prepare Eagle's minimum essential medium with GlutaMAX (i.e., serum-free medium) containing a final concentration of 15 μM MSDH and incubate the cultures at 37 °C with 5% of CO_2 in air.
4. Aspire the cell culture media after 2 h.
5. Beside your samples prepare the following controls:
   (a) Sample to analyze unspecific labeling and autofluorescence
   (b) Samples for single labeled antibodies

3.2 Staining of Cells for Immunofluorescence

1. Wash cells with PBS, RT (room temperature) 2 × 2 min (*see* **Note 11**).
2. Fix the cells in 4% paraformaldehyde (PFA) for 20 min at 4 °C.
3. Remove PFA (*see* **Note 12**) and wash in PBS, RT, 2 × 5 min.
4. Incubate in incubation buffer for 20 min, RT.
5. Dilute the first antibody in incubation buffer (*see* **Note 13**).
6. Add primary antibody and incubate overnight in 4 °C in a moist box.
7. Wash in incubation buffer 2 × 5 min, RT.
8. Dilute the second antibody in incubation buffer.

9. Incubate with second antibody in the dark for 1 h, RT (*see* **Note 14**).
10. Wash in incubation buffer 2 × 5 min, RT.
11. Wash in PBS 5 min and then in MilliQ water 5 min.
12. The methodological controls should be prepared as described above but
    (a) Add no first antibody to sample for analysis of unspecific labeling and autofluorescence
    (b) Add only one first antibody and the corresponding second antibody to samples for single labeled antibodies
13. Add mounting medium and cover the cells with a coverslip (*see* **Notes 5** and **15**).
14. Place the specimens in refrigerator.

3.3 Microscopy

3.3.1 Confocal Laser Scanning Microscopy

1. Select an appropriately magnifying objective with high numerical aperture and position a slide with coverslip (*see* **Note 10**) facing the objective lens. If using an oil immersion objective, add a small drop of immersion oil on top of the coverslip facing the objective lens.
2. Set coarse focus manually on sample via the eyepiece using brightfield (halogen lamp) or widefield illumination (e.g., mercury arc discharge lamp) with the preferred filter set for the most intense labeling.
3. Switch to confocal mode and fine tune focus during a fast, low resolution, confocal live scan.
4. Select preferred scanning and acquisition settings for one wavelength at a time using live scan to set the channel-specific acquisition parameters and to focus on the optical volume with most dense labeling.
5. Set laser power. Keep laser power as low as feasible in order to avoid photobleaching (*see* **Note 16**).
6. Set pinhole size, the larger the pinhole the thicker the optical section, and vice versa. The resulting section thickness depends on both wavelength and the numerical aperture of the objective (*see* **Note 17**).
7. Set detector offset (black level) carefully not to filter out true, but weak signals.
8. Set detector voltage (gain) using a range or intensity level indicator or histogram, so that no saturated pixels are present in the image.
9. In the confocal acquisition mode, select frame size (*see* **Note 18**).
10. Leave scan speed unaltered or increase speed to faster acquisition (*see* **Note 19**).

11. To reduce image distortion resulting from electronic and stochastic noise in the system, it is generally advisable to use averaging. Normally two or four images for mean are sufficient.
12. Check for autofluorescence and unspecific labeling by analyzing the sample stained with only second antibodies, i.e., detected emission for these samples should be negligible in either channel.
13. Make sure no fluorescence is bleeding through by analyzing the single labeled samples, i.e., detected emission in the red channel should be negligible for the single green labeled sample and vice versa.
14. To compare samples, the acquisition parameters need to be exactly the same for all samples (*see* **Note 20**).
15. To acquire optical sections, activate the Z-stacking or sectioning option.
16. Using only one wavelength and live scanning, focus manually clockwise until the sample is just about visible. Set this focal plane as starting point for the Z-stack.
17. Then focus counter-clockwise to the opposite side of the sample and set this opposite focal plane as stopping point for the Z-stack.
18. All reasonably modern confocal systems have a software option that allows setting the Nyqvist criteria for appropriate sampling frequency (*see* **Note 21**).
19. As resolution is dependent on wavelength the optical volume may vary markedly between the shortest and the longest wavelength. Therefore, some confocal systems allow users to match pinhole. However, adjusting pinhole will also alter the amount of emission reaching the detector and therefore it is advisable to re-check the levels after matching pinholes.
20. Activate scanning (Fig. 1) (*see* **Note 22**).
21. To compare samples, the acquisition parameters (illumination intensity and exposure time) need to be exactly the same for all samples (*see* **Note 20**).

3.3.2 Spinning Disc

Most confocal laser scanning microscopes have a similar hard- and software setup irrespective of manufacturer. However, the setup of spinning disk confocals seems to vary more, as different alternative designs are commercially available. The description below is thus somewhat superficial.

1. Select an appropriately magnifying objective with high numerical aperture and position slide with coverslip (*see* **Note 10**) facing the objective lens.
2. If using an inverted microscope with an oil immersion objective, add a small drop of immersion oil directly on the objective lens.

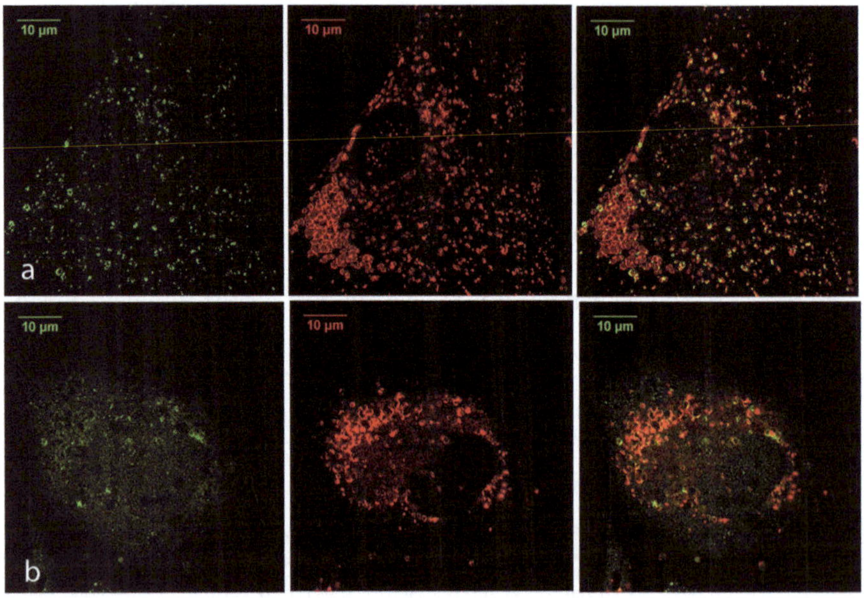

Fig. 1 Single z-plane images of human fibroblasts immunolabeled for cathepsin B (*green*) and LAMP-2 (*red*) acquired using a confocal laser scanning microscope. (**a**) Untreated cells, (**b**) cells treated with 15 μM MSDH for 2 h to induce lysosomal membrane permeabilization

3. Observe the sample via the eyepiece and set coarse focus using brightfield (halogen lamp) or widefield illumination (mercury arc or xenon discharge lamp, or LED illumination) with the preferred filter set for the most intense labeling.
4. Start live scanning and activate the confocal alternative for the preferred fluorescent channels available in the sample (*see* **Note 23**).
5. Since spinning disk confocals use digital cameras, set camera integration time (exposure) so that the resulting image contains no clipped pixels (saturation). Use the histogram option to visualize the distribution of intensities. When a suitable exposure time is set, use the autoscale function or manually adjust the look-up-table (LUT) to adjust and optimize the outgoing intensity levels.
6. Check for autofluorescence and unspecific labeling by analyzing the sample stained with only secondary antibodies, i.e., detected emission for these samples should be negligible in either channel.
7. Make sure no fluorescence is bleeding through by analyzing the single labeled samples, i.e., detected emission in the red channel should be negligible for the single green labeled sample and vice versa.
8. If intensities are too low or too high, adjust camera exposure time in the camera settings or, optionally, if using a LED-based illumination, adjust the illumination intensity. For imaging in real-time mode using sCMOS camera, one frame or less is

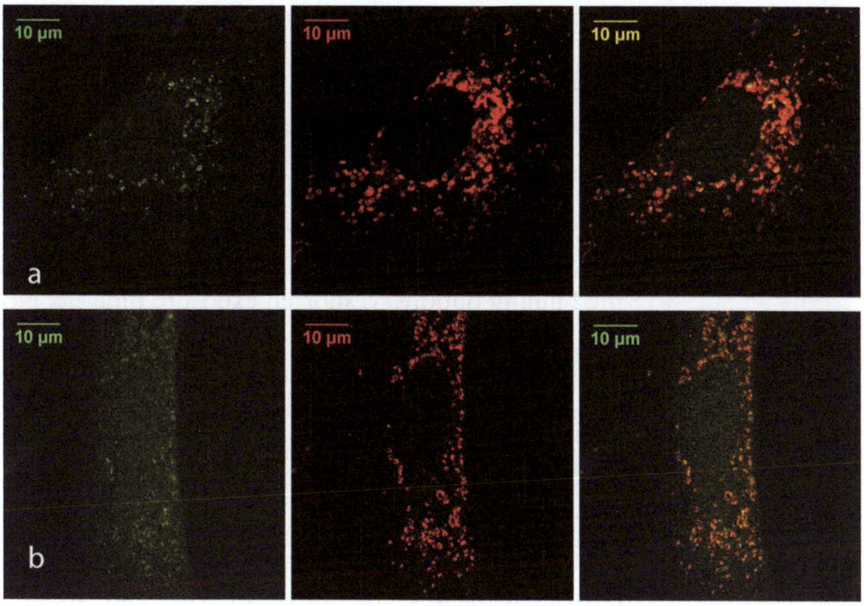

Fig. 2 Single z-plane images of human fibroblasts immunolabeled for cathepsin B (*green*) and LAMP-2 (*red*) acquired using a spinning disk confocal microscope. (**a**) Untreated cells, (**b**) cells treated with 15 μM MSDH for 2 h to induce lysosomal membrane permeabilization

normally sufficient exposure for reasonably dense fluorescence distribution and widefield imaging. However, transmission through the spinning disk will be slightly reduced, and therefore the confocal mode usually requires longer exposure times.

9. Set Z-stack starting and stopping as described for confocal laser scanning microscopy (*see* Subheading 3.3.1, **steps 16** and **17**). Due to the slightly larger out-of-focus contribution in spinning disk compared with the point scanning confocal, the number of sections becomes larger than in the point scanning confocal microscope.
10. For adequate sampling interval, select the Nyqvist optimization option (*see* **Note 21**).
11. Activate scanning (Fig. 2) (*see* **Note 24**).
12. To compare samples the acquisition parameters (illumination intensity and exposure time) need to be exactly the same for all samples (*see* **Note 20**).

3.3.3 VivaTome Aperture Correlation Spinning Disk Confocal

The VivaTome is an aperture correlated confocal microscope and although it is equipped with a spinning disk it is not similar to a spinning disk confocal. Rather it is a form of structured illumination as the disk is a grid. However, like spinning disk it is a live confocal.

1. Select an appropriately magnifying objective with high numerical aperture and position slide with coverslip (*see* **Note 10**) facing the objective lens.

2. If using an inverted microscope with an oil immersion objective, add a small drop of immersion oil on top of the coverslip facing the objective lens.
3. Observing the sample via the eyepiece, set coarse focus using brightfield (halogen lamp) or widefield illumination (mercury arc or xenon discharge lamp, or LED illumination) with the preferred filter set for the most intense labeling.
4. Set camera acquisition at maximum frame size, or use one of the binning options to shorten exposure time (*see* **Note 25**).
5. Work fast not to bleach or harm the sample, close shutter in front of the excitation light source when not viewing the sample.
6. Activate the aperture correlation confocal. The VivaTome spinning grid has a low and a high magnification option depending on the objective used for image acquisition. Focus quickly, not to bleach or harm the sample.
7. In the live mode, adjust camera integration time and check intensity levels to avoid saturation. When running the aperture correlation confocal the live view shows both the widefield and the confocal image, side by side. Center sample in the viewing area. The confocal part of the image will be clearly dimmer than the widefield part as it only contains the in-focus emission.
8. Check for autofluorescence and unspecific labeling by analyzing the sample stained with only second antibodies, i.e., detected emission for these samples should be negligible in either channel.
9. Make sure no fluorescence is bleeding through by analyzing the single labeled samples, i.e., detected emission in the red channel should be negligible for the single green labeled sample and vice versa.
10. Activate the Z-stacking/sectioning option and set start and stop position as described for confocal laser scanning microscopy (*see* Subheading 3.3.1, **steps 16** and **17**).
11. Set appropriate sampling interval according to Nyqvist (*see* **Note 21**).
12. Activate scanning (Fig. 3) (*see* **Note 26**).
13. To compare samples the acquisition parameters (frame size, illumination intensity, and exposure time) need to be exactly the same for all samples (*see* **Note 20**).

3.3.4 Huygens

Below we present a very brief introduction to the Deconvolution wizard and the colocalization analysis.

1. Open image file (*see* **Note 27**).
2. Right-click on the image and select Parameter Editor, check settings and correct for embedding media, channel wavelengths, objective numerical aperture and quality, and coverslip. Also check that no parameter is indicated in red (*see* **Note 28**).
3. Click on Set all verified, and accept the settings.

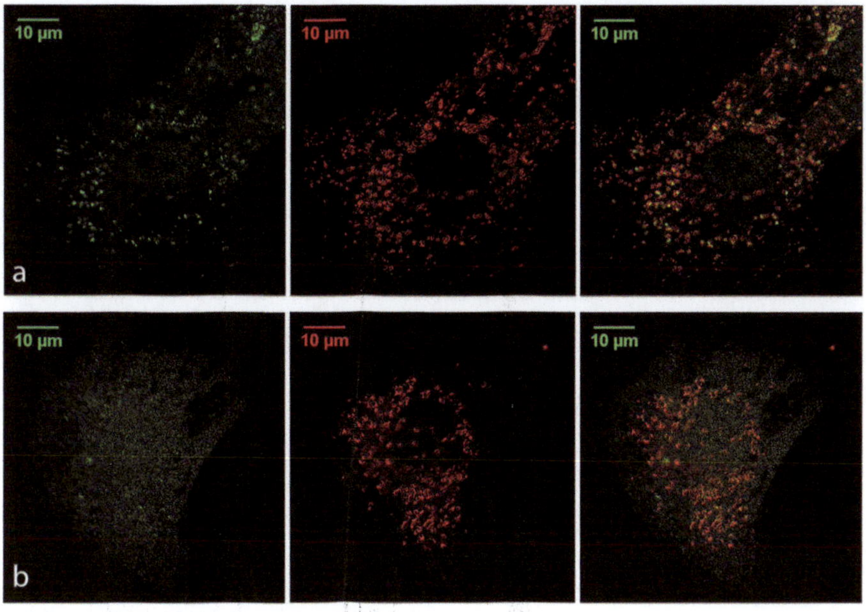

Fig. 3 Single z-plane images of human fibroblasts immunolabeled for cathepsin B (*green*) and LAMP-2 (*red*) acquired using an aperture correlation spinning disk confocal microscope. (**a**) Untreated cells, (**b**) cells treated with 15 μM MSDH for 2 h to induce lysosomal membrane permeabilization

4. Highlight the image and select the Deconvolution wizard under the Deconvolution menu or use the Decon icon.
5. Enter wizard, unless a measured PSF (Point Spread Function) is available, use a theoretical PSF.
6. Crop your image if necessary, otherwise proceed to next step.
7. Select channel to deconvolve and proceed. Select compute and check if image contains clipped voxels (saturation).
8. Proceed to background estimation. If background is known, proceed with Manual settings, otherwise select Auto to allow the computer to estimate the background (*see* **Note 29**).
9. Select deconvolution algorithm. For most uses the Classic Maximum Likelihood Estimation (CMLE) is sufficient.
10. Decide if you wish to proceed with default settings or alter number of iterations, signal to noise ratio, quality threshold, iteration mode, or brick layout (*see* **Note 30**).
11. Start deconvolution of the selected channel. During the iterations the original image is shown to the left and the deconvolution progress to the right (*see* **Note 31**).
12. When deconvolving widefield images an option to correct for bleaching and unstable illumination will appear before the iterations start.
13. When iterations are done, select Accept and proceed to next channel. Repeat settings as above.
14. When all channels are deconvolved, click All done (Figs. 4 and 5).

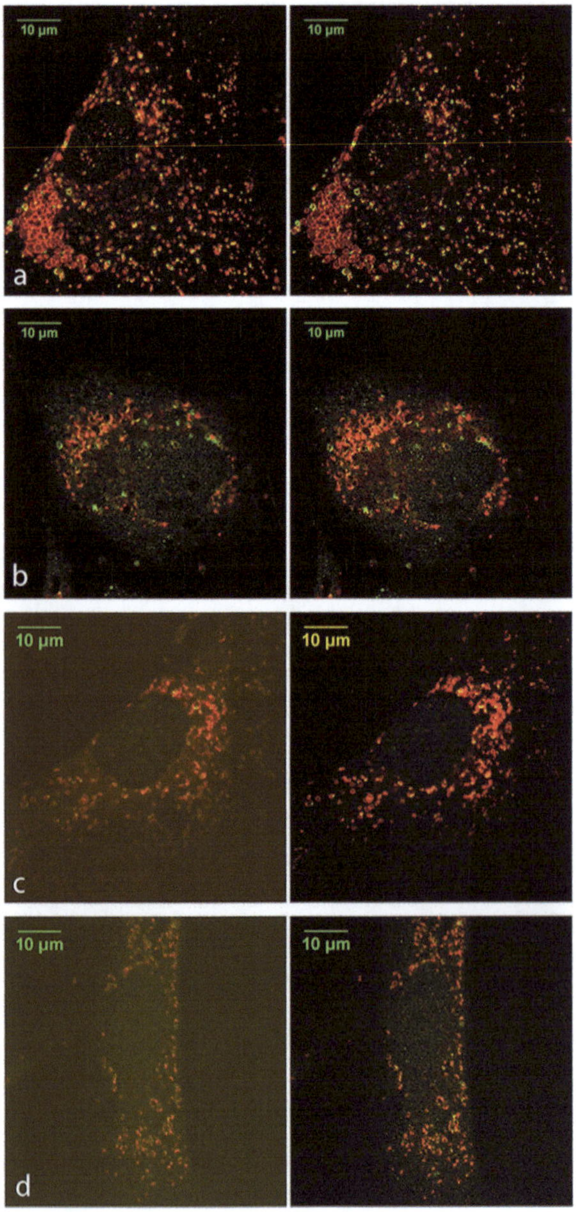

Fig. 4 Single z-plane images before (*left*) and after (*right*) deconvolution using Huygens professional software. (**a**) Untreated and (**b**) LMP-induced cells imaged using a confocal laser scanning microscope. (**c**) Untreated and (**d**) LMP-induced cells imaged using a spinning disk confocal microscope. (**e**) Untreated and (**f**) LMP-induced cells imaged using an aperture correlation spinning disk confocal microscope

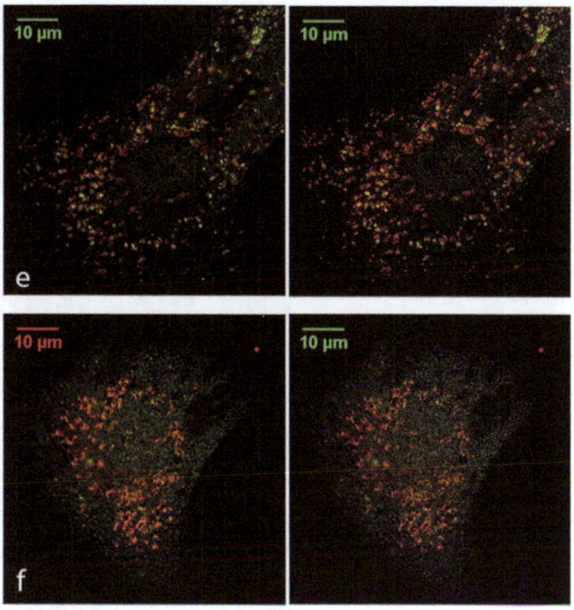

Fig. 4 (continued)

15. In the following steps it is possible to select images to merge and to rename the file. Proceed to next option and Done.
16. The deconvolved image (ending with _decon, if not changed in previous step) will appear next to the original image in the Thumbnail overview.
17. To view the original and the deconvolved image side by side, highlight one of the images in the Thumbnail overview and select Twin slicer under the Visualization menu.
18. The highlighted image appears to the left in a new window. To compare, select the other image using the option above the right-hand image field.
19. For Colocalization, highlight image in Thumbnail overview, select the option Colocalization analyzer under the Analysis menu.
20. Select channels to colocalize and click on Compute (Fig. 6) (*see* **Notes 32** and **33**).
21. Save images in the Huygens professional hdf5-format (.h5).
22. In order to open images in Image J (Fiji), save images as TIFF (*see* **Note 34**).

4 Notes

1. The protocol is described for human fibroblasts, but can be modified to fit any cell type. The cells should be cultured in their regular cell culture medium.

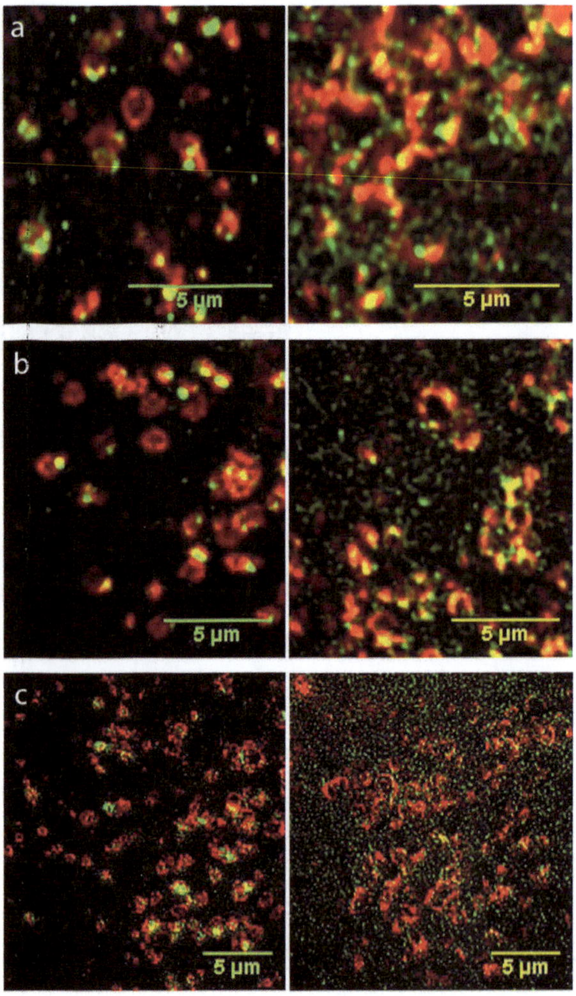

Fig. 5 Detailed images of untreated (*left*) or LMP-induced (*right*) human fibroblasts immunolabeled for cathepsin B (*green*) and LAMP-2 (*red*) using (**a**) a confocal laser scanning microscope, (**b**) a spinning disk confocal microscope, and (**c**) an aperture correlation spinning disk confocal microscope

2. PFA is a chemical crosslinker that preserves the morphology well, do not fix for longer time than 20 min and keep the sample cold to save the epitopes. PFA can be kept in the refrigerator for max 1 month.
3. To analyze membrane proteins it is important to use very mild permeabilization, here we use saponin, which mainly permeabilizes the plasma membrane, leaving the intracellular membranes largely intact.
4. Depending on which antibodies used the blocking substance can be varied. The background might be reduced with 1% milk powder, 1% BSA, or 5% FCS. It is important that the blocking proteins

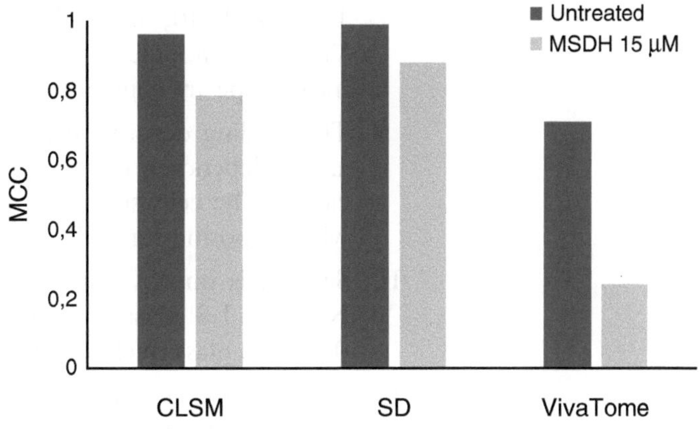

Fig. 6 Manders' colocalization coefficient (MCC) calculated for untreated or LMP-induced cells after deconvolution of single z-plane images acquired using confocal laser scanning microscope (CLSM), spinning disk confocal microscope (SD) and aperture correlation spinning disk confocal microscope (VivaTome)

do not originate from the species in which the primary antibody was raised since the second antibody will then lose its specificity.

5. We recommend not to use nuclear stains such as DAPI since it is might interfere with the detection of longer wavelength fluorescence. Moreover, with time DAPI tends to detach from dsDNA and diffuses into the cytosol. Also, UV excitation may photoconvert DAPI so that it can be detected in the FITC/Alexa Fluor 488 channel [22].
6. The confocal laser scanning microscope used in the protocol is a LSM700 attached to a Zeiss AxioImager.Z2. For our image acquisitions, we used a Plan-Apochromat 63×/1.4 oil DIC.
7. Spinning disk confocal microscope is performed using an inverted Nikon Ti-E (Tokyo, Japan) equipped with a CREST Optics X-Light V2 (Rome, Italy) spinning disk module with 60 µm pinholes illuminated by a Lumencor Spectra X (Beaverton, OR) LED-based light source. Images are captured using a 16-bit Hamamatsu Photonics Orca-Flash4 LT sCMOS (Shizuoka, Japan) with an effective pixel area of 2048 × 2048. The software used in the setup is Nikon NIS Elements AR 4.5.
8. The VivaTome module used is mounted on an inverted Zeiss Axiovert 200. Illumination for the VivaTome is done using a metal halide (HXP 120C) illumination source. The Semrock filter wheel in the VivaTome contains filters for DAPI, FITC, and Texas Red. Two 12-bit charged coupled devices (CCD) are used to acquire images of widefield fluorescence and VivaTome confocal, respectively. A Zeiss AxioCam MRm is used for widefield and an AxioCam MRm Rev. 3 is used to acquire images using the VivaTome. Both have a standard mono setting of

1388 × 1040, i.e., when running without binning. The VivaTome-module and the peripherals associated with the microscope are controlled via Zeiss AxioVision Rel. 4.8.

9. The seeding density needs to be optimized to cell type used and is dependent on, e.g., proliferation rate and plating efficiency. The cells should be at approximately 70% confluence when exposing for LMP-inducing drugs.
10. Seed cells on coverslips or in dish with glass bottom. Use Number 1.5 glass, which most closely corresponds to the 0.17 mm glass thickness that the objectives are adapted for.
11. Wash gently, especially when the cells have been exposed to cell damaging agents.
12. PFA is toxic and potentially carcinogenic. Consult local regulations for handling of the waste.
13. Always use sterile pipet tips when pipetting from the stock with antibody to avoid contamination.
14. Keep fluorophores protected from light.
15. We recommend the use of hard-set mounting medium to reduce the photobleaching and preserve the specimen. Several different brands are commercially available, e.g., Vechtashield and ProLong Gold/Platinum.
16. Low laser power is even more critical for live cells as the high intensity beam causes phototoxicity. For live cells it is therefore advisable to keep laser power at a minimum.
17. The smallest, or generally most practical, pinhole is 1 Airy Unit, which yields a thin optical volume together with an adequate transmission of emission from the sample. In some systems the pinhole size is indicated with millimeters or other system-specific indication. It is thus important to carefully consult the manual or the local technical staff responsible for the equipment.
18. In most systems it is advisable to use 512 × 512 or preferably 1024 × 1024 to speed up acquisition time of analysis images, but the acquisition of high-resolution publication images benefits from using a higher frame rate such as 2048 × 2048, if available.
19. A too slow scan will cause unnecessary photobleaching. A too fast scan will introduce higher noise but if using image restoration software such as deconvolution this slight increase in noise is usually of no major concern. Some additional noise may even be beneficial for the deconvolution algorithms calculating the out-of-focus contribution.
20. In most systems the option to reuse specific settings from a previous image is readily available. Therefore, acquire an image

with the preferred settings and save the image in the software-specific file format.

21. It is critical to avoid under-sampling, as information may be lost. Oversampling, on the other hand, gives no additional information but risk generating very large data files.
22. The setup in our acquisitions was: Widefield observation using the mercury arc lamp was done using the following filter sets for green and red fluorescence: Zeiss #10, BP 450-490/FT510/BP515-565; and Zeiss #20, BP546/12/FT560/BP575-640. For setting up confocal acquisition we used the Smart Setup settings for Alexa Fluor 488 and Alexa Fluor 546, and the option Best signal in Zen 2012. Laser power was set at 2% and pinholes were 1.17 Airy Units for the green channel and 1.01 Airy Units for the red channel, resulting in an equal section thickness of 0.8 μm. PMT voltage (gain) for Alexa Fluor 488 required 393 and the Alexa Fluor 546 required 398 in gain. Digital Offset was left unaltered. In our acquisitions, we used the highest Frame Size (2048 × 2048) and also increased scanning speed one step up resulting in a pixel dwell time of 0.79 μs. We used two images for averaging. Images were acquired as 12-bit. Using the 63× objective, we left zoom unaltered at one yielding a pixel size of 0.05 μm and image size of 101.6 × 101.6 μm. As the cells in our samples were very thin, Z-stack setting resulted in 5–8 sections per cell, with a thickness of 0.33 μm. To ensure appropriate sampling, we added one slice to slightly oversample our image stacks
23. To avoid bleaching and phototoxicity, live imaging and Z-scan setup should preferably be done using 2 × 2 binning or more together with a fast scan mode to speed up settings.
24. Images using the Nikon Ti-E at the Microscopy Unit, Linköping University were acquired using a Nikon Plan Apo 100×/1.40 oil DIC H. We set exposure at 100 ms for both green and red acquisition using the spinning disk-mode. The LED-based illumination for green nm and red was set at 50% and 38%, respectively. There are six available excitation filters in the Spectra X at the Microscopy Unit, Linköping University. We used the 470/24 and 550/15. In the X-Light spinning disk module, we currently have a "quad-band" DAPI/FITC/TRITC/Cy5 emission and dichroic mirror available, CrEST EM and CrEST Di, respectively. The quad-band filter means that there is a risk of shorter wavelength fluorophore excitation of longer wavelength fluorophores. All filters in the spinning disk set up are from Chroma Technology Corp., (Bellows Farms, VT). Both the dichroic mirror and the emission filter in the X-Light V2 spinning disk module are quad-band. In our spinning disk confocal imaging, we only used acquisition in

the FITC- and TRITC-channels. For Z-scanning, step size was set to oversample slightly, meaning 7–19 slices, depending on cell morphology, with 0.2 μm for each optical slice.

25. For live cell imaging it is very important to reduce the exposure time and thus at least a 2 × 2 binning should be used. This will obviously reduce the resolution, but will also prolong the survival of the cells.
26. For our VivaTome image acquisitions, we used FITC exc. 494/20–25 emi. 536/40–25 nm, and Texas Red exc. 575/25–25 emi. 628/40–25 nm. In the module there is a filter wheel with a common dichroic F436/514/604 and a blocking filter 457/530/628–25. The objective was an alphaPlan-Apochromat 63×/1.46 oil Correction collar. Widefield imaging was done using FITC (Zeiss #17, BP485/20/FT510/BP515-565), and Texas Red (Zeiss #14, BP510-560/FT580/LP590) filter setup. We used only green and red fluorescence with 1 and 2 s exposure, respectively. Full frame, standard mono, no binning. Sectioning was set by manually focusing on the approximate middle of the cell using the red channel in VivaTome live mode, and then setting start at 1.7 μm below the focus and stop at 1.7 μm above the sample. This yielded a step size of 0.246 μm for each of the 15 acquired slices.
27. We saved our point scanning confocal images as .czi, the spinning disk images as .nd2, and the VivaTome images as .zvi, which all can be opened in Huygens Professional 16.05. Single z-plane images in which labeled LAMP-2 made the lysosomal membrane clearly visible as enclosing a compartment were selected for further analysis.
28. In all images, we changed embedding media from oil (refractive index 1.515) to glycerin (refractive index 1.474). For point scanning laser images no other changes were considered necessary. For spinning disk we changed channel 1 to emission 600 nm, as it was set on 520 nm for both channel 0 and 1. For VivaTome-images we changed objective quality from good to perfect as the 63×/1.46 objective was considered to be of higher quality than the ones used in the other techniques.
29. Background estimation depends on the specific microscopy method. For confocal, use Lowest, otherwise select Widefield.
30. For point scanning laser the default settings are: number of iterations 40, signal to noise ratio 20, quality threshold 0.05, iteration mode optimized, and brick layout auto. For widefield the default settings are: number of iterations 50, signal to noise ratio 40, quality threshold 0.01, iteration mode optimized, and brick layout auto. Widefield also has an additional option, Bleaching correction, in which the default is: If possible.

In our deconvolutions, we used the default settings. However, as the software at the Microscopy Unit at Linköping University does not include the spinning disk option, we used confocal settings for both Spinning Disk and the VivaTome images.

31. The deconvolution will either iterate all selected steps or interrupt when quality threshold is fulfilled.
32. Although the colocalization analyzer gives an assortment of colocalization coefficients, we selected to use only the Manders' overlap coefficient in our measurements. Background was set using the single labeled images, i.e., the background for the emission in the red channel was set using the single green labeled sample and vice versa.
33. Upon LMP, MCC calculated for (green-labeled) cathepsin B follows the same trend in all three confocal methods, i.e., the fraction of this protein colocalizing with LAMP-2 decreases as expected in a situation where cathepsin B is leaking out from the lysosomes. However, the magnitude of this decrease differs among the different techniques. Our interpretation is that the amount of colocalization likely reflects the resolution in each technique. The confocal laser scanning microscope based on the Zeiss LSM700 and the spinning disk confocal based on the X-Light V2 module have corresponding, theoretical axial resolutions, especially after deconvolution. This is also apparent in the obtained, corresponding coefficients before and after treatment. The aperture correlation-based confocal, i.e., the VivaTome, shows a lower amount of overlap between the fluorescent channels, both in controls as well as after treatment. The VivaTome, with its grid-based structured illumination-type of beam path, yields a slightly increased axial resolution. The thinner optical volume acquired in accordance will result in less overlap. This is a likely outcome, as the labeled proteins do not interact physically. Rather, the green fluorescence associated with cathepsin B is observed as filling a void formed up by a red circle of membrane-associated LAMP-2, being an optical section through a small membrane enclosed organelle. The thinner the optical section, the less out-of-focus light is colocalizing with the inner green fluorescence. The higher axial resolution will also affect the overlap after permeabilization of the lysosomal structure. In all methods the red and green disperses in 3D after lysosome permeabilization, but it becomes increasingly apparent at higher axial resolution, hence leaving less overlapping pixels in the thin optical section.
34. We selected to save images as TIFF 16-bit, one per channel, and with the linked scale option for deconvolved images and keep values for the original.

Acknowledgement

This work was supported by grants from the Swedish Research Council, Swedish Cancer Society, County Council of Östergötland, Konung Gustav V och Drottning Victorias Frimurarestiftelse and Stiftelsen Olle Engkvist Byggmästare.

References

1. De Duve C (1959) Lysosomes, a new group of cytoplasmic particles. In: Hayashi T (ed) Subcellular Particles. The Ronald Press Co, New York, pp 128–159
2. Turk B, Turk V (2009) Lysosomes as 'suicide bags' in cell death: myth or reality? J Biol Chem 284:21783–21787
3. Roberg K, Öllinger K (1998) Oxidative stress causes relocation of the lysosomal enzyme cathepsin D with ensuing apoptosis in neonatal rat cardiomyocytes. Am J Pathol 152(5):1151–1156
4. Boya P, Kroemer G (2008) Lysosomal membrane permeabilization in cell death. Oncogene 27(50):6434–6451. doi:10.1038/onc.2008.310
5. Bivik CA, Larsson PK, Kågedal KM, Rosdahl IK, Öllinger KM (2006) UVA/B-induced apoptosis in human melanocytes involves translocation of cathepsins and Bcl-2 family members. J Invest Dermatol 126(5):1119–1127
6. Roberg K, Kågedal K, Öllinger K (2002) Microinjection of cathepsin d induces caspase-dependent apoptosis in fibroblasts. Am J Pathol 161(1):89–96
7. Schestkowa O, Geisel D, Jacob R, Hasilik A (2007) The catalytically inactive precursor of cathepsin D induces apoptosis in human fibroblasts and HeLa cells. J Cell Biochem 101(6):1558–1566
8. Scheffler JM, Schiefermeier N, Hube LA (2014) Mild fixation and permeabilization protocol for preserving structures of endosomes, focal adhesions, and actin filaments during immunofluorescence analysis. Methods Enzymol 535:93–102
9. Stadler C, Skogs M, Brismar H, Uhlén M, Lundberg E (2010) A single fixation protocol for proteome-wide immunofluorescence localization studies. J Proteomics 73(6):1067–1078
10. Donaldson J (2001) Immunofluorescence staining. Curr Protoc Cell Biol 4. doi:10.1002/0471143030.cb0403s00
11. Jamur MC, Oliver C (2010) Permeabilization of cell membranes. Methods Mol Biol 588: 63–66
12. Bauer CR (2014) Labeling and use of monoclonal antibodies in immunofluorescence: protocols for cytoskeletal and nuclear antigens. Methods Mol Biol 1131:543–548
13. Yamanaka M, Smith NI, Fujita K (2014) Introduction to super-resolution microscopy. Microscopy 63:177–192
14. Cox S, Jones GE (2013) Imaging cells at hte nanoscale. Int J Biochem Cell Biol 45: 1669–1678
15. Conchello JA, Lichtman JW (2005) Optical sectioning microscopy. Nat Methods 2: 920–231
16. Sanderson MJ, Smith I, Parker I, Bootman MD (2014) Fluorescence microscopy. Cold Spring Harb Protoc 2014(10):pdb.top071795. doi:10.1101/pdb.top071795
17. Oreopoulos J, Berman R, Browne M (2014) Spinning-disk confocal microscopy: present technology and future trends. Methods Cell Biol 123:153–175
18. Karlsson T, Lagerholm BC, Vikström E, Loitto VM, Magnusson KE (2013) Water fluxes through aquaporin-9 prime epithelial cells for rapid wound healing. Biochem Biophys Res Commun 430(3):993–998
19. Karlsson T, Bolshakova A, Magalhães MA, Loitto VM, Magnusson KE (2013) Fluxes of water through aquaporin 9 weaken membrane-cytoskeleton anchorage and promote formation of membrane protrusions. PLoS One 8(4):e59901
20. Wang YL (2007) Computational restoration of fluorescence images: noise reduction, deconvolution, and pattern recognition. Methods Cell Biol 81:435–445
21. Dunn WD, Kamocka MM, McDonald JH (2011) A practical guide to evaluating colocalization in biological microscopy. Am J Physiol Cell Physiol 300:C723–C742
22. Jež M, Bas T, Veber M, Košir A, Dominko T, Page R, Rožman P (2013) The hazards of DAPI photoconversion: effects of dye, mounting media and fixative, and how to minimize the problem. Histochem Cell Biol 139: 195–204

Chapter 6

Quantitative Co-Localization and Pattern Analysis of Endo-Lysosomal Cargo in Subcellular Image Cytometry and Validation on Synthetic Image Sets

Frederik W. Lund and Daniel Wüstner

Abstract

Late endosomes and lysosomes (LE/LYSs) play a central role in trafficking of endocytic cargo, secretion of exosomes, and hydrolysis of ingested proteins and lipids. Failure in such processes can lead to lysosomal storage disorders in which a particular metabolite accumulates within LE/LYSs. Analysis of endocytic trafficking relies heavily on quantitative fluorescence microscopy, but evaluation of the huge image data sets is challenging and demands computer-assisted statistical tools. Here, we describe how to use SpatTrack (www.sdu.dk/bmb/spattrack), an imaging toolbox, which we developed for quantification of the distribution and dynamics of endo-lysosomal cargo from fluorescence images of living cells. First, we explain how to analyze experimental images of endocytic processes in Niemann Pick C2 disease fibroblasts using SpatTrack. We demonstrate how to quantify the location of the sterol-binding protein NPC2 in LE/LYSs relative to cholesterol-rich lysosomal storage organelles (LSOs) stained with filipin. Second, we show how to simulate realistic vesicle patterns in the cell geometry using Markov Chain Monte Carlo and suitable inter-vesicle and cell-vesicle interaction potentials. Finally, we use such synthetic vesicle patterns as "ground truth" for validation of two-channel analysis tools in SpatTrack, revealing their high reliability. An improved version of SpatTrack for microscopy-based quantification of cargo transport through the endo-lysosomal system accompanies this protocol.

Key words Cholesterol, Diffusion, Endocytosis, Fluorescence, Lysosomal storage disorder, Monte Carlo simulation, Single particle tracking, Sorting, Time-lapse

1 Introduction

1.1 Niemann Pick Disease Type C, A Prototypic Lysosomal Storage Disorder

Neurodegenerative diseases and other disorders of the central nervous system represent worldwide an increasing health problem. Many neurodegenerative diseases can be tracked down to traffic jams in the degradative pathway of cells, comprised of late endosomes and lysosomes (LE/LYSs). These cellular organelles play a central role in maintenance of intracellular cholesterol homeostasis. LE/LYSs receive low-density lipoprotein (LDL) after endocytosis from the plasma membrane (PM) and house the machinery for

Karin Öllinger and Hanna Appelqvist (eds.), *Lysosomes: Methods and Protocols*, Methods in Molecular Biology, vol. 1594, DOI 10.1007/978-1-4939-6934-0_6, © Springer Science+Business Media LLC 2017

hydrolysis of LDL-derived cholesteryl esters (CEs). Export of cholesterol liberated from LDL in LE/LYSs is not understood. Niemann-Pick disease type C (NPC) is a rare neurodegenerative disorder caused by a mutation in either protein NPC1 or protein NPC2 [1]. NPC1 is a large transmembrane protein mostly located in LE/LYSs but also found in the trans-Golgi network (TGN) [1]. NPC2 is a small-soluble glycoprotein found in the lumen of LE/LYSs, but being also present in several body fluids including milk, bile, and epididymal fluid [1, 2]. Loss of function in either NPC1 or NPC2 has been shown to cause severe lysosomal accumulation of cholesterol, glycosphingolipids, sphingosine, and sphingomyelin (SM) [1, 3]. The resulting cellular phenotype is characterized by large amounts of lipid-laden crescent-shaped LE/LYSs, sometimes called lysosomal storage organelles (LSOs) [1]. Fibroblasts lacking functional NPC1 or NPC2 hydrolyze LDL-derived cholesteryl-esters (CEs) normally but have a strongly reduced ability to elicit normal regulatory responses. This includes a lack of stimulation of cholesterol esterification and diminished suppression of cholesterol synthesis by ingested LDL [1, 2].

1.2 Intracellular Distribution of Endo-Lysosomes—Location Matters

It is increasingly acknowledged that the mobility and intracellular steady-state distribution of endosomes and other vesicles are intimately linked to their function. While sorting endosomes (SEs) are mostly located in the periphery of cells, the endocytic recycling compartment (ERC) is often found in the perinuclear para-Golgi region, depending on the structural organization and orientation of microtubules and the centrioles [4–6]. LE/LYSs often reside slightly scattered in proximity to the ERC, and their movement depends on motor proteins, such as dynein and kinesin. Similarly, iterative geometric-based sorting of endocytic receptors, as the transferrin (Tf) receptor, is driven by syntaxin-mediated tubulation of SEs, which at the same time controls positioning of these organelles relative to the ERC [7–9]. Not only are endosomes actively translocated by motor-assisted movement along cytoskeleton tracks, a process which is directly involved in endocytic sorting [10]. There is also ample evidence for a direct involvement of lysosome positioning in the nutrient status of the cell [11]. In particular, cholesterol enrichment in LE/LYSs due to defective rab7 GTPase or NPC1 can cause dynein-dependent perinuclear accumulation and reduced mobility of LE/LYSs [12–15]. Similarly, the ion composition of LE/LYSs is tightly regulated and intimately linked to ligand receptor sorting and cargo degradation [16]. A recent study reported that the intraluminal pH of LE/LYSs depends on their position in the cell with more acidic lysosomal organelles in the cell center compared to the cell periphery [17]. Interestingly, depletion of functional rab7 and its effector, RILP, accounted for the more alkaline LE/LYSs in the cell periphery providing a further link between the composition, function, and intracellular position of LE/LYSs. The intracellular distributionof organelles in the endocytic pathways can be manipulatedby optogenetic control of the associated motor proteins [18, 19]. This is a promising tool for further elucidation of the interrelationship of endosome location, mobility, and function. Importantly, quantitative image analysis tools are instrumental for interpreting such experiments, not only for assessing cargo co-localization but also for determining spatial relationships of various endosome populations.

1.3 Quantitative Measures of Co-Localization in the Endocytic Pathway

Function and pathological alterations of endocytic and lysosomal membrane trafficking are often studied by quantitative fluorescence microscopy. Modern imaging platforms, however, generate a huge amount of multidimensional image data, either as z-stacks, time-lapse sequences or multicolor data sets. In order to obtain objective evidence for co-trafficking of endocytic cargo, changes of endosome motility or for clustering/dispersion of organelles upon a particular treatment, computational image analysis tools are required. A particularly demanding task is to determine to which extent two differently labeled molecular species co-localize in living cells at a given time point. The easiest but least reliable approach is to make an overlay of a red and green image and judge by eye the extent of yellow pixels as indication for co-localization. This should be avoided, as visual inspection can only be used as a guide for later unbiased computational analysis of the image sets. Most current quantification methods assess co-localization on a pixel-by-pixel basis from two-channel image sets. The intensity of the green and red signals at each pixel in an image can be plotted as a two-dimensional histogram (also called cytofluogram, scatter plot, or correlation plot) after setting appropriate thresholds in each channel. Visual inspection of such scatter plots can sometimes reveal important features about relative cargo content in endosomes [20]. The degree of co-localization can be quantified from such plots via a correlation coefficient, the Pearson correlation coefficient (PCC), a common measure in statistics [20, 21]. Alternative measures of pixel-based co-localization are Mander's coefficients, the fraction of overlapping pixels or variants of pixel correlation measures, as Costes method or image correlation spectroscopy [21–23]. These and related methods have been reviewed recently, and a wide variety of software is available for this type of analysis [17, 20–31]. Pixel-based co-localization analysis of fluorescence images including optimal confocal imaging of labeled cells for that purpose has been recently described in this protocol series [24]. An alternative approach is to consider the individual fluorescent entities as separate objects and calculate the distance between such objects between both channels. Here, the critical step is to define the fluorescent entities unequivocally, i.e., to segment the images in a reliable and reproducible way. This can also accompany the pixel-based co-localization analysis but is mandatory for object-based quantification of co-localization. Since LE/LYSs are punctate structures, a variety of algorithms can be employed for spot detection and segmentation of round objects, as recently reviewed [32, 33]. From the distance of the centers of mass of detected objects in both channels, distance maps are generated in two or three dimensions for object-based analysis, and a criterion of co-localization based on the microscope resolution is used to quantify co-localization [20, 31, 34–36]. Importantly, object-based but not pixel-based co-localization analysis also allows for determining spatial relationships between tagged molecules, as their tendency to cluster in a particular area of the cell. This approach provides useful information also in the absence of co-localization and is particularly useful for endo-lysosomes. Here, one cannot only assess the content of particular cargo in LE/LYSs but also determine their spatiotemporal relationships within a given cell geometry.

1.4 Analysis of Endocytic Trafficking and Lysosomal Storage Using SpatTrack

SpatTrack is a user-friendly and freely available software package which we developed for determining trafficking and disposition of endocytic cargo in living mammalian cells based on two-color images or time-lapse fluorescence image sequences. Particle tracking in SpatTrack is based on the method described by Crocker and Grier [37]. We showed that this method allows for determining vesicle trajectories with high accuracy and quantified the localization error based on various particle filtering methods [38]. SpatTrack calculates statistical characteristics of the obtained trajectories, as the mean square displacement (MSD) or the radius of gyration (RoG) [39, 40]. By fitting physical models of the underlying dynamics to the MSD, SpatTrack can determine diffusion constants, anomaly coefficients, and flow components. The RoG analysis allows for classifying individual trajectories as likely originating in normal diffusion/anomalous subdiffusion or directed motion. The accuracy of these image processing and analysis tools in SpatTrack can be validated by the user on synthetic image data generated by Monte Carlo (MC) simulations. In this way, one can simulate various diffusion scenarios of particles and determine the tracking accuracy. One can also assess the reliability of various diffusion models to fit the measured MSD of tracked vesicles. These features of SpatTrack have been reviewed and explained in depth in a separate article [41].

Once the particle coordinates have been determined, these can be imported into another subroutine for object-based co-localization analysis. Here, the particle sizes are determined by fitting a 2D Gaussian function to each particle. From that data, a mean particle diameter, M, is estimated for both particle populations (i.e., in the red and green channel). To avoid redundant assignments, the population with fewer vesicles is called reference and the other vesicle population is named candidate. Co-localization is judged based on a predetermined overlap criterion (i.e., alpha times M, with alpha ~0.5). From that, the degree of overlap between particles in two different channels may be determined (*see* Fig. 6 in [39]). Apart from that "particle detection-based co-localization" (PDBCA), several measures have been introduced in SpatTrack for quantification of the relative proximity of vesicles detected in both channels, as the nearest neighbor (NN) analysis and the calculation of a two-channel radial distribution function (two-channel-RDF). These tools will be discussed below. We used simple MC simulations to generate random vesicle patterns in the 2D-cell geometry extracted from microscope images of fibroblasts. This was essential for comparing the experiment-derived vesicle patterns with a given random particle distribution. It allows for assessing the degree of vesicle clustering or dispersion using spatial pattern analysis based on the NN measure or the two-channel-RDF, respectively.

Together, the different subroutines in SpatTrack allow for detection and tracking of fluorescent vesicles in single images and time-lapse sequences, respectively. SpatTrack calculates a variety of statistical properties from the particle coordinates including motility characteristics of the vesicles, spatial vesicle distribution in the cell geometry, and potential co-localization in multicolor image sets. Here, we show how SpatTrack's co-localization tools can be used to quantify trafficking of NPC2 to cholesterol-rich LSOs visualized by filipin staining in NPC2 disease fibroblasts. Finally, we show how the co-localization tools implemented in SpatTrack can be validated on synthetic image sets. Simulation techniques rooted in statistical equilibrium mechanics can be used to generate realistic benchmark vesicle patterns in synthetic cell images. By choosing suitable interaction potentials between the vesicles and between vesicles and cells, one can perform simulations in the real cell geometry. From such simulations, one can generate a variety of vesicle patterns, which closely resemble those detected in living cells. The synthetic vesicles form a thermodynamic ensemble from which configurations are sampled by Markov Chain Monte Carlo (MCMC) simulations. The generated vesicle patterns can be applied to validate image analysis toolboxes, such as the PDBCA and two-channel-RDF subroutines in SpatTrack. In this protocol, we quantify the precision of SpatTrack's object-based co-localization tools and show that PDBCA and two-channel-RDF gather complementary information about co-trafficking events.

2 Materials

1. Culture medium: DMEM supplemented with 1% glutamine, 1% penicillin, and 10% FBS.
2. Buffer medium: 150 mM NaCl, 5 mM KCl, 1 mM $CaCl_2$, 1 mM $MgCl_2$, 5 mM glucose, and 20 mM HEPES (pH 7.4).
3. Simulations of intracellular vesicle patterns and image analysis via SpatTrack were run on a 64-bit Windows 7 desktop computer having a 3.07 GHz Intel Xeon W3550 processor and 12 GB RAM. On such a machine, the simple MC simulations

take about 15 s per image analyzed, MCMC simulations take about 3 min and batch processing of a stack of 10 images with 20 random distributions per analysis takes between 7 and 15 min depending on the types of analyses included (*see* **Note 1**).

3 Methods

3.1 Co-Labeling of Human Fibroblasts with Fluorescent NPC2 and Filipin

1. Seed fibroblasts on microscope dishes in culture medium with glass coverslip inlets and let them settle for 24 h.
2. Exchange medium to DMEM medium containing human lipoprotein-depleted serum (LPDS) instead of FCS, in the absence or presence of 100 nM NPC2 or Alexa546-NPC2 and incubate for 48 h (*see* **Note 2**).
3. Wash the cells with ice-cold buffer medium to prevent further membrane trafficking.
4. Fix the cultures in 3% PFA in PBS for 1 h at room temperature and quench with 50 mM glycine in PBS.
5. Incubate cells for 2 h at room temperature in buffer medium containing 50 μg/ml filipin added from a stock solution in DMSO (final concentration 0.2% v/v).
6. Rinse cells several times with buffer medium to remove non-specifically adhered filipin.
7. Image cells in buffer medium at room temperature on a DMIRBE inverted widefield microscope equipped with a 63 × 1.4 NA oil immersion Fluotar objective (Leica Lasertechnik GmbH) with a Lambda SC smartshutter (Sutter Instrument Company) as illumination control. Images were acquired with an Andor Ixon^{EM} blue EMCCD camera operated at −75 °C and driven by the Solis software supplied with the camera. The final pixel size for these settings is 0.25 × 0.25 μm. Filipin was imaged using a Leica A-cube having a 360-nm (20-nm bandpass) excitation filter, 400-nm dichroic and a 425-nm longpass emission filter (*see* **Note 3**). Alexa546-NPC2 was imaged using a standard rhodamine filter set [535-nm, (50-nm bandpass) excitation filter, 565-nm longpass dichromatic filter and 610-nm (75-nm bandpass) emission filter].

3.2 Analysis of Spatial Patterns of Cholesterol-Rich LSOs Using SpatTrack (www.sdu.dk/bmb/spattrack)

3.2.1 Isolating the Cell from the Background

In order to analyze the spatial pattern of endosomes with respect to the cell geometry SpatTrack requires a mask of the cell on a dark background. For that purpose, we extract the cell outline from images of filipin-stained fibroblasts. As a significant amount of cellular cholesterol resides in the plasma membrane, outlines are easily extracted from the filipin images of the cells. This may be obtained by the cell segmentation utility in SpatTrack as shown in Fig. 1. Initially, the image is divided into a number of tiles in each direction, and the contrast of the image is enhanced

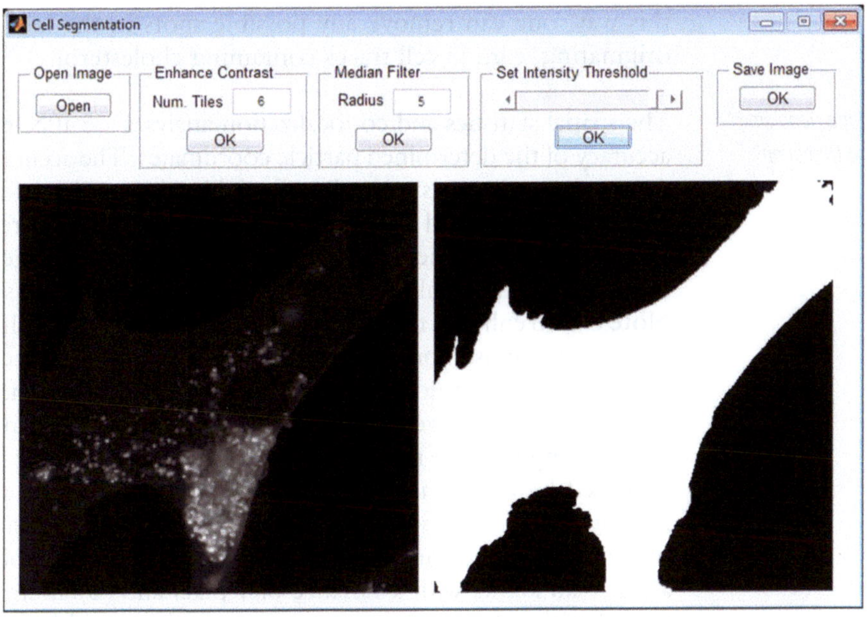

Fig. 1 GUI for cell segmentation. Initially, the image is divided into a number of tiles and the contrast of the image is enhanced locally in each tile using the contrast-limited adaptive histogram equalization (CLAHE) algorithm. Here the image is divided into 6 × 6 = 36 tiles. Subsequently, low frequency background signal is diminished by a median filter. Finally, the cell can be isolated from the background by an intensity threshold

locally in each of these tiles using the contrast-limited adaptive histogram equalization (CLAHE) algorithm [42]. A potential problem in fluorescence microscopy is uneven illumination which leaves some areas dimmer than other areas. In that case one can typically not apply a local threshold to the whole image. Dividing the image into tiles has the advantage that the contrast is enhanced locally in each tile. Thus, potential artifacts from uneven illumination will even out (*see* **Note 4**). In Fig. 1, the image is divided into 6 tiles in the x- and y-direction giving a total of 36 tiles. Second, the image is applied a median filter, with a radius of 5 pixels, for an easier separation of the cell from the background and finally, the cell may be isolated by adjusting the intensity threshold. For a given filipin image, one can first adjust the number of tiles and click OK in the Cell Segmentation menu. The result is shown in the right side panel of the graphical user interface (GUI) in Fig. 1, while the left side panel always shows the raw image. Usually using fewer tiles makes it easier to isolate the cell. Subsequently, one can try adjusting the size of the median filter to see the effect on the processed image. Finally, the user can adjust the intensity threshold to isolate the cell from the background. At this stage several smaller spots outside the cell may be included in the segmentation. However, once "OK" is clicked, only the largest region, which is the cellular area, is preserved in the resulting image. In this way, one can remove any possible spots in the background, originating, e.g., in cell traces containing cholesterol.

3.2.2 Image Denoising and Particle Detection

The spatial statistics and co-localization analyses are affected by the accuracy of the determined particle coordinates. The accuracy of the detected coordinate is primarily affected by the number of collected photons, the standard deviation of the objective's point spread function (PSF) and by the pixel size of the final image. Image noise is unavoidable but should be minimized during image acquisition (*see* **Note 5**). To enhance the particle detection performance, the images should be denoised prior to the spatial analyses. Subsequently, the location of the endosomes should be determined using the particle detection GUI. Image denoising in SpatTrack is based on a bandpass filtering method first described by Crocker and Grier [37] and also used by Lund and Wüstner [38] which requires an estimated particle radius. Briefly described, to generate the denoised image two low pass images are first generated (1) an image convolved with a Gaussian kernel with a variance of 1 pixel and (2) an image convolved with a boxcar (or square) kernel with a width of $2w + 1$, where w is an integer larger than the average particle radius. Next, the denoised image is generated by subtracting the boxcar filtered image from the Gaussian filtered image. Finally, an intensity threshold is applied to remove remaining background noise. This threshold is given by a user-defined value multiplied by the average intensity of the denoised image. For example, the image in Fig. 2a was denoised with a particle radius (Object Size) of 4 pixels and an intensity threshold of 5×, the average intensity of the denoised image. Open the "Image Denoising" GUI from the main SpatTrack GUI and open the "NPC2_Image.tif" image from the folder containing the example files (*see* **Note 6**). Selecting the appropriate intensity threshold and particle size is crucial, as the size of the particles in the image is affected by these two parameters. Therefore, the denoising GUI includes an optimization function. Click the "Optimize" button and then load the image into the new GUI and set the "Object Size" to 4 pixels and the "Intensity Threshold" to 5. Subsequently, click the "Optimize" button which opens a dialog box asking for a number of particles to include in the analysis. Select a number between 5 and 10 and click "OK". This opens a new GUI where the user can now select a given number of particles in the raw image. For ease of use the image in this GUI is inverted so one can easily see the difference between the particles and the crosshair. Once the particles have been selected, the program calculates the radius of each particle given by the variance of a 2D Gaussian function fitted to the particle. Next, the image is denoised with a constant intensity threshold but different particle sizes. For example, the image in Fig. 2 was denoised with particle radii of 2, 3, 4, 5, or 6 pixels. After denoising the sizes of the pre-selected particles are determined as the variance of a fitted 2D Gaussian function. Finally, the average size of

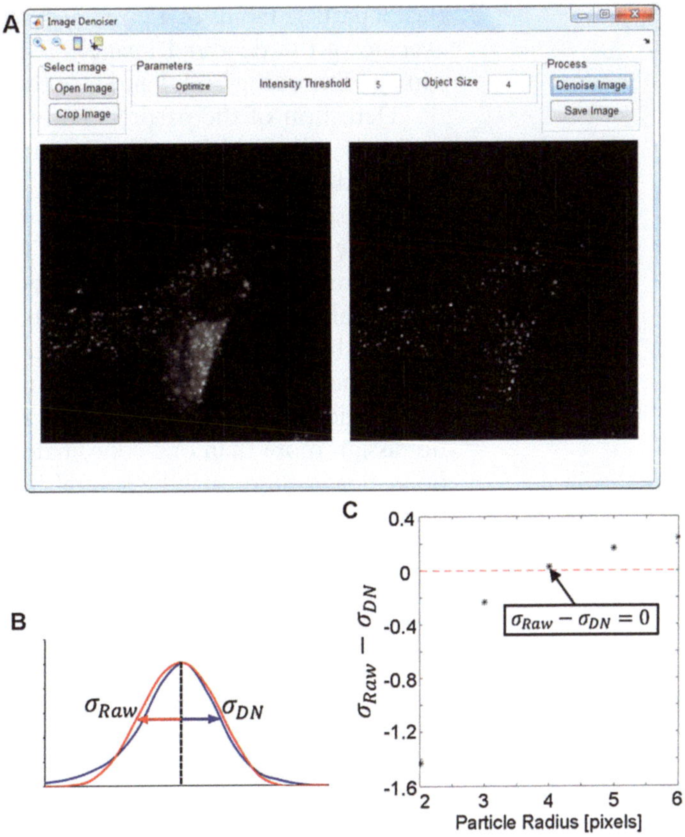

Fig. 2 Image denoising in SpatTrack. (**a**) Gui for image denoising. The image is denoised using the Boxcar algorithm which requires a particle radius (Object Size) and an intensity threshold. Here the image was denoised with a particle radius of 4 pixels and an intensity threshold of 5× the average intensity of the denoised image. (**b**) The denoising settings are optimized manually by comparing the size of the particles in the raw and denoised image. The particle radius is determined from the average variance (σ) of a 2D Gaussian function fitted to a number of particles in the raw and denoised image. Panel B exemplifies this process with a 1D Gaussian function where the red and blue curves show the Gaussian functions fitted to a particle in the raw and denoised images, respectively. To determine the optimum settings, the image is denoised with different particle radii, but with the same intensity threshold the size of the denoised particles is determined for each radius. Panel C shows the difference between the average particle radius for the particles in the raw image (σ_{Raw}) and the average particle radii in the denoised and thresholded images (σ_{DN}) from which it can be seen that for the NPC2-containing LE/LYSs the optimum particle radius for denoising is 4 pixels

the denoised particles is compared to the average size of the particles in the raw image. This is illustrated in Fig. 2b where the red and blue curves show the 1D Gaussian profiles of a particle in the raw image and the denoised image, respectively. Finally, Fig. 2c shows that $\langle\sigma_{Raw}\rangle - \langle\sigma_{DN}\rangle \sim 0$, for the selected denoising parameters (i.e., a particle radius of 4 pixels and an intensity threshold of 5× the average of the denoised image). Thus, apply these settings in the "Image Denoising" GUI and save the denoised image.

Detection of the particles requires two parameters, which are set in the "Particle Detection" panel, (1) a fluorescence intensity threshold and (2) a particle location diameter. The intensity threshold is based on the intensity values of the image. For example, in Fig. 3 the intensity threshold is 93. Thus, if a histogram was generated from all pixel intensities in the image, pixels with intensity lower than the 93rd percentile will be ignored. To easily determine the threshold click "Find Threshold" which will open a new GUI where the threshold can be determined by adjusting a slider bar. The particle "Location Diameter" ensures that the algorithm does not assign more than one coordinate per particle. For example in a noisy image there may be several high intensity pixels in a single particle. The optimum setting depends on the image. For example, if some particles are partially overlapping the "Location diameter" should be set such that the particles can still be separated without false positives. Typically, the "Location Diameter" should be set to a value slightly larger than the particles but smaller than the distance

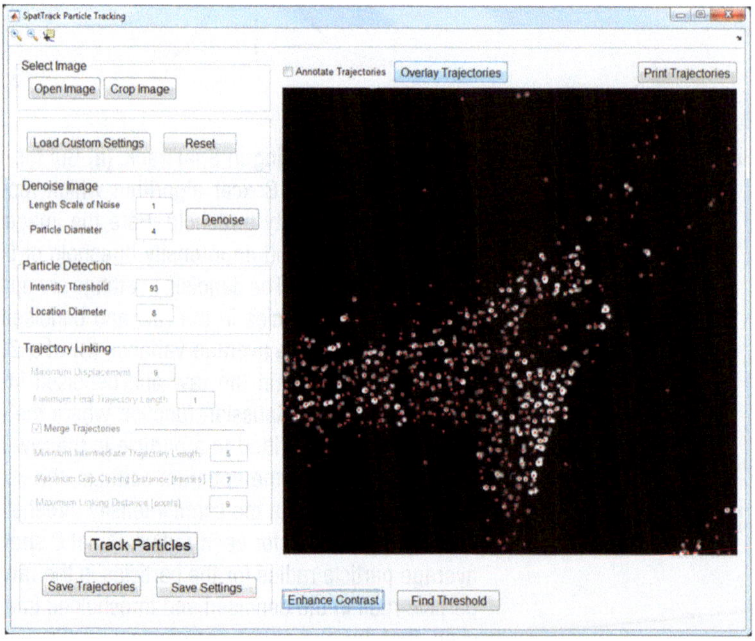

Fig. 3 Particle detection GUI in SpatTrack. Particle detection in single images requires two parameters, (1) an intensity threshold and (2) a minimum distance between two detected particles. Here the Intensity threshold is set to 93, corresponding to a cut-off at the 93rd percentile of the image intensity histogram. The "Location Diameter" gives the minimum distance between two particles. Here it is set to 8 pixels corresponding to 2× the estimated particle radius during image denoising

between two neighboring particles. In Fig. 3 it was set to 8 pixels corresponding to two times the radius of 4 pixels used in the denoising step. Thus, no two coordinates can be found with a distance of two particle radii from each other. The "Location Diameter" is also used for refining the particle coordinate. Initially, the particles are located at a pixel-wise precision by identifying the brightest pixel in each blob making up a particle. Subsequently, the coordinate is refined to sub-pixel precision by calculating the center of mass in a box centered around the initially detected coordinate. For this operation the side length of the box is given by the "Location Diameter" plus one pixel. In the following paragraphs, we will refer to the particles found by particle detection as the detected particles to distinguish these from randomly generated particles.

3.2.3 Determining the Distance of LSOs from the Plasma Membrane

Given the location of endosomes or LSOs in a living cell and the shape of the cell, one can determine the distance of the particles to the PM. Figure 4 shows the GUI for this analysis in SpatTrack. Here the minimum, maximum, and average distances may be calculated. Furthermore, one can generate a random particle distribution and compare the detected and random distributions. To perform this analysis, open the "Distance to Membrane" GUI from the main SpatTrack menu. Then click "From Memory" which opens a new GUI. In this GUI load the "NPC2_xCoords.

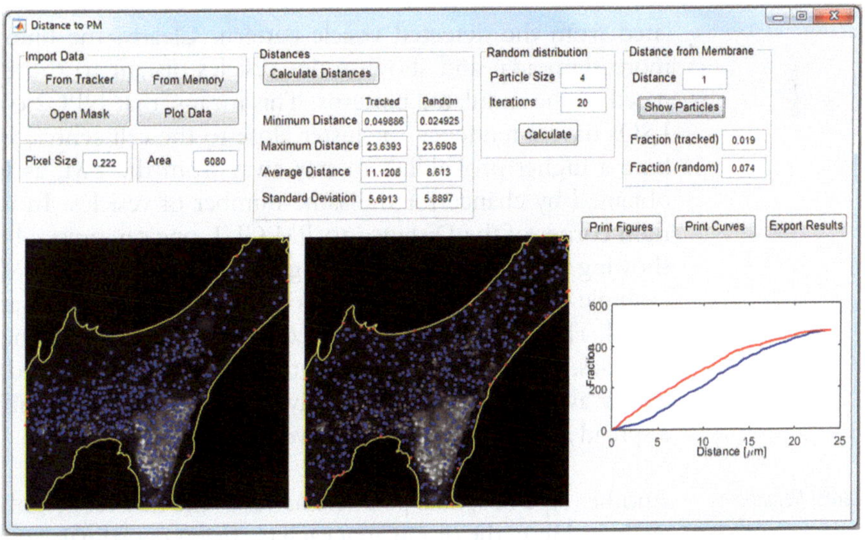

Fig. 4 GUI for analysis of the particles' distance to the plasma membrane (PM). Detected particles in the cell geometry, as determined from the PM signal of the filipin image (*yellow*) lines, are shown as blue dots on top of the filipin image in the left panel. In the middle panel, the same number of particles but randomly distributed within the same cell are shown. Most right panel in the GUI shows the number of particles in a given distance from the PM for the random particle population (*red curve*) and the experimental vesicle distribution (*blue curve*). Clearly the experimental distribution is more clustered, as fewer filipin-labeled LSOs are found close to the PM (e.g., in a distance of up to 15 μm from the cell edge) as expected by chance

txt" and "NPC2_yCoords.txt" as the x- and y-coordinates for population 1. Additionally, open the "Mask.tif" image. Leave the "File Manager" GUI open and return to the "Distance to Membrane" GUI and click "From Memory" again. Finally, click "Plot Data". Now you should see the outline of the cell in yellow and the detected particles in blue. Notice that any particles outside of the cell geometry are automatically excluded from the calculation. Set the pixel size to 0.222 (μm/pixel) and click "Calculate Distances" to determine the minimum and maximum distances from the detected particles to the membrane as well as the mean and standard deviation of all particle-to-membrane distances. Notice that the distances in the second column under "Random" have not yet been calculated. To compare the detected particle locations to a random distribution set the "Particle Size" to 4 pixels corresponding to the radius of the particles used during image denoising. Set the "Iterations" to 20 and click "Calculate" to generate 20 random particle distributions to compare to the detected particles. Importantly, the distance between two points in the random distribution cannot be closer to each other than two times the estimated radius of the detected particles.

Performing 20 rounds of MC simulations gives us a 95% confidence interval for subsequent calculations [39]. Now you can see, as in Fig. 4, two curves for the number of particles as a function of the distance to the PM; in red the distance according to the average from 20 random vesicle distributions, and in blue, the curve calculated from the detected vesicle pattern. Clearly, the blue curve is more sigmoidal and shows more vesicles distant to the PM compared to the simulated patterns. This means, that NPC2 containing LSOs have a tendency to cluster close to the cell center, since they have a higher probability to stay away from the PM, as would be obtained by chance for the same number of vesicles. In the upper right corner of the Distance-to-PM GUI, one can press a button for showing the particles within a given distance to the PM. In the example given in Fig. 4, particles within 1 μm of the PM are shown in red. This is 1.9% of the detected particles and 7.4% of the random particles, supporting visually the tendency of filipin-stained LSOs to cluster away from the PM. Finally, pressing "Export Results" opens a spread sheet from which the two curves may be copied.

3.2.4 Nearest Neighbor Analysis and Fractional Area Occupied by LE/ LYSs

Another type of spatial pattern analysis is the nearest neighbor (NN) analysis. Here, the distance of each particle's coordinate to its closest neighbor particle, also called its nearest neighbor, is calculated. Again, the coordinates of all detected vesicles are considered as a set of points which have an associated physical dimension (the average vesicle size estimated from fitting a 2D Gaussian). The distance to all points in the plane for a given coordinate pair (x, y) is calculated, and the smallest distance (i.e., the NN) is stored. This is repeated for all detected particles in the image. Subsequently, the average NN of the detected particles is calculated. In parallel, the average NN is determined for a number of random vesicle patterns in the same cell geometry, as described above. Finally, the NN of the detected particles is divided by the average NN of a number of random particle distributions in the same cell geometry, and this measure can reveal whether the particles are randomly distributed, accumulating or repulsive to each other. Figure 5a shows the GUI for NN analysis. In the GUI click "From Memory" to import the particle coordinates and the mask image and click "Plot Data". Set the pixel size to 0.222 μm/pixel, as described above for this particular image, and set the particle size to 4 pixels. Notice that any particles located outside of the boundary are not included in the calculation. To calculate the minimum, maximum and mean NN distances for the detected and the random particles press "Calculate". This yields an ANN score of 0.95. The ANN is the average NN distance of the detected particles divided by the average NN distance of the randomly generated particles. Thus, for ANN = 1 the particles are randomly distributed while ANN < 1 or ANN > 1 indicates accumulation or repulsion, respectively. Thus, for this example the particles are randomly distributed on the NN length scale. This, however, is not in contradiction to the results of the distance-to-PM analysis described above. The NN measure ignores the overall cellular context and assesses only the local neighborhood of a given particle.

In addition to the NN analysis, this GUI lets you calculate the fractional area occupied by the particles in the cell. This calculation starts from the fact that each triangle connecting three dots can be circumscribed by a circle (the circumcircle). By using a method from computational geometry called Delaunay triangulation, SpatTrack uses the circumcircles to connect the given coordinates such that the minimum angle of all triangulations is maximized. Each edge of the Delaunay triangulation is subsequently associated with a specific radius (the radius of the smallest empty circle containing the edges of the triangle). The union of all these radii and triangles forms a shape covering the area occupied by the vesicles within the cell geometry. This is shown in the left panel of the GUI in Fig. 5a by the blue lines connecting the particles. However, this triangulation alone tends to overestimate the particle-occupied area, since the outer boundary of the Delaunay triangulation is the convex hull and endosome patterns do not need to fill convex areas (i.e., the vesicle area can also contain concave regions). To account for this, the lines connecting the particle coordinates are refined by a method called alpha-shaping. In this process we place a circle with a radius given by $r = \alpha\langle \mathrm{NN_{obs}} \rangle$, where α is a user-defined parameter, and NN_{obs} is the average NN distance of the detected particles. This circle is placed at the center of each line segment in the triangulation, and circles which do not contain any of the tracked coordinates are deleted. Thus, for a sufficiently chosen α, one can remove any empty convex areas (i.e., regions inside the triangulation, which do not contain vesicles). Thereby, the algorithm achieves tight fitting of the alpha-shape around all vesicles. If α is chosen too large, no line

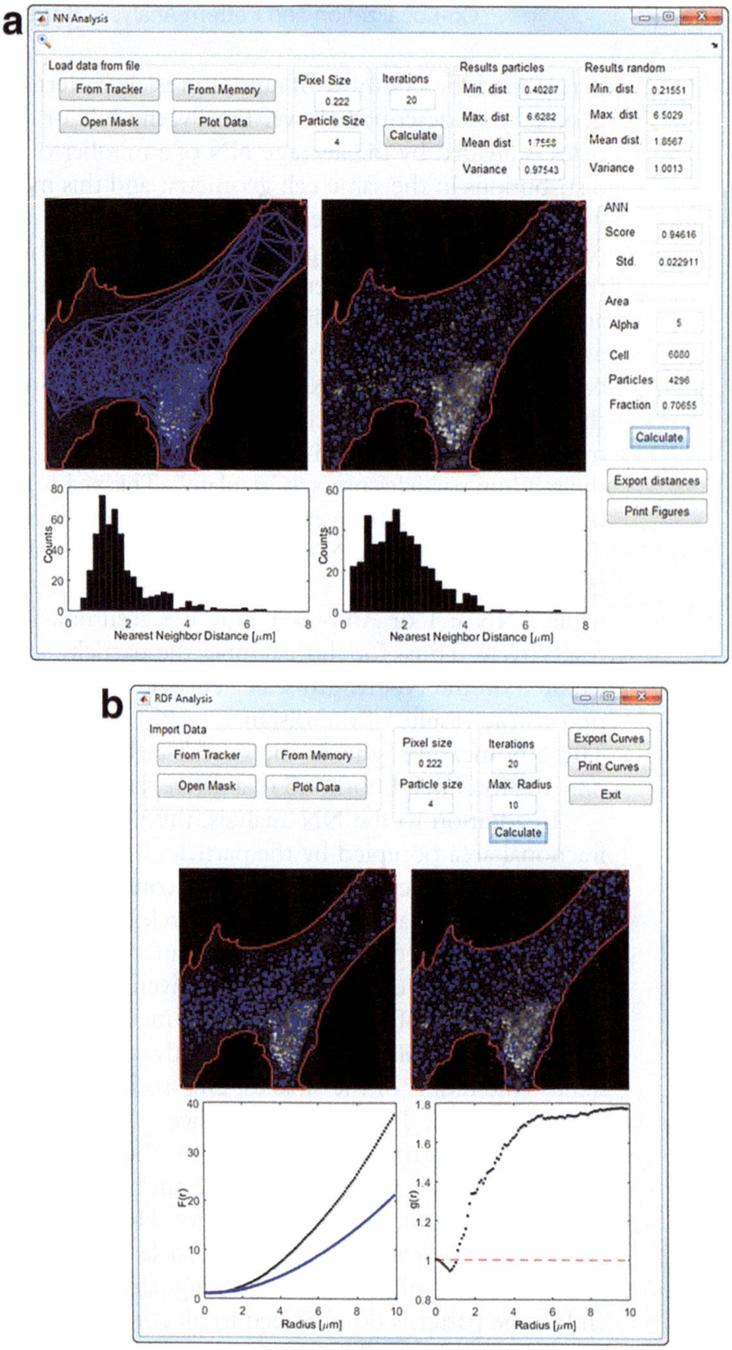

Fig. 5 GUIs for one-channel spatial analysis. (**a**) NN analysis. The detected particles are shown in the left panel while the random particles are shown in the right panel. From the ANN = 0.94616 ± 0.022911 one can see that the endosomes are randomly distributed with respect to each other on the NN length scale. (**b**) GUI for analysis by the radial distribution function (RDF). The detected particles are shown in the top left side panel while the random particles are shown in the right side panel. The lower left hand panel shows the number of particles as a function of distance (i.e., the RDF) for the detected and random particle distributions in *black* and *blue* respectively. In the lower right hand panel the ratio g(r) of the RDF for the detected particles with respect to the RDF for the randomly generated particles is shown. Careful inspection of the g(r) curve shows that for distances smaller than ~1 μm the particles repel each other while for distances larger than ~1 μm the particles are accumulating. Interestingly, 1 μm is only slightly larger than the estimated particle radius of 0.888 μm

segments are deleted, while for α being too small, all line segments are deleted. For further explanations of this method we refer to reference [39] and the supplementary material therein. In the open image set $\alpha \sim 5$ and press "Calculate". Now we can see that the area of the cell is 6080 μm^2 while the area occupied by the detected particles is 4296 μm^2 which yields a fractional area occupied by 0.71.

3.2.5 Analysis by the Radial Distribution Function (RDF)

While the NN analysis may reveal if the particles are randomly distributed, accumulating or repulsive, it is limited to the nearest neighbor length scale. One can calculate the radial distribution function (RDF), as shown in Fig. 5b, to analyze spatial patterns on a large range of distances, thereby accounting for the cellular context of endosome distributions. Calculation of the RDF starts by placing a given particle's coordinates at the center of a circle of defined radius and counting the number of neighboring particles within this radius. This procedure is repeated for increasing radii of concentric rings surrounding the particle's coordinate. Specifically, the range is divided into 100 equally spaced bins from 0 μm to the maximum radius defined by the user. This calculation is carried out for each particle detected in the images of a given channel, and the average of all particle RDFs is determined. Next a separate average RDF is generated from a user-defined number of random particle distributions. Typically, we set the number of random distributions to 20 to obtain a 95% confidence of the result. Finally, we calculate the ratio g(r) given by the average RDF for the detected particles divided by the average RDF for the random particle distributions, *see* Eqs. 13 and 14 in [39]. The GUI for RDF analysis, *see* Fig. 5b, may be opened from the main SpatTrack GUI. In the GUI click "From Memory" to open the particle coordinates and the mask image. Then click "Plot Data" to display the particle coordinates in the cell geometry. Set the pixel size to 0.222 μm/pixel and the particle size to 4 pixels. Keep the number of iterations to 20 and the maximum radius for calculating the RDF for the simulated pattern to "Max. Radius = 10". When you press "Calculate" two plots are generated. On the left, the average number of particles is shown as a function of radius, both, for the detected vesicle coordinates (black dots) and the random particle sets (blue dots), respectively. The plot in the lower right panel of the GUI in Fig. 5b shows g(r), i.e., the ratio of the experimental and simulated average RDF, as a function of the radius. Here, $g(r) = 1$ indicates a random distribution while $g(r) < 1$ or $g(r) > 1$ indicate repulsion or accumulation, respectively. Clearly, this analysis shows clustering of cholesterol-rich LSOs in NPC2 disease fibroblasts, thereby confirming the outcome of the distance-to-PM analysis presented in Fig. 4.

3.3 Object-Based Co-Localization of Fluorescent NPC2 with Cholesterol-Rich LSOs

Before starting the following analyses go to the "file manager" GUI and load the x- and y-coordinates for the filipin-stained LSOs as coordinates for particle population 2. Denoising of the filipin image was done with a particle radius of 5 pixels and a threshold of

8 times the average of the denoised image. Thus, in the following analyses the particle size for channel 2 should be set to 5 pixels.

3.3.1 Two-Channel NN Analysis

Two-channel NN analysis reveals how the particles in one population relate to the particles in another population, both recorded in the same cell (e.g., endosomes containing Alexa546-NPC2 relate to filipin containing LSOs). In this analysis, the NN distance for a particle in population 1 is the distance to the nearest particle of type 2. Subsequently, this is compared to the average NN distance for the particles of type 1 to a population of randomly distributed particles of the same size as population 2. Open the two-channel NN analysis GUI from the SpatTrack main GUI and click "From Memory" followed by "Plot Data". Once the image mask and the particle coordinates are loaded, set the pixel size to 0.222 μm/pixel and the number of iterations to 20. Additionally, set the particle size of population 1 and 2 to 4 and 5 pixels, respectively. These are the radii we used for denoising of the NPC2 and filipin images, respectively. Figure 6a shows the GUI for two-channel NN analysis where the fibroblast cell co-labeled with filipin and NPC2 described above is analyzed. In Panel A the coordinates for NPC2 containing endosomes are shown in red while the coordinates for filipin containing LSOs are shown in yellow. The distribution of particles in population 1 (NPC2) with respect to particles in population 2 (filipin) is calculated by clicking the "Ch. 1" button while the opposite calculation is performed by clicking the "Ch. 2" button. The right side panel of Fig. 6a shows the distribution of NPC2 containing endosomes compared to the distribution of a random particle population containing the same number of endosomes as the population of filipin containing vesicles. From this it can be seen that the average NN distance from NPC2 containing endosomes to filipin containing vesicles is 1.6172 μm while the average NN distance to the randomly distributed particles is 2.1415 μm. This yields an average NN distance ratio (ANN) of 1.6172 μm/2.1415 = 0.7559. From that one can conclude that LE/LYSs containing NPC2 and cholesterol-rich LSOs stained with filipin have a tendency to form clusters in the fibroblasts, since they are more closely packed, as would be expected by chance. The same analysis for the filipin vs. NPC2-containing vesicles (i.e., by changing the assignment of populations) gives a value for the ANN of 0.54, confirming the non-random but close distribution of cholesterol-rich and NPC2-containing vesicles in disease fibroblasts.

3.3.2 Two-Channel RDF Analysis

As described above for the one-population analyses, the NN measure provides information about particle aggregation or repulsion on short spatial scales. Thus, NN analysis is restricted to relatively short distances. On the other hand, analysis by the RDF provides information about the distribution of the particles over a range of distances. The two-channel RDF for the detected particles is a

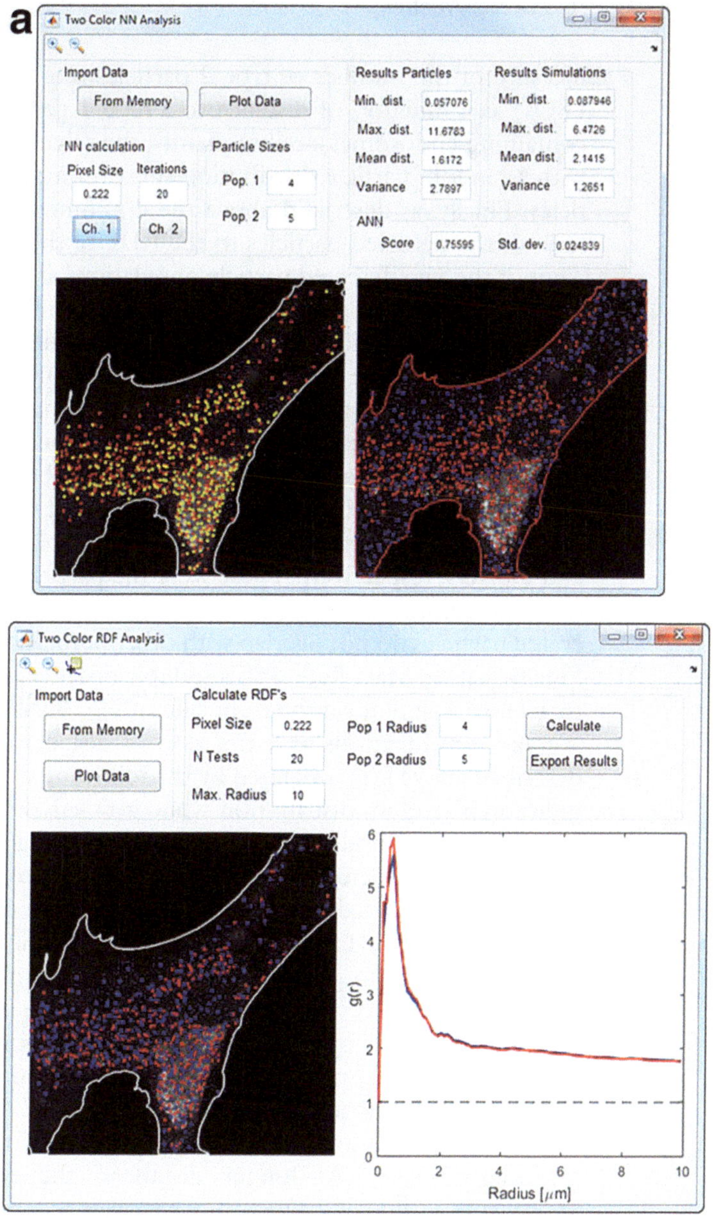

Fig. 6 GUIs for two-channel spatial analysis. (**a**) GUI for two-channel NN analysis. The two-channel NN analysis provides a measure of whether two particle populations are randomly distributed with respect to each other. Data are imported at the top left corner of the GUI and the required parameters can be entered below. In the lower left panel the coordinates of NPC2 containing endosomes and filipin-stained LSOs are plotted as *red* and *yellow* dots, respectively. In the bottom right panel, the distribution of NPC2-containing endosomes (*red*) is plotted against a random particle distribution (*blue*). To ensure that the random particles are simulated with a physical size the user must enter the size of each type of particle. Subsequently, the ANN for particles of one type with respect to the other can be calculated. Here the ANN is 0.75595 indicating that the NPC2 and filipin-containing vesicles are slightly accumulating with respect to each other. (**b**) GUI for two-color RDF analysis. Data import and calculations are performed at the top of the GUI. In the left panel the particles are plotted on top of the cell where particles of type 1 and 2 are shown as *blue* and *red* dots, respectively. The plot in the right hand panel shows g(r) as a function of radius in the same colors. Notice the peak at ~0.4 μm which indicates partial co-localization of the endosomes containing NPC2 and the filipin-stained LSOs

measure of the number of type 2 particles (e.g., filipin containing LSOs) as a function of distance to a type 1 particle (e.g., NPC2-containing endosomes). Subsequently, the calculation is performed with a random particle distribution substituting for the type 2 particles. Finally, as described above the g(r) function is given by the ratio of the detected particles to the RDF of the random particles. Thus, if the two detected particle populations are randomly distributed with respect to each other $g(r) = 1$. On the other hand, $g(r) < 1$ or $g(r) > 1$ indicate repulsion or accumulation, respectively. Figure 6b shows the GUI for two-channel-RDF analysis. Load the mask image and particle coordinates by clicking "From Memory" and "Plot Data". Again population 1 and 2 contain the coordinates of NPC2-positive endosomes and of filipin-labeled LSOs, respectively. Set the pixel size to 0.222 μm/pixel and the number of tests to 20. Additionally, set the particle size for population 1 and 2 to 4 and 5 pixels, respectively. These were the particle radii used during image denoising and are required to ensure that the randomly generated particles do not overlap with other particles of the same type. Finally, set the Max. Radius to a suitable distance, in Fig. 6b it is set to 10 μm. The plot on the right side of Fig. 6b shows a plot of the average g(r)-functions as a function of distance to a particle. As described above g(r) is defined by the following properties; $g(r) = 1$ indicates a random distribution while $g(r) < 1$ or $g(r) > 1$ indicate repulsion or accumulation, respectively, where g(r) is the ratio of the RDF for the detected particles with respect to the RDF for the randomly distributed particles. Thus, from Fig. 6b it can be seen that at radii below 1.5–2 μm the g(r) is much larger than 1 which shows that the two particle populations are accumulating on this length scale. Notice that g(r) peaks at ~0.4 μm which is approximately half the size of the detected particles (4 pixel radius × 0.222 μm/pixel = 0.888 μm radius). Thus, the location of the peak indicates that not only do the NPC2-containing LE/LYSs and filipin-stained LSOs accumulate but some fraction of these vesicles are co-localized [39]. At longer distances, the two-channel-RDF approaches a value of 2 suggesting a weaker accumulation over longer distances. Within the cells, the NPC2- and cholesterol-rich vesicles never become randomly distributed relative to each other, which would mean $g(r) = 1$ (see above). This is in line with the one-color RDF analysis, which revealed for both vesicle populations a tendency to cluster in the perinuclear region [39].

3.3.3 Particle Detection-Based Co-localization Analysis (PDBCA)

Object-based co-localization in SpatTrack can be carried out using the PDBCA subroutine. Figure 7a shows the GUI for PDBCA. Here the x- and y-coordinates for the detected particles in each channel as well as the denoised images used for particle tracking are first loaded in the top of the GUI. Then the number and fraction of co-localized particles are determined on basis of the pixel size of the images and the maximum allowed displacement between the identified particles in the two channels. The GUI operates in two

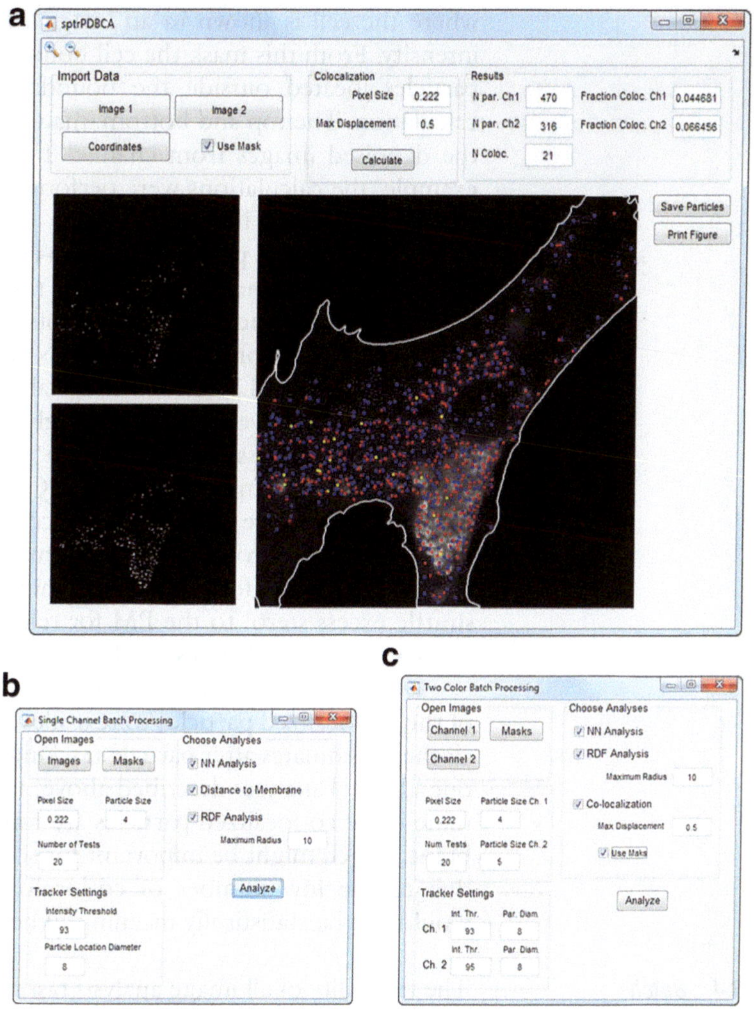

Fig. 7 GUIs for particle detection-based co-localization analysis (PDBCA) and batch processing. (**a**) GUI for PDBCA. Initially, the data for the two-particle populations are loaded. Then the number of co-localized particles is determined based on the pixel size of the images and the maximum allowed displacement between co-localized particles in the two channels. (**b**) GUI for single channel batch processing. (**c**) GUI for two-channel batch processing. The batch processing GUIs take stacked Tiff-files of denoised images for one or two channels as generated in Fig. 2 and a stack of mask images as generated in Fig. 1. Subsequently, the batch processing first performs particle detection followed by the analyses described above. The types of analyses are determined by the user by clicking one or more check boxes in the right side panel. Finally, the results are saved to text files which can be analyzed further by the user

modes. In the first mode, the co-localization between all particles in the two-channel images is calculated. This is possible, as the PDBCA is independent of a particular cellular geometry, such that it could for example be applied to images containing more than one cell. Alternatively, the user may choose to load a mask image where the cell is shown in an image with a background of zero intensity. From this mask the cell boundary is determined and any particles located outside the boundary are excluded from the calculation. The top and bottom images to the left in Fig. 7a show the denoised images from channel 1 and 2, respectively. In this example, the calculations were performed with a mask. The larger panel to the right in Fig. 7a shows the cell with the boundary (white line) and the particles found in channel 1 (Alexa546-NPC2) and 2 (filipin) as blue and red points, respectively. Finally, co-localized particles are shown as yellow dots. As shown in the top right corner of Fig. 7a only ~5% of the NPC2-containing endosomes and filipin-stained LSOs were co-localized. This shows that while the two types of vesicles are strongly clustered they are almost exclusive. One possible explanation for this could be that the endosome-containing functional NPC2 rapidly efflux free cholesterol and, thus, have weak fluorescence intensity in the filipin channel. The cholesterol-rich LSOs serve in this scenario as inert cholesterol donors for NPC2-containing endocytic vesicles, which shuttle excess sterol to the PM for release from the cells. Further studies are required to test such a model carefully.

Once the analysis is completed one can export the coordinates of the co-localized particles to text files similar to those containing all the coordinates after particle tracking. Thus, one can repeat the one-channel analyses described above. For example, one can determine if the co-localized particles are found near the plasma membrane, which might be important for some applications. However, due to the low number of co-localized particles such analyses would not be statistically meaningful here.

3.4 Batch-Processing of Large Image Sets in SpatTrack

The reliability of all image analysis tasks increases with the number of biological replicates acquired for each condition. Therefore, the microscopist typically acquires several images per condition and experiment and repeats a given experiment several times. Analysis of such image data can be automated, since a given analysis protocol is identically carried out several times. For that purpose, SpatTrack contains an option called batch processing. Assuming one has acquired many images for two channels under identical conditions and stored them as one stack per acquired channel for subsequent image processing and analysis. One can carry out all the described analyses automatically in all images, and Fig. 7b and c shows the GUIs for single- and two-channel batch processing, respectively. Batch processing requires a stack of images which would normally be used for particle tracking as well as a corresponding stack of mask images. Notice that the batch-processing GUIs do not include image denoising. This must be done separately if required. When the user has set the correct parameters and chosen the type(s) of analysis the program runs the following for each image (1) detect coordinates for particles in the given image,

(2) using the mask and the detected coordinates calculate the spatial statistics for the image, (3) save the results in memory until all images have been analyzed. Finally, the program outputs a number of comma separated txt files.

3.4.1 Output from Single-Channel Analysis

1. Particle detection: x- and y-coordinates and intensities for each image (in three separate files).
2. NN-analysis: A single file containing the following columns; col. 1–4: min NN, max NN, mean NN and standard deviations for experimentally detected particles, col. 5–8: min NN, max NN, mean NN and standard deviations for random particles (these are the mean values obtained from numTests simulations) and cols 9 and 10: ANN and standard deviation of ANN.
3. Number of particles as a function of the distance to the plasma for each cell and for random particle populations in each cell.
4. Radial distribution function g(r) for the particles in each cell.

3.4.2 Output from Two-Channel Analysis

1. x- and y-coordinates and intensities for particles detected in each cell in each channel in 6 individual files.
2. Results of NN-analysis in one file for each channel, e.g., for channel 1 the following is returned.
   - min NN for channel 1 with respect to channel 2.
   - max NN for channel 1 with respect to channel 2.
   - mean NN for channel 1 with respect to channel 2.
   - Standard deviation of NN for channel 1 with respect to channel 2
   - min NN for channel 1 with respect to random particles.
   - max NN for channel 1 with respect to random particles.
   - mean NN for channel 1 with respect to random particles.
   - Standard deviation of NN for channel 1 with respect to random particles.
3. RDF analysis for each channel in the same format as for single-channel analysis.
4. Results of co-localization in the following format

   col1: number of particles detected in channel 1.

   col2: number of particles detected in channel 2.

   col3: number of co-localized particles.

   col4: fraction of co-localized particles in channel 1.

   col5: fraction of co-localized particles in channel 1.

3.5 Markov Chain Monte Carlo Simulation of Intracellular Vesicle Patterns

The validity of co-localization measures is often assessed by Monte Carlo simulations, in particular to generate randomized images for assessing the extent of co-localization by chance [21, 26, 43]. We have made heavy use of simulations for generating random vesicle patterns, as described above, but also for defining diffusion modalities of mobile vesicles tracked in SpatTrack [38, 39]. Image simulations can serve other purposes as for mimicking realistic cellular patterns of organelles, thereby allowing for assessing the impact of microscope optics or imperfect alignment or detection noise on the co-localization analysis [33, 44]. In such cases, the basic idea is that the generated organelle patterns or diffusion can serve as "ground truth", as particle coordinates are known before the co-localization or motion analysis. In a realistic scenario, intracellular vesicle patterns are a consequence of mutual vesicle interactions as well as of binding of vesicles to cytoskeleton elements via motor proteins. One can generate such patterns by including physical interactions into Monte Carlo simulations. By including an interaction potential into the simulation, the vesicles are considered as a statistical ensemble of interacting particles from which representative configurations must be sampled to ensure thermodynamic equilibrium. In Markov Chain Monte Carlo (MCMC) simulations originally proposed by Metropolis et al. (1953), configurational sampling is performed along a Markov chain [45]. Let $\mathbf{X}^N$ be the configuration vector comprising positional (x, y-) coordinates of an N-particle system in two dimensions, i.e., $\mathbf{X}^N = (X_1, X_2, \ldots, X_N)$, average thermodynamic properties like the internal energy, U, can be expressed as average according to:

$$\mathrm{U} = \int \ldots \int E\left(X^N\right) \cdot P\left(X^N\right) dX^N = \langle E\left(X^N\right) \rangle \tag{1}$$

Here, $\mathbf{E}(\mathbf{X}^N)$ is the configuration energy, while the probability of finding the system in state $\mathbf{X}^N$ is given by

$$\mathbf{P}\left(X^N\right) = \frac{\exp\left(-\beta \cdot \mathrm{E}\left(X^N\right)\right)}{\int \ldots \int \exp\left(-\beta \cdot \mathrm{E}\left(X^N\right)\right) \cdot \mathrm{d}X^N} \tag{2}$$

This configuration integral is high-dimensional (e.g., for 250 particles in two dimensions, N = 62.500) and therefore, thermodynamic average properties cannot be calculated directly. Instead, the configuration integral is evaluated by generating an irreversible Markov chain, whose limiting distribution is the Boltzmann distribution:

$$\mathbf{P}\left(X^N\right) = \left\{\mathbf{P}\left(X_1^N\right), \mathbf{P}\left(X_2^N\right), \ldots\right\} \tag{3}$$

Here, $\mathbf{P}\left(\mathbf{X}_{\mathbf{i}}^{\mathrm{N}}\right)$ is the Boltzmann probability of the ith configuration of the system. For a detailed discussion of the underlying statistical theory, see refs. [45–47]. For the concrete purpose of this study, 250 particles with a distribution of diameters, obtained from a Gaussian distribution were generated in the red and green channel. The MCMC simulation was performed by displacing them sequentially according to:

$$x_i(n+1) = x_i(n) + \xi_1 \tag{4a}$$

$$y_i(n+1) = y_i(n) + \xi_2 \tag{4b}$$

Here, x_i and y_i are the (x, y) coordinates of the ith particle, n is the Monte Carlo cycle and ξ_1, ξ_2 are two random numbers drawn from a Gaussian probability density function (PDF) of mean equal to zero and variance equal to one. In MatLab, random numbers following a normal distribution were generated by the inherent function randn. The inter-particle distances, rij, as well as the distance of each particle to a reference point, r_c, in the cell were calculated before and after each attempted move and entered an energy function of the form:

$$E = E_{\text{particle-particle}} + E_{\text{particle-cell}} \tag{5}$$

Here, the first term can be a classical Lennard-Jones interaction:

$$E_{\text{particle-particle}} = \sum_{i<j} \frac{a_{ij}}{r_{ij}^{12}} - \frac{b_{ij}}{r_{ij}^{6}} \tag{6}$$

This is implemented for each particle pair as

$$E_{PP} = \varepsilon \left[\left(\frac{r_{\text{m}}}{r} \right)^{12} - \left(\frac{r_{\text{m}}}{r} \right)^{6} \right] \tag{7}$$

where ε is a measure of the strength of the interaction, r_{m} is the particle radius, and r is the inter-particle distance (given in distance units of pixels). The interaction term, E_{PP}, makes sure that particles will propel each other when they come into contact (given by the first term in Eq. 7), while they attract each other on a longer scale (given by the second term in Eq. 7). To simulate partially overlapping particles, as it takes place in co-localization events, the expression for E_{PP} can be modified by using a Buckingham instead of a Lennard-Jones potential, which is given by:

$$E_{PP} = \varepsilon \left[\frac{6}{\alpha - 6} \exp\left(-\alpha\left(r / r_{\text{m}} - 1\right)\right) - \left(\frac{\alpha}{\alpha - 6} + 1 \right) \cdot \left(\frac{r_{\text{m}}}{r} \right)^{6} \right] \tag{8}$$

Here, α is a parameter defining the nature and strength of the interaction, allowing us to tune the potential for simulating partially overlapping particles. For example, if α is set to slightly lower than 6, one gets a positive potential at $r = 0$, which means that the particles would repel each other. In contrast, if $\alpha = 6.1$ we have a strongly attractive potential for small distances, which results in particle overlap (or "fusion" from a physics perspective). In

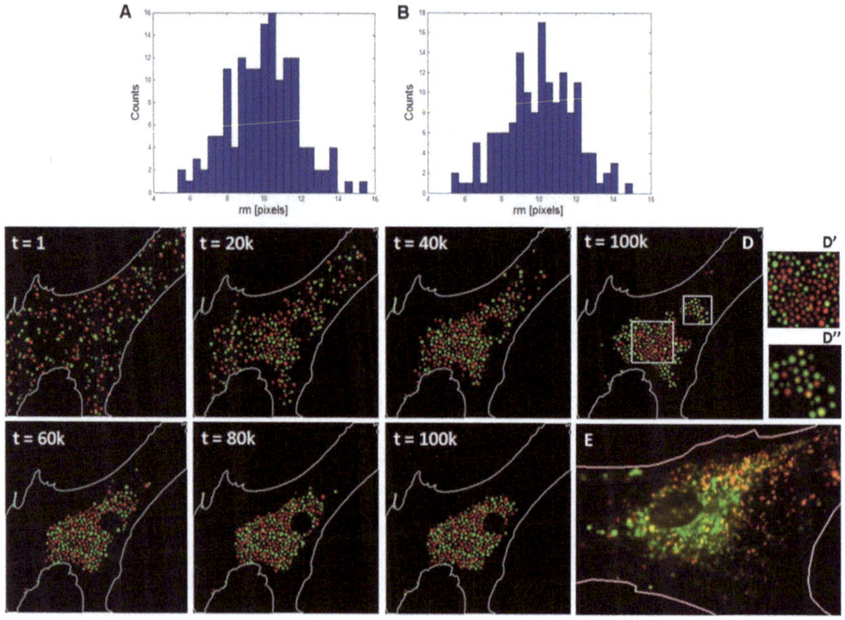

Fig. 8 MCMC simulation of two vesicle populations in the cell geometry. (**a**), (**b**) show particle $r_{m.}$ values for particles of type 1 and type 2, respectively. (**c**) Evolution of a MCMC simulation of two-particle populations with 250 particles of each type (i.e., time-lapse images of the simulation). The number of iterations is indicated as "t" but does not mean physical time. The two-particle populations pack closely in the perinuclear area, but they do not overlap. (**d**) Snapshot of simulation of two-particle populations with partial co-localization. The initial particle distribution at $t = 0$ for that simulation was the final configuration of a simulation with LJ-potentials performed over 100,000 iterations (as the one shown in panel **c**). The attraction toward the MTOC for the green particles in panel **d** was 10 times smaller than the attraction of the red particles. Thus, the red particles are more clustered in this simulation than in the one shown in Panel C. Subsequently, the simulation was run with both the LJ and the Buckingham potentials to simulate partially overlapping particles, **d** and **d**″. (**e**) Human skin fibroblast labeled with 20 μg/ml rhodamine-dextran (70 kD, pseudo-colored *green*) overnight followed by incubation with 10 μg/ml Alexa647-Transferrin for 1 h (pseudo-colored red), wash and imaging on a widefield microscope. Partial co-localization is apparent as *yellow-orange* spots

practice, we kept a given number of interacting vesicles in a list by making a vector of α-values. A list of α-values is kept for all simulated particles. For non-overlapping particles, the LJ-potential as described above and in Eq. 7 would be used instead (i.e., there is a conditional statement in the code such that if $\alpha(\mathrm{i}) = 0$ the energy is calculated with the LJ potential and if $\alpha(\mathrm{i}) > 0$ the energy is calculated with the Buckingham potential) (*see* **Note 7**). We varied the size of particles by drawing $r_{m.}$ values from a Gaussian distribution (Fig. 8a, b). This was important for generating realistic vesicle packing and thereby spatial distributions. The second term of Eq. 5, i.e., $E_{\text{particle-cell}}$, contains two terms: the interaction with the nucleus, whose position is defined by the user and the interaction with the microtubule organization center (MTOC), which is given by the center of mass of the detected vesicle distribution (i.e., the vesicle distribution detected by fluorescence microscopy as described above) from which the cell geometry was extracted from real cell micrographs. The particles were allowed to move in the cell geometry only, by rejecting all attempted moves from inside to outside of the cell. Inside the cell, a move was accepted according to the Metropolis criterion [46]. The start configuration was a random vesicle distribution of 250 particles assigned to the green channel and 250 particles assigned to the red channel (Fig. 8c). Equilibration was assessed by recording the system energy which dropped rapidly and reached equilibrium after about 100.000 MC steps (not shown). One can clearly see that the particles change their spatial arrangement from an almost random distribution at the beginning of the simulation to a clustered organization surrounding the "nucleus" after 100.000 iterations. The particles "avoid" the "nucleus" due to the second interaction term $E_{\text{particle-cell}}$. Selected snapshots of the trajectory during equilibration of the system will represent different degrees of particle clustering as it might be found for intracellular vesicles (e.g., compare simulated images after 20.000 and 80.000 MCMC steps in Fig. 8c). To simulate partial co-localization and an eventual differing spatial distribution of the two vesicle populations in the cell, we used the Buckingham instead of the Lennard-Jones potential (compare Eqs. 7 and 8) and additionally varied the attraction to the MTOC for both particle populations. This not only gives overlapping vesicles (yellow in Fig. 8d) but also results in partial demixing of the vesicle populations (Fig. 8d, only snapshot after 100.000 MCMC steps is shown). The latter was often found in experimental images of filipin vs. Alexa546-NPC2 but also, for example, for fluorescent transferrin (*see* **Note 8**) and rhodamine-dextran, two markers of recycling endosomes and LE/LYSs, respectively (Fig. 8e).

3.6 Validation of Two-Channel NN and RDF Analysis as well as PDBCA Using Simulation Vesicle Patterns

Having established a valid method for simulating realistic vesicle patterns, we use such images as benchmark for assessing the accuracy of SpatTrack for determining the spatial relationship of two vesicle populations in the cell geometry. The purpose of this section is additionally a pedagogic one, as we can learn from known vesicle patterns, how we should interpret the two-channel NN distance, two-channel RDF and PDBCA. In our belief, this increases the intuition about the meaning of these spatial measures among microscopists, who are often not familiar with these concepts. Directly from the simulations, we save the coordinates of the particle's coordinate and can determine the two-channel NN distance, two-channel-RDF and PDBCA as ground truth from such simulations. Next, we load these images into SpatTrack for determining the same properties from the simulated images (i.e., including finding the particles, estimating their size and coordinates, etc.; see above). As a first check, we simulate randomly distributed vesicles. In contrast to the naive MC simulations used for generating random point patterns in SpatTrack, the MCMC simulations

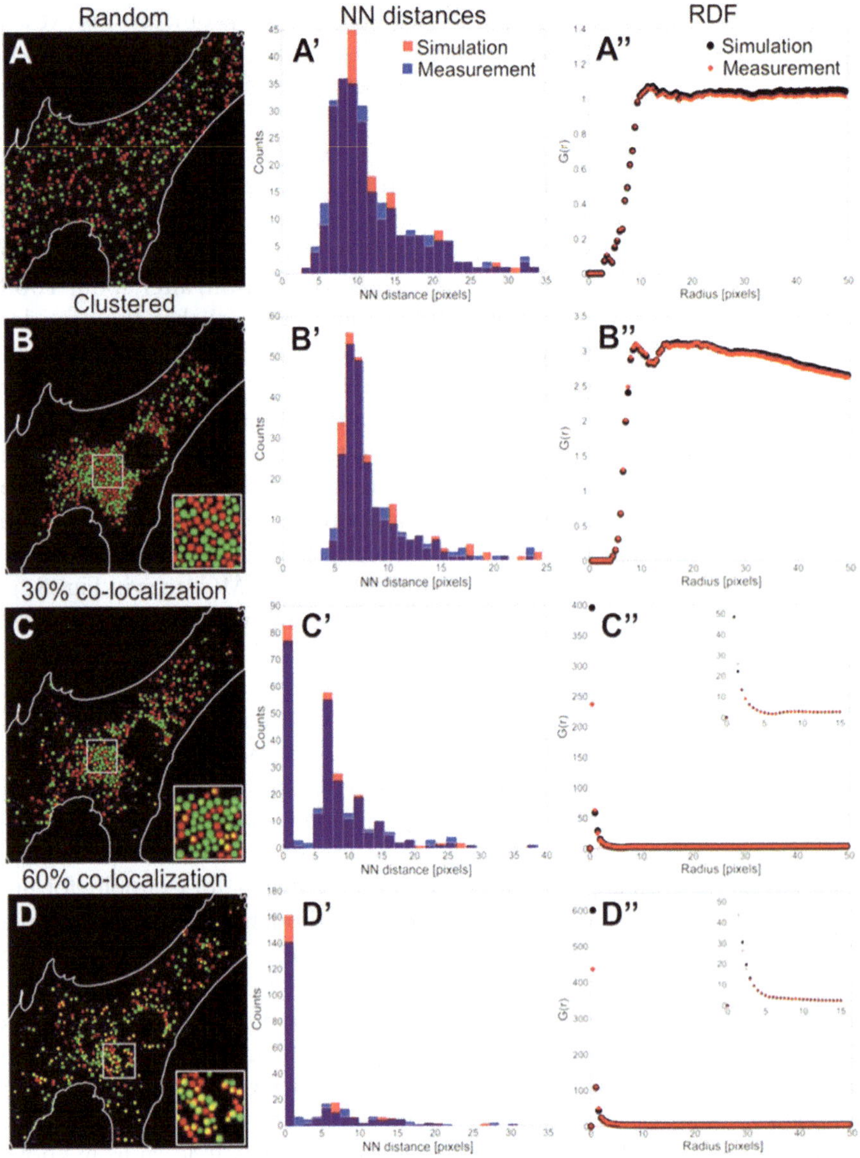

Fig. 9 Validation of NN distance and two-channel-RDF subroutines in SpatTrack on synthetic vesicle images. Simulated images are shown, with increasing clustering and co-localization. In (**a**) the particles are randomly distributed which is reflected in the RDF (**a**″) which is ~1 for distances larger than the average particle radius. In (**b**) the particles are strongly accumulated as seen by RDF > 1 for distances larger than the particle radius. Finally, (**c**) and (**d**) show simulations with 30% and 60% co-localization, respectively. Notice how co-localization results in a sharp peak in g(r) at distances much smaller than the particle radius, (**c**″) and (**d**″). Furthermore, notice how the distributions of NN distances in panels (**c**′) and (**d**″) are split into a population with a NN distance close to zero, reflecting the co-localized particles, and another population with NN distances larger than zero

performed here allow the particles to interact via a Lennard-Jones potential (*see* Eqs. 6 and 7). This method is often used in simulation of gases or liquids and represents the simplest physical interaction via van der Waals forces [46]. SpatTrack recovers precisely the NN distances and two-channel RDF for such images (Fig. 9a). The NN distances peak at around seven to eight pixels corresponding to 1.554–1.776 μm, both in the simulations and the SpatTrack result. This resembles the average density of the altogether 500 particles in the area available within the cell borders. In both, simulation and SpatTrack analysis, the distribution of NN distances between both channels is not symmetrical but skewed to higher values (Fig. 9a). The two-channel g(r) shows a steep increase and peaks around 8–10 pixels, confirming the result of the NN analysis. The two-channel g(r) plateaus at 1, indicating a completely random distribution between the red and green particles for inter-particle distances >15–20 pixels. Next, we simulate perinuclear clustering of vesicles, this time by including the $E_{\text{particle-cell}}$ terms into the MCMC simulation, as described in **Note 7** (Fig. 9b). While both, simulation and SpatTrack analysis find similar peaks around 6–7 pixels in the two-channel NN analysis, the distribution of NN distances becomes less skewed compared to the simulation of a random distribution (compare Fig. 9b 'and Fig. 9a'). The two-channel g(r) for the simulation of clustered vesicles and for its analysis using SpatTrack shows a characteristic dip between two peaks at around 8 and 15 pixels, corresponding to 1.776 μm and 3.33 μm, respectively. The absolute values of g(r) are threefold higher than for the first simulation (of random particles), illustrating the clustering of particles (Fig. 9b"). Again, simulations and their analysis using SpatTrack coincide very well. Next, we go on and simulate clustering particles which additionally co-localize (overlap) to either 30% or 60% (Fig. 9c and d). For the later, we use the modified Buckingham potential in the MCMC simulation, as given in Eq. 8. Here, the histogram of the two-channel NN analysis of simulated and analyzed particles show two peaks; one high peak between 0 and 2 pixels, corresponding to the co-localizing fraction, and a lower broader peak between 5–7 pixels (Fig. 9c' and d'). For the higher fraction of co-localizing particles, the first peak increases on expense of the second, exactly mirroring the increased overlap between the particles (Fig. 9d'). Notably, the analysis performed in SpatTrack (blue bars) coincides nicely with the ground truth of simulated images (red bars). When looking at the two-channel RDF for these two simulations of partially overlapping particles, we find that the peak is shifted to lower values of around 1.5 pixels (i.e., 0.33 μm). Given that the simulated particles have a physical diameter between 5 and ca. 15 pixels (corresponding to 1.11 and 3.33 μm, *see* Fig. 9a and b), we can conclude that this peak in the two-channel-RDF resembles the fraction of co-localizing particles. This reminds us of the result found for the vesicles containing Alexa546-NPC2 and overlapping with vesicles labeled with filipin (*see* Fig. 6b above and [39]). Thus, the extent of co-localization can be directly inferred from the two-channel RDF, if the peak is found at distances smaller than the average particle diameter. Importantly, the particle analysis in SpatTrack exactly

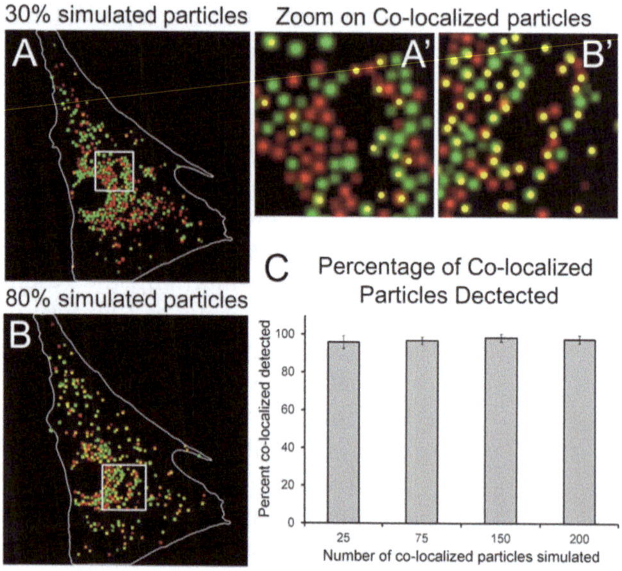

Fig. 10 Validation of the co-localization subroutine PDBCA in SpatTrack on synthetic vesicle images. The PDBCA algorithm was tested on images with increasing co-localization. (**a**) and (**b**) show examples of 30% and 80% co-localization, respectively, while (**a′**) and (**b′**) show a zoom of the boxed area in A) and B) with particles determined as co-localized indicated by yellow dots. C) Images were simulated with 25, 75, 150, or 200 co-localized particles out of 250 particles in each channel. In all cases SpatTrack detected ~95% of these particles as co-localized

recovers the two-channel-RDF calculated directly from the simulations (compare black and red symbols in Fig. 9c" and d").

Simulated vesicle patterns with a predefined percentage of overlapping particles can also be employed for validating the accuracy of the object-based co-localization, as performed in the PDBCA subroutine in SpatTrack (Fig. 7a). For that purpose, we have performed a number of simulations with increasing percentage of particle overlap. Next, we have used such simulated images for analysis in SpatTrack and quantified the accuracy of this analysis by determining the percentage of particles correctly assigned as being co-localized (Fig. 10c). SpatTrack correctly identifies about 95% of all co-localizing particles as such, independent of percentage of "true" co-localization. Thus, for high-quality images without significant noise, SpatTrack is highly accurate in finding co-localizing particles. This analysis underlines the value of realistic image simulations as benchmark for validation of image analysis algorithms. The MCMC simulations presented here could be also combined with simulating various degrees of noise (additive and multiplicative), spatially varying illumination or various degrees of blurring to simulate the effect of diffraction in the microscope optical train. This approach has been used recently for assessing the accuracy of particle tracking in SpatTrack versus image correlation spectroscopy in the MatLab suite developed by the Wiseman group [38].

In this chapter, we have presented a detailed protocol for using SpatTrack in co-localization experiments for endocytic cargo. We have demonstrated how the individual subroutines of SpatTrack play together and can be applied for two-channel co-localization and spatial pattern analysis. Other potential applications of SpatTrack lie in ratiometric measurement of the ion composition of LE/LYSs and direct linking of luminal pH, calcium or chloride concentration to the intracellular location and mobility of these organelles [17, 48]. The presented analysis is limited to diffraction-limited far-field fluorescence microscopy. However, with the increasing availability of super-resolution microscopy, as stimulated emission depletion (STED), stochastic optical reconstruction microscopy (STORM) or photoactivatable localization microscopy (PALM), the current methodology can be easily applied to images generated by such techniques [49–51]. This is, because the criteria used in SpatTrack for assessing co-localization or vesicle clustering can be easily scaled down to images with resolution beyond the diffraction limit. Additional features of SpatTrack, as the particle tracking suite and diffusion analysis have not been discussed, since they can be found elsewhere [38, 39, 41]. Future extensions of SpatTrack might include multi-channel analysis, as there is an increasing demand for quantifying co-localization of more than two markers at the same time [26]. In addition, one could expand imaging into the third dimension for object-based co-localization of endo-lysosomes in 3D stacks [34]. In its current form, SpatTrack could be useful for automated batch processing of large image sets, as they are obtained in high-content drug screening assays of cholesterol-binding fluorescent polyenes, as filipin [52–55], or of tagged proteins, as perifringolysin derivatives [56]. (*see* **Note 9**). Currently, image processing in such applications of automated microscopy is limited to simple thresholding operations [52, 53], and one could imagine, that for increased resolution, subcellular cytometry with SpatTrack becomes feasible. For that purpose, one would need to extend the routines for extracting the cellular geometry and generation of random vesicle patterns for several cells in an image field in parallel. This is computationally more demanding, but with the increasing availability of fast processors, should be doable in the near future. We emphasize that SpatTrack is not the only program, which can perform spatial analysis of vesicle patterns, nor is it the only program for object-based co-localization analysis (other programs for such purposes are the MosaicIA and Squassh plugins for ImageJ [57–59]). Thus, the image analyst can choose from a variety of computer programs. However, SpatTrack provides a user-friendly option which assembles a variety of algorithms dedicated to spatiotemporal analysis of LE/LYSs under one main program. Along with this publication comes an improved version of SpatTrack compared to our original publication [39]. In SpatTrack 2.0, we have implemented an improved tracking algorithm, which implements the simpletracker frame linking algorithm developed by Jean-Yves Tinevez and freely available at the MatLab File Exchange (http://www.mathworks.com/matlabcentral/

fileexchange/34040-simple-tracker) Simpletracker uses the Hungarian algorithm to optimize the nearest-neighbor distances between coordinates in one frame and the next and is approximately 10 times faster than the linking algorithm implemented in SpatTrack 1.0 [38, 39]. Additional features in the latest version of SpatTrack include i) an improved two-channel batch GUI (*see* Fig. 7b and c), which includes options for defining particle size and other settings for both particle populations; ii) the ability to detect particle merging and splitting, and iii) the option of calculating the MSD from trajectories with missing entries. Often one will use some kind of memory during tracking to deal with particles which disappear for one or more frames. During that time there will be no information about the particle location, and thus the trajectories will contain one or more NaN (not a number) entries. The MSD algorithm can now calculate the average MSD of several trajectories where some of the particle coordinates are missing. You can download SpatTrack 2.0 together with an updated User Guide and sample images described in the text as stand-alone version at www.sdu.dk/bmb/spattrack. Please note that MatLab is not required to run the program, but the MatLab Runtime environment (version 8.5 2015a 64-bit for Windows available at http://www.mathworks.com/products/compiler/mcr/) must be installed in the same folder as the main program. Further download instructions can be found in the User Guide.

4 Notes

1. The computer used for running should have 4 GB RAM as minimum. For running the MCMC simulations, at least 8GB is advised.
2. Purification of NPC2 was performed as described by Larsen and co-workers, and its labeling with Alexa546 succimidyl ester followed the protocol described by the manufacturer [39, 60].
3. 3 Filipin is a polyene antibiotic with fluorescence properties similar to the nuclear stain DAPI. Filipin binds cholesterol and other sterols with a 3′-hydroxy group. It is often used for detecting cholesterol and cholesterol-rich LSOs in lysosomal storage disorders [61]. Filipin can be imaged using a DAPI filter on any standard widefield microscope equipped with a CCD camera sensitive in the spectral range centered around 400 nm. However, other light efficient instrumentation, as Airy scan or spinning disk confocal microscopes could be used as well. As other polyenes, filipin bleaches rapidly, so it is not suitable for long-term observations. We advise, focusing cells in multicolor experiments in another channel, as the rhodamine channel in case of Alexa546-NPC2, take an image and switch to the DAPI filter for acquiring the corresponding image in the filipin channel afterwards.

4. This procedure is only intended here to optimize the segmentation procedure of the studied cell from background. Other procedures are available to correct for uneven illumination for subsequent quantifications of intensity from the images, as calculation of the time-integrated emission (TiEM) in bleach stacks of labeled cells [62].

5. Thompson et al. described the error of the determined particle coordinates as [63]:

$$(x^2) = \frac{s^2}{N} + \frac{a^2}{12N} + \frac{8\pi s^4 b^2}{a^2 N^2},$$

where N is the number of photons collected, s is the standard deviation of the PSF, a is the pixel size, and b is the background noise including autofluorescence and noise of the detector. The PSF of the objective is inversely correlated with the numerical aperture (NA) and furthermore, a high NA typically improves photon collection. Hence, increasing the NA of the objective is the single most important parameter for improved particle detection. Finally, due to pixelation, it is generally recommended that the pixel size is about equal to the standard deviation of the PSF. Here, we have described the analysis of widefield images acquired with a 63× objective with a 1.4 NA. However, SpatTrack is not limited to this particular configuration and could in principle be used even with super-resolution microscopy such as stimulated emission depletion (STED) microscopy.

6. SpatTrack only reads Tiff files but if required the user could convert .jpg images to tiffs in ImageJ (https://**imagej**.nih.gov/ij/) prior to analysis. To our knowledge a lot of the newer microscopes save the files in vendor-specific formats. Thus, it might be necessary to convert the files anyway. However, we cannot comment about any changes in image properties upon conversion, so saving images as Tiff files is advised.

7. Energy terms for MCMC simulation of two types of particles interacting via LJ-potentials only

   For particle population 1 with particle population 1:

$$E_{par1-par1} = \varepsilon_1 \left[\left(\frac{r_{m1}}{r_{11}} \right)^{12} - 2 \left(\frac{r_{m1}}{r_{11}} \right)^{6} \right],$$

where ε_1 is the depth of the potential well for interactions between particles of type 1, r_{m1} is a distribution of $r_{m.}$-values for particle population 1, and r_{11} is the distance between a particle of type 1 and another particle of type 1. To calculate the total system energy this term is summed over all type 1 particles and their interactions with other particles of type 1.

For particle population 1 with particle population 2:

$$E_{par1-par2} = \varepsilon_1 \left[\left(\frac{r_{m1}}{r_{12}} \right)^{12} - 2 \left(\frac{r_{m1}}{r_{12}} \right)^{6} \right],$$

where ε_1 is the depth of the potential well for interactions between particles of type 1, r_{m1} is a distribution of $r_{m.}$-values for particle population 1, and r_{12} is the distance between a particle of type 1 and a particle of type 2. To calculate the total system energy, this term is summed over all particles of type 1 and their interactions with particles of type 2.

For particle population 1 with the cell nucleus:

$$E_{par1-Nuc} = \varepsilon_N \left[\left(\frac{r_{mN}}{r_N} \right)^{12} - 2 \left(\frac{r_{mN}}{r_N} \right)^{6} \right],$$

where ε_N is the depth of the potential well for interactions between particles of any type with the nucleus, and r_{mN} is the $r_{m.}$-value for interactions between particles of any type and the nucleus (r_{mN} is the same value for all particles of both types). To calculate the total system energy this term is summed over all particles of type 1 and their interaction with the nucleus.

For particle population 1 with the MTOC:

$$E_{par1\text{-MTOC}} = \beta_1 r_{1\text{MTOC}},$$

where β is the strength of the quadratic potential between particles of type 1 and the MTOC and $r_{1\text{MTOC}}$ is the distance between a type 1 particle and the MTOC.

For particle population 2 with particle population 2:

$$E_{par2-par2} = \varepsilon_2 \left[\left(\frac{r_{m2}}{r_{22}} \right)^{12} - 2 \left(\frac{r_{m2}}{r_{22}} \right)^{6} \right],$$

where ε_2 is the depth of the potential well for interactions between particles of type 2, r_{m2} is a distribution of $r_{m.}$-values for particle population 2, and r_{22} is the distance between a particle of type 2 and another particle of type 2. To calculate the total system energy this term is summed over all type 2 particles and their interactions with other particles of type 2.

For particle population 1 with particle population 2:

$$E_{par2-par1} = \varepsilon_2 \left[\left(\frac{r_{m2}}{r_{21}} \right)^{12} - 2 \left(\frac{r_{m2}}{r_{21}} \right)^{6} \right],$$

where ε_2 is the depth of the potential well for interactions between particles of type 2, r_{m2} is a distribution of $r_{m.}$-values for particle population 2, and r_{21} is the distance between a particle of type 2 and a particle of type 1. To calculate the total system energy this term is summed over all particles of type 2 and their interactions with particles of type 1.

For particle population 1 with the cell nucleus:

$$E_{par2-Nuc} = \varepsilon_{\mathrm{N}}\left[\left(\frac{r_{mN}}{r_{\mathrm{N}}}\right)^{12} - 2\left(\frac{r_{mN}}{r_{\mathrm{N}}}\right)^{6}\right],$$

where ε_{N} is the depth of the potential well for interactions between particles of type 1 or 2 with the nucleus, and r_{mN} is the $r_{\mathrm{m.}}$-value for interactions between type 1 or 2 particles and the nucleus (r_{mN} is the same value for all particles of both types). To calculate the total system energy this term is summed over all particles of type 2 and their interaction with the nucleus.

For particle population 1 with the MTOC:

$$E_{par2-\mathrm{MTOC}} = \beta_2 r_{2\mathrm{MTOC}},$$

where β_2 is the strength of the quadratic potential between particles of type 2 and the MTOC and $r_{2\mathrm{MTOC}}$ is the distance between a type 2 particle and the MTOC.

8. Transferrin (Tf) was iron-loaded and labeled with Alexa647 to get Alexa647-Tf as previously described [4, 64].
9. SpatTrack is not affected by the choice of fluorophore. In principle, it could even be used with antibody staining as long as the organelles of interest are vesicle like.

References

1. Mukherjee S, Maxfield FR (2004) Lipid and cholesterol trafficking in NPC. Biochim Biophys Acta 1685:28–37
2. Storch J (2009) Niemann-Pick C2 (NPC2) and intracellular cholesterol trafficking. Biochim Biophys Acta 1791(7):671–678
3. Lloyd-Evans E, Platt FM (2010) Lipids on trial: the search for the offending metabolite in Niemann Pick type C disease. Traffic 11:419–428
4. Yamashiro DJ, Tycko B, Fluss SR, Maxfield FR (1984) Segregation of transferrin to a mildly acidic (pH 6.5) para-Golgi compartment in the recycling pathway. Cell 37:789–800
5. Lin SX, Gundersen GG, Maxfield FR (2002) Export from pericentriolar endocytic recycling compartment to cell surface depends on stable, detyrosinated (glu) microtubules and kinesin. Mol Biol Cell 13(1):96–109
6. Bennett EM, Lin SX, Towler MC, Maxfield FR, Brodsky FM (2001) Clathrin hub expression affects early endosome distribution with minimal impact on receptor sorting and recycling. Mol Biol Cell 12(9):2790–2799
7. Dunn KW, McGraw TE, Maxfield FR (1989) Iterative fractionation of recycling receptors from lysosomally destined ligands in an early sorting endosome. J Cell Biol 109:3303–3314
8. Traer CJ, Rutherford AC, Palmer KJ, Wassmer T, Oakley J, Attar N et al (2007) SNX4 coordinates endosomal sorting of TfnR with dynein-mediated transport into the endocytic recycling compartment. Nat Cell Biol 9(12):1370–1380. doi:10.1038/ncb1656
9. Skanland SS, Walchli S, Brech A, Sandvig K (2009) SNX4 in complex with clathrin and dynein: implications for endosome movement. PLoS One 4(6):e5935. doi:10.1371/journal.pone.0005935
10. Murray JW, Wolkoff AW (2003) Roles of the cytoskeleton and motor proteins in endocytic sorting. Adv Drug Deliv Rev 55:1385–1403

11. Korolchuk VI, Saiki S, Lichtenberg M, Siddiqi FH, Roberts EA, Imarisio S et al (2011) Lysosomal positioning coordinates cellular nutrient responses. [Research Support, N.I.H., Extramural Research Support, Non-U.S. Gov't]. Nat Cell Biol 13(4):453–460. doi:10.1038/ncb2204
12. Bucci C, Thomsen P, Nicoziani P, McCarthy J, van Deurs B (2000) Rab7: A key to lysosome biogenesis. Mol Biol Cell 11:467–480
13. Rocha N, Kuijl C, van der Kant R, Janssen L, Houben D, Janssen H et al (2009) Cholesterol sensor ORP1L contacts the ER protein VAP to control Rab7-RILP-p150 Glued and late endosome positioning. J Cell Biol 185(7):1209–1225
14. Ganley IG, Pfeffer SR (2006) Cholesterol accumulation sequesters Rab9 and disrupts late endosome function in NPC1-deficient cells. J Biol Chem 281(26):17890–17899
15. Neufeld EB, Wastney M, Patel S, Suresh S, Cooney AM, Dwyer NK et al (1999) The Niemann-Pick C1 protein resides in a vesicular compartment linked to retrograde transport of multiple lysosomal cargo. J Biol Chem 274(14):9627–9635
16. Luzio JP, Bright NA, Pryor PR (2007) The role of calcium and other ions in sorting and delivery in the late endocytic pathway. Biochem Soc Trans 35(5):1088–1091
17. Johnson DE, Ostrowski P, Jaumouille V, Grinstein S (2016) The position of lysosomes within the cell determines their luminal pH. J Cell Biol 212(6):677–692. doi:10.1083/jcb.201507112
18. Duan L, Che D, Zhang K, Ong Q, Guo S, Cui B (2015) Optogenetic control of molecular motors and organelle distributions in cells. Chem Biol 22(5):671–682. doi:10.1016/j.chembiol.2015.04.014
19. van Bergeijk P, Adrian M, Hoogenraad CC, Kapitein LC (2015) Optogenetic control of organelle transport and positioning. Nature 518(7537):111–114. doi:10.1038/nature14128
20. Bolte S, Cordelieres FP (2006) A guided tour into subcellular colocalization analysis in light microscopy. [Research Support, Non-U.S. Gov't]. J Microsc 224(Pt 3):213–232. doi:10.1111/j.1365-2818.2006.01706.x
21. Costes SV, Daelemans D, Cho EH, Dobbin Z, Pavlakis G, Lockett S (2004) Automatic and quantitative measurement of protein-protein colocalization in live cells. Biophys J 86(6):3993–4003. doi:10.1529/biophysj.103.038422
22. van Steensel B, van Binnendijk EP, Hornsby CD, van der Voort HT, Krozowski ZS, de Kloet ER et al (1996) Partial colocalization of glucocorticoid and mineralocorticoid receptors in discrete compartments in nuclei of rat hippocampus neurons. J Cell Sci 109(Pt 4):787–792
23. Comeau JWD, Costantino S, Wiseman PW (2006) A guide to accurate fluorescence microscopy colocalization measurements. Biophys J 91:4611–4622
24. Adler J, Parmryd I (2013) Colocalization analysis in fluorescence microscopy. Methods Mol Biol 931:97–109. doi:10.1007/978-1-62703-056-4_5
25. Landmann L (2002) Deconvolution improves colocalization analysis of multiple fluorochromes in 3D confocal data sets more than filtering techniques. J Microsc 208:134–147
26. Fletcher PA, Scriven DR, Schulson MN, Moore ED (2010) Multi-image colocalization and its statistical significance. Biophys J 99(6):1996–2005. doi:10.1016/j.bpj.2010.07.006
27. Adler J, Parmryd I (2010) Quantifying colocalization by correlation: the Pearson correlation coefficient is superior to the Mander's overlap coefficient. Cytometry A 77(8):733–742. doi:10.1002/cyto.a.20896
28. Adler J, Parmryd I (2014) Quantifying colocalization: thresholding, void voxels and the H(coef). PLoS One 9(11):e111983. doi:10.1371/journal.pone.0111983
29. Li Q, Lau A, Morris TJ, Guo L, Fordyce CB, Stanley EF (2004) A syntaxin 1, Galpha(o), and N-type calcium channel complex at a presynaptic nerve terminal: analysis by quantitative immunocolocalization. J Neurosci 24(16):4070–4081. doi:10.1523/JNEUROSCI.0346-04.2004
30. Jaskolski F, Mulle C, Manzoni OJ (2005) An automated method to quantify and visualize colocalized fluorescent signals. J Neurosci Methods 146(1):42–49. doi:10.1016/j.jneumeth.2005.01.012
31. Lagache T, Sauvonnet N, Danglot L, Olivo-Marin JC (2015) Statistical analysis of molecule colocalization in bioimaging. Cytometry A 87(6):568–579. doi:10.1002/cyto.a.22629
32. Smal I, Loog M, Niessen W, Meijering E (2010) Quantitative comparison of spot detection methods in fluorescence microscopy. IEEE Trans Med Imaging 29(2):282–301. doi:10.1109/TMI.2009.2025127
33. Ruusuvuori P, Aijo T, Chowdhury S, Garmendia-Torres C, Selinummi J, Birbaumer M et al (2010) Evaluation of methods for detection of fluorescence labeled subcellular objects in microscope images. BMC Bioinformatics 11. doi:10.1186/1471-2105-11-248
34. Woodcroft BJ, Hammond L, Stow JL, Hamilton NA (2009) Automated organelle-

based colocalization in whole cell imaging. Cytometry A 75:941–950
35. Demandolx D, Davoust J (1997) Multicolour analysis and local image correlation in confocal microscopy. J Microsc (Oxford) 185:21–36. doi:10.1046/j.1365-2818.1997.1470704.x
36. Morrison IEG, Karakikes I, Barber RE, Fernandez N, Cherry RJ (2003) Detecting and quantifying colocalization of cell surface molecules by single particle fluorescence imaging. *Biophys J 85*(6):4110–4121. doi:10.1016/S0006-3495(03)74823-1
37. Crocker JC, Grier DG (1996) Methods of digital video microscopy for colloidal studies. J Colloid Interface Sci *179*(1):298–310. doi:10.1006/jcis.1996.0217
38. Lund FW, Wüstner D (2013) A comparison of single particle tracking and temporal image correlation spectroscopy for quantitative analysis of endosome motility. J Microsc 252(2):169–188
39. Lund FW, Jensen MLV, Christensen T, Nielsen GK, Heegaard CW, Wüstner D (2014) SpatTrack, an imaging toolbox for analysis of endosome motility and distribution. Traffic 15(12):1406–1429
40. Saxton MJ (1983) Lateral diffusion in an archipelago. Biophys J 64:1766–1780
41. Modzel M, Lund FW, Wüstner D (2016) Synthesis and live-cell imaging of fluorescent sterols for analysis of intracellular cholesterol transport. Methods Mol Biol (in press).
42. Zuiderveld K (1994) Contrast-limited adaptive histogram equalization. *Graphics Gems IV*:474–485
43. Lachmanovich E, Shvartsman DE, Malka Y, Botvin C, Henis YI, Weiss AM (2003) Co-localization analysis of complex formation among membrane proteins by computerized fluorescence microscopy: application to immunofluorescence co-patching studies. J Microsc (Oxford) *212*:122–131. doi:10.1046/j.1365-2818.2003.01239.x
44. Ghaye J, De Micheli G, Carrara S (2012) Simulated biological cells for receptor counting in fluorescence imaging. *BioNanoSci 2*:94–103
45. Metropolis N, Rosenbluth AW, Rosenbluth MN, Teller E (1953) Equation of state calculations by fast computing machines. J Chem Phys 21:1087–1092
46. Allen MP, Tildesley DJ (1987) Computer simulation of liquids. Oxford University Press
47. Frenkel D (2004) Introduction to Monte Carlo Methods. In: Attig N, Binder K, Grubmüller H, Kremer K (eds) *Computational soft matter: from synthetic polymers to proteins*, vol 23. NIC Series, Lecture Notes, Jülich, pp 29–60
48. Majumdar A, Cruz D, Asamoah N, Buxbaum A, Sohar I, Lobel P et al (2007) Activation of microglia acidifies lysosomes and leads to degradation of Alzheimer amyloid fibrils. Mol Biol Cell 18(4):1490–1496
49. Donnert G, Keller J, Wurm CA, Rizzoli SO, Westphal V, Schönle A et al (2007) Two-color far-field fluorescence nanoscopy. Biophys J 92(8):L67–L69
50. Hess ST, Gould TJ, Gudheti MV, Maas SA, Mills KD, Zimmerberg J (2007) Dynamic clustered distribution of hemagglutinin resolved at 40 nm in living cell membranes discriminates between raft theories. Proc Natl Acad Sci U S A 104(44):17370–17375
51. Hell SW (2007) Far-field optical nanoscopy. Science 316(5828):1153–1158
52. Pipalia NH, Huang A, Ralph H, Rujoi M, Maxfield FR (2006) Automated microscopy screening for compounds that partially revert cholesterol accumulation in Niemann-Pick C cells. J Lipid Res 47:284–301
53. Bartz F, Kern L, Erz D, Zhu M, Gilbert D, Meinhof T et al (2009) Identification of cholesterol-regulating genes by targeted RNAi screening. Cell Metab 10(1):63–75
54. Rosenbaum AI, Rujoi M, Huang AY, Du H, Grabowski GA, Maxfield FR (2009) Chemical screen to reduce sterol accumulation in Niemann-Pick C disease cells identifies novel lysosomal acid lipase inhibitors. Biochim Biophys Acta 1791(12):1155–1165
55. Wehrmann ZT, Hulett TW, Huegel KL, Vaughan KT, Wiest O, Helquist P et al (2012) Quantitative comparison of the efficacy of various compounds in lowering intracellular cholesterol levels in Niemann-Pick type C fibroblasts. PLoS One 7(10):e48561. doi:10.1371/journal.pone.0048561
56. Ishitsuka R, Saito T, Osada H, Ohno-Iwashita Y, Kobayashi T (2011) Fluorescence image screening for chemical compounds modifying cholesterol metabolism and distribution. J Lipid Res 52(11):2084–2094. doi:10.1194/jlr.D018184
57. Shivanadan A, Radenovic A, Sbalzarini IF (2013) MosaicIA: an ImageJ/Fiji plugin for spatial pattern and interaction analysis. BMC Bioinformatics 14:349
58. Davies MJ (2004) Reactive species formed on proteins exposed to singlet oxygen. Photochem Photobiol Sci 3(1):17–25. doi:10.1039/b307576c
59. Davies M (2014) Long-lived reactive species formed on proteins induce changes in protein and lipid turnover. Free Radic Biol

Med 75(Suppl 1):S6–S7. doi:10.1016/j.freeradbiomed.2014.10.841

60. Larsen LB, Ravn P, Boisen A, Berglund L, Petersen TE (1997) Primary structure of EPV20, a secretory glycoprotein containing a previously uncharacterized type of domain. Eur J Biochem 243(1-2):437–441
61. Maxfield F, Wüstner D (2012) Analysis of cholesterol trafficking with fluorescent probes. Methods Cell Biol 108:367–393
62. Wustner D, Christensen T, Solanko LM, Sage D (2014) Photobleaching kinetics and time-integrated emission of fluorescent probes in cellular membranes. Molecules 19(8):11096–11130. doi:10.3390/molecules190811096
63. Thompson RE, Larson DR, Webb WW (2002) Precise nanometer localization analysis for individual fluorescent probes. Biophys J 82(5):2775–2783. doi:10.1016/s0006-3495(02)75618-x
64. Wüstner D, Solanko LM, Sokol E, Lund FW, Garvik O, Li Z et al (2011) Quantitative assessment of sterol traffic in living cells by dual labeling with dehydroergosterol and BODIPY-cholesterol. Chem Phys Lipids 164(3):221–235

Chapter 7

Preparation of a Two-Photon Fluorescent Probe for Imaging H_2O_2 in Lysosomes in Living Cells and Tissues

Mingguang Ren, Beibei Deng, Xiuqi Kong, Yonghe Tang, and Weiying Lin

Abstract

Hydrogen peroxide (H_2O_2) plays important roles in many physiological and pathological processes. At the cellular organelle level, the abnormal concentrations of H_2O_2 in the lysosomes may cause redox imbalance and the loss of the critical functions of the lysosomes. Herein, we describe the preparation of a potent lysosome-targeted two-photon fluorescent probe (Lyso-HP) for the detection of H_2O_2 in the lysosomes in the living cells. This unique fluorescent probe can also be employed to effectively detect H_2O_2 in the living tissues using two-photon fluorescence microscopy.

Key words Fluorescent Probe, H_2O_2, Lysosome, Two photon, Fluorescence imaging

1 Introduction

As a type of reactive oxygen species (ROS), hydrogen peroxide (H_2O_2) plays key roles in mediating various physiological processes, for instance, serving as an oxidative stress marker and a messenger in cellular signal transduction [1–4]. However, aberrant production of H_2O_2 is associated with many diseases including cancers, diabetes, neurodegenerative disorders, and cardiovascular diseases [5]. Lysosomes containing digestive enzymes are the major degradation machinery in cells to recycle damaged organelles as well as to digest nucleic acids, polysaccharides, fats, and proteins. Both endogenous and exogenous molecules can be delivered to lysosomes [6]. There is a close relationship between the lysosomes and hydrogen peroxide. For example, the accumulation of peroxidized lipids and proteins in the lysosomes of the brain cells can result in Alzheimer's disease. In 1993, Zdolsek and coworker found that abnormal endogenous or exogenous production of H_2O_2 through a primary interaction with lysosomal ferrous iron may induce the formation of hydroxyl radicals (intralysosomal Fenton reaction) to lead to lysosomal membrane damage and leakage of lysosomal contents such as hydrolytic lysosomal enzymes [7].

Karin Öllinger and Hanna Appelqvist (eds.), *Lysosomes: Methods and Protocols*, Methods in Molecular Biology, vol. 1594,
DOI 10.1007/978-1-4939-6934-0_7,

Therefore, the efficient monitoring and detection of H_2O_2 in lysosomes are significant in physiology and clinical diagnosis.

Small molecule-based fluorescent probes for detecting ROS in subcellular organelles have been developed [8–12]. However, detecting H_2O_2 with the capability of two-photon fluorescence imaging in lysosomes is still very challenging. This protocol describes a recipe-style preparation of Lyso-HP, a potent H_2O_2-specific fluorescent probe for monitoring H_2O_2 within the lysosomes of the living systems, as well as a procedure for imaging lysosomal H_2O_2 in the cultured cells and tissues.

Lyso-HP is a bifunctional molecule, which combines a boronate-masked naphthalimide fluorescent scaffold for selective H_2O_2 detection and a morpholine targeting group for lysosomal localization. The naphthalimide dye was chosen as the fluorescent scaffold in this probe owing to its excellent photochemical and photophysical properties [13–18]. The boronate group has been known as the robust reaction site for H_2O_2 to generate phenol with a high selectivity [19]. The targeting group, morpholine, can significantly increase the probe distribution in lysosomes and is relatively difficult to be washed out [20, 21]. Combined with the above-mentioned elements, we designed the compound Lyso-HP as the lysosome-targeted two-photon fluorescent H_2O_2 probe (Scheme 1) [22].

Lyso-HP selectively and efficiently localizes to the lysosomes in HeLa and RAW264.7 cells, where it can respond to local

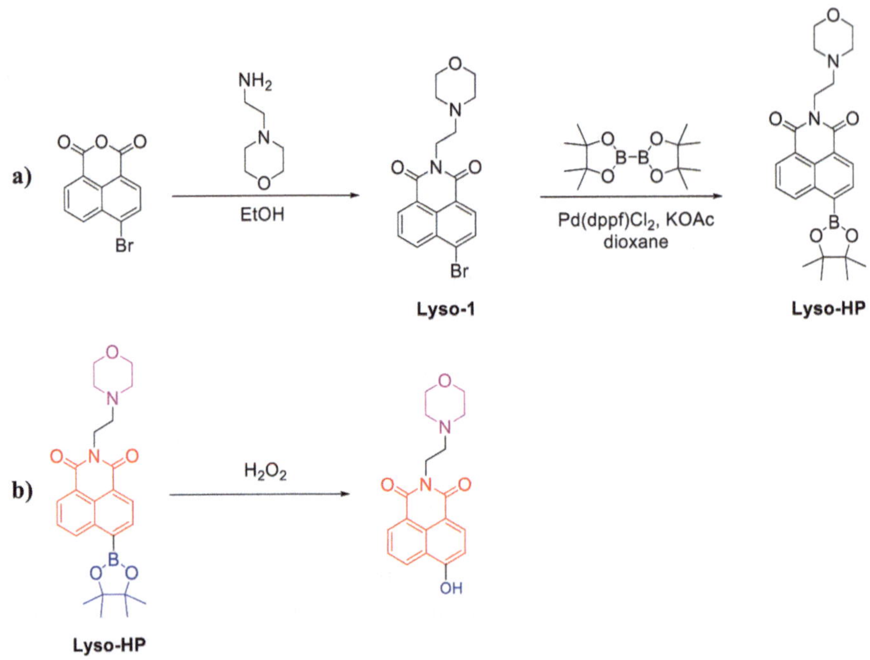

Scheme 1 (**a**) Synthesis of the fluorescent probe Lyso-HP. (**b**) H_2O_2-mediated activation of the probe Lyso-HP

changes in H_2O_2 levels with a turn-on fluorescence increase. The advantages of the probe include fast response, large turn-on fluorescence signal changes, organelle-specific detection, and two-photon properties.

In the procedure described in the following text, we describe the synthetic procedure of the probe Lyso-HP and a typical experiment using the probe to image exogenous H_2O_2 in HeLa cells and the endogenous H_2O_2 in RAW264.7 cells, as well as to detect H_2O_2 in mice tissue. We recommend that each experiment should be repeated at least in triplicate to obtain the average fluorescent intensities. We demonstrate that the lysosome indicator, Lysotracker, can be added simultaneously with Lyso-HP in order to confirm the expected localization of the probe in cells.

2 Materials

2.1 Reagents and Equipment

All chemicals and reactions used in this protocol are potentially harmful, and thus a lab coat, gloves, and eye protection should be used. The solid and liquid waste should be appropriately disposed of according to the relevant local and institutional guidelines.

1. Ethanol, 99.9% (EtOH).
2. 1,1′-Bis(diphenylphosphino)ferrocene]dichloropalladium(II) in complex with dichloromethane (Pd(dppf) $Cl_2 \bullet CH_2Cl_2$).
3. Silica gel (200–300 mesh).
4. Ethyl acetate, petroleum ether, methanol, and dichloromethane all of chromatography grade.
5. *N,N*-Dimethyl formamide (DMF; anhydrous).
6. Dulbecco's phosphate-buffered saline (DPBS).
7. HeLa and RAW 264.7 cell lines.
8. Analytical thin-layer chromatography (TLC) plates.
9. Desiccator with vacuum manifold.

2.2 Cell Culturing and Exposures

1. Cell culture medium: DMEM (Dulbecco's Modified Eagle Medium), 10% FBS (fetal bovine serum), 100 U/ml penicillin, 100 μg/ml streptomycin. The medium should be stored at 4 °C.
2. 100 mM hydrogen peroxide: Dissolve 10.9 μl of 30% (w/v) hydrogen peroxide solution in 1 ml of distilled water (DW). Use freshly prepared hydrogen peroxide each time.
3. Lipopolysaccharide (LPS) 1 mg/ml: Dissolve 1 mg of LPS in 1 ml of DW and store at −20 °C for up to 6 months.
4. 100 μM phorbol 12-myristate 13-acetate (PMA): To make stock solution of 10 mM, dissolve 6.2 mg of PMA in 1 ml of DMSO. To make 100 μM PMA, dilute 1:100 in DMSO and store it at −20 °C for up to 6 months.

5. Cell isolation medium: RPMI-1640, 2.5% FBS, without penicillin and streptomycin. This medium can be stored at 4 °C for up to 6 months.

3 Methods

3.1 Synthesis of the Compound Lyso-1

1. Connect a 100-ml one-necked round-bottomed flask containing a jujube nucleus-shaped Teflon-coated magnetic stirring bar to a reflux condenser and ensure that the magnetic stir bar can rotate freely in the flask.
2. To the flask, add 277 mg (1 mmol, 1 eq) of 4-bromo-1,8-naphthalic anhydride and 130 mg (1 mmol,1 eq) of 2-morpholinoethanamine.
3. Add 30 ml of anhydrous EtOH to the mixture in the flask.
4. Securely attach the flask with a clamp into an RT (20–25 °C) oil bath placed on a temperature-controlled hot plate with a magnetic stirrer.
5. Heat the oil bath to 85 °C while stirring on the magnetic stirrer.
6. After 5 h, turn off the heating element and allow the reaction to cool to room temperature.
7. Once the mixture is completely cool, the flask is removed from the oil bath and the water supply to the condenser is switched off.
8. Remove the solvent via rotary evaporation by heating the mixture to 30 °C under reduced pressure and then add 80 ml ethyl acetate to the flask.
9. Transfer the resulting solution to a 250 ml reparatory funnel, and rinse with 20 ml ethyl acetate.
10. Shake and separate the layers and wash the organic layer three times with water (80 ml in each wash), shaking and separating as before.
11. Take the organic layer, dry it over ~10 g of sodium sulfate for 10–15 min, gravity filter it through the fluted filter paper, and remove the solvent by rotary evaporation.
12. Purify the product by silica gel column chromatography (column diameter: 4–5 cm; silica gel column height: 30–35 cm) using a 1:40 (v/v) mixture solution of methanol and dichloromethane as the eluent. Monitor the collected fractions by silica gel TLC (thin layer chromatography) with a 1:40 (v/v) solution of methanol and dichloromethane eluent. The R_f (retardation factor) value of the product on TLC is 0.35 (methanol/dichloromethane 1:50 (v/v) as extending solvent). The compound can be visualized as a dark spot ($R_f = 0.35$) in silica gel TLC with UV absorbance at 254 nm. Collect those fractions

containing the desired product and remove the solvent using a rotary evaporator at room temperature under reduced pressure to get 310 mg of the compound Lyso-1 as a white solid (80% yield).

13. The identity and purity of the compound can be established by ^1H NMR, ^{13}C NMR, and ESI-MS (electrospray ionization-mass spectrometry). Lyso-1 can be stored in the dark at 4 °C for at least 12 months.

3.2 Synthesis of the Compound Lyso-HP

1. Dry a 50-ml two-necked round-bottom flask and a magnetic stir bar overnight in an electric oven at 130 °C.
2. After the glassware has cooled to room temperature, connect the 50-ml two-necked round-bottomed flask containing a jujube nucleus-shaped Teflon-coated magnetic stirring bar to a reflux condenser and ensure that the magnetic stir bar can rotate freely in the flask.
3. Weigh out 500 mg (1.28 mmol, 1.0 eq.) of the compound Lyso-1, 94.4 mg of Pd(dppf) $Cl_2{\cdot}CH_2Cl_2$ (0.12 mmol, 0.1 eq.), 688 mg of bis(pinacolato)diboron (1.93 mmol, 1.5 eq.), and 378 mg (3.85 mmol, 3.0 eq.) of potassium acetate.
4. Seal the apparatus with rubber septa, and fill the flask with argon gas using the standard Schlenk techniques (*see* **Note 1**).
5. Add 20 ml of anhydrous 1,4-dioxane to the mixture in the flask using a 50 ml syringe.
6. Heat the oil bath to 110 °C while stirring on the magnetic stirrer, and stir it for 10 h in the dark by covering it in aluminum foil and switching off the light in the fume hood.
7. Turn off the heating plate and allow the reaction to cool to room temperature.
8. After the mixture is completely cool, remove the flask from the oil bath, carefully remove the cap, and switch off the water supply to the condenser.
9. Dilute the reaction with 20 ml of ethyl acetate, and gravity filter the mixture through the fluted filter paper.
10. Transfer the filtrate to a 100 ml separator funnel. Shake and separate the water and organic layers three times with water (20 ml in each wash), shaking and separating as before.
11. Take the organic layer, dry it over ~10 g of sodium sulfate for 10–15 min, gravity filter it through the fluted filter paper, and remove the solvent by rotary evaporation.
12. Purify the product by silica gel column chromatography (column diameter: 2–3 cm; silica gel column height: 20–25 cm) using a 1:1 (v/v) mixture of ethyl acetate and petroleum ether as the eluent. Monitor the collected fractions by silica gel TLC with a 1:1 (v/v) solution of ethyl acetate and petroleum ether

eluent (*see* **Note 2**). The R_f value of the compound Lyso-HP on TLC is 0.45 (methanol/dichloromethane 1:40 (v/v) as extending solvent). The compound can be visualized as a dark spot (R_f = 0.45) in silica gel TLC with UV absorbance at 254 nm. Collect those fractions containing the desired product and remove the solvent using a rotary evaporator at room temperature under reduced pressure to get 401 mg of the compound Lyso-HP as a white solid (72% yield).

13. Establish the identity and purity of the compound Lyso-HP by ^{1}H NMR, ^{13}C NMR, and ESI-MS.
14. The solid Lyso-HP can be stored in the dark at −20 °C for 3–4 months.

3.3 Preparation of the Lyso-HP Stock Solution

1. Remove the solid compound (Lyso-HP) from the freezer and allow it to warm to room temperature.
2. After the probe has warmed to room temperature, prepare a stock solution (1.0 mM) of Lyso-HP (MW 436 g/mol) by dissolving 4.4 mg of Lyso-HP in 10 ml dry DMF (*see* **Note 3**).

3.4 Experiments in the Cells

3.4.1 Seeding of the Test Cells

Seed three groups of cells (HeLa or RAW 264.7) in 35-mm glass-bottom dishes at a density of 2 × 10^5 cells per dish in 2 ml of culture medium and incubate them in an incubator containing 5% CO_2 for 24 h at 37 °C. The cells will attach to the glass surface during this time.

3.4.2 Imaging the Exogenous H_2O_2 in the HeLa Cells and Colocalization Experiment

1. Treat the first group of the HeLa cells with 5 μM Lyso-HP probe for 20 min at 37 °C in a CO_2 incubator. This group is considered as the control group.
2. Treat the second group of the HeLa cells with 5 μM Lyso-HP probe for 20 min, wash twice with DPBS, and then add 30 μM H_2O_2 and incubate for 20 min at 37 °C. This group is considered as the stimulated group.
3. Treat the third group of the HeLa cells with 5 μM Lyso-HP and 50 nM Lysotracker Red for 20 min, rinse twice with DPBS, and then add 30 μM H_2O_2 and incubate for 20 min at 37 °C. This group is considered as the colocalization experiment group.
4. Wash the first and second group of cultures with DPBS, and visualize by confocal microscopy. The fluorescence images can be acquired by confocal microscopy with excitation at 488 nm using a diode laser as the light source (*see* **Note 4**) and detecting the emission wavelengths at 500–550 nm (Fig. 1). The fluorescence images of the third group can be acquired by confocal microscopy with excitation at 488 nm and 561 nm and emission wavelengths 500–550 nm (the green channel) and 570–620 nm (the red channel), respectively (Fig. 2).

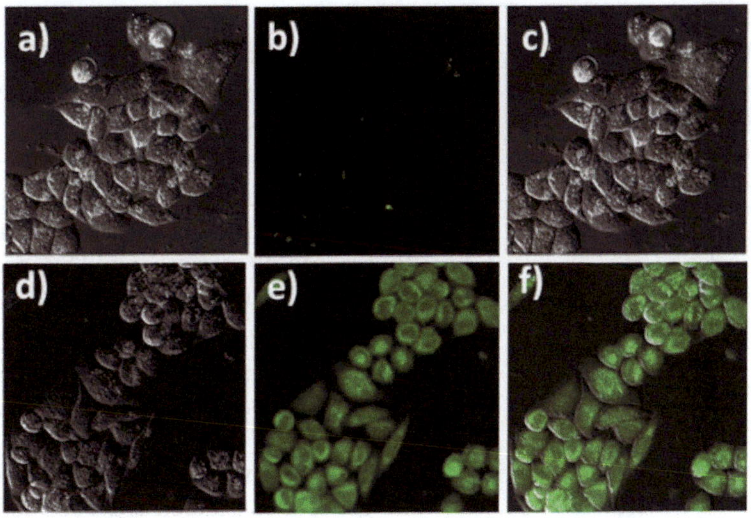

Fig. 1 Images of exogenous added H_2O_2 in HeLa cells stained with the probe Lyso-HP. (**a**) Brightfield image of control HeLa cells treated with the probe Lyso-HP (5 μM); (**b**) fluorescence image (a) from the *green* channel; (**c**) overlay of (a) and (b); (**d**) brightfield image of HeLa cells incubated with Lyso-HP (5 μM) for 30 min and then further treated with H_2O_2 (30 μM); (**e**) fluorescence image of (d) from the *green* channel; (**f**) overlay of (d) and (e). (Reprinted from ref. [22] Copyright (2016), with permission from Elsevier)

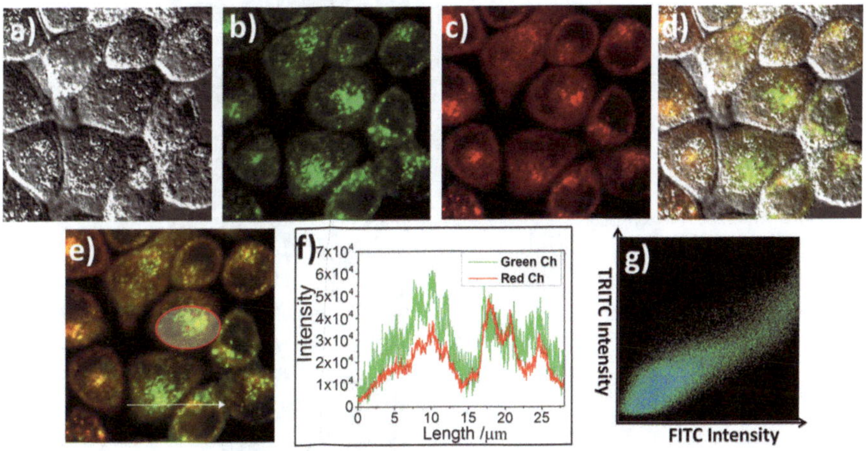

Fig. 2 Brightfield and fluorescence images of the HeLa cells exposed to H_2O_2 stained with the probe Lyso-HP and Lysotracker Red. (**a**) Brightfield image; (**b**) fluorescence image from the *green* channel (Lyso-HP); (**c**) from the *red* channel (Lysotracker Red staining); (**d**) overlay of the brighfield, *green* and *red* channels; (**e**) overlay of the *green* and *red* channels; **(f)** intensity profile of linear region of interest across in HeLa cell costained with Lysotracker Red and Lyso-HP; (**g**) intensity scatter plot of the green and red channels. (Reprinted from ref. [22] Copyright (2016), with permission from Elsevier)

3.4.3 Imaging the Endogenous H_2O_2 in the RAW 264.7 Cells and Colocalization Experiment

1. Treat the first group of the RAW 264.7 cells with 5 μM Lyso-HP probe for 20 min at 37 °C in a CO_2 incubator. This group is considered as the control group.
2. Treat the second group of the RAW 264.7 cells with 4.8 μM PMA for 24 h, wash with DPBS, and incubate with 5 μM Lyso-HP probe for 20 min at 37 °C in a CO_2 incubator. After washing with DPBS, the fluorescence images are acquired using confocal microscopy. This group is considered as the stimulated group.
3. Treat the third group of the RAW 264.7 cells with 4.8 μM PMA for 24 h, wash with DPBS, and incubate with 5 μM Lyso-HP probe and 50 nM Lysotracker Red for 20 min at 37 °C in a CO_2 incubator. After washing twice with DPBS, the fluorescence images are acquired using confocal microscopy. This group is considered as the colocalization experiment group.
4. After washing with DPBS, the first and second groups can be visualized by confocal microscopy. The fluorescence images can be acquired by confocal microscopy with excitation at 488 nm using a diode laser as the light source and detection the emission wavelengths 500–550 nm (the green channel) (Fig. 3). The fluorescence images of the third group can be acquired by

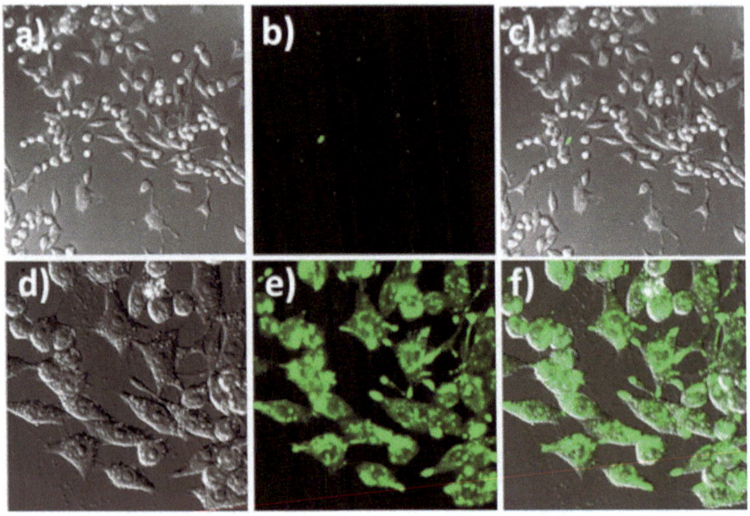

Fig. 3 Images of the endogenous H_2O_2 production in RAW 264.7 cells stained with the probe Lyso-HP. (**a**) Brightfield image of control RAW 264.7 macrophages cells stained with Lyso-HP; (**b**) fluorescence images of (a) from the *green* channel; (**c**) overlay of (a) and (b). (**d**) Brightfield image of RAW 264.7 macrophages stimulated with PMA (4.8 μM) and stained with Lyso-HP, (**e**) fluorescence images of (d) from the *green* channel; (**f**) overlay of the brightfield image (d) and *green* channels (e). (Reprinted from ref. [22] Copyright (2016), with permission from Elsevier)

confocal microscopy with excitation at 488 nm and 561 nm and the emission wavelengths are 500–550 nm (the green channel) and 570–620 nm (the red channel), respectively.

3.5 Experiments in Liver Tissue

Obtain liver tissue from mice (20–30 day old) in accordance with the institutional and governmental regulations and guidelines.

1. Remove the liver from mice and cut it to obtain two slices of 400 μm thickness using a vibrating-blade microtome in DPBS (pH 7.4).
2. Place the two tissue slices in two 35-mm glass-bottom dishes.
3. For the control group, incubate the slice with 30 μM Lyso-HP in DPBS buffer in an incubator bubbled with 95% air and 5% CO_2 for 30 min at 37 °C.
4. For the stimulated group, treat the slice with 100 μM H_2O_2 for 20 min and then wash three times with DPBS. After that, incubate the slice with 30 μM Lyso-HP in DPBS for 30 min in an incubator bubbled with 95% air and 5% CO_2 for 30 min at 37 °C, and then wash three times with DPBS.
5. The control and stimulated groups can be visualized by confocal microscopy. The fluorescence images are acquired by confocal microscopy with excitation at 780 nm using a femtosecond pulse as the light source and emission wavelengths 500–550 nm (Fig. 4).

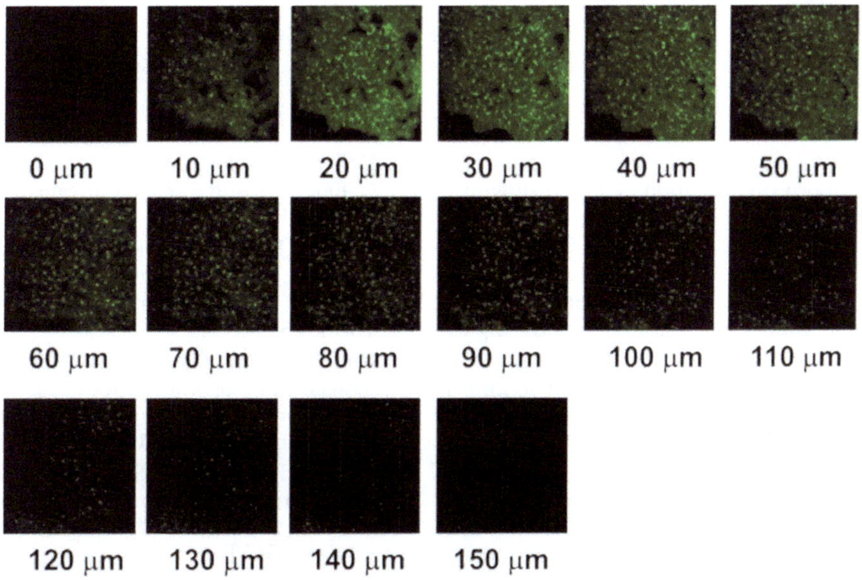

Fig. 4 Two-photon fluorescence images of a fresh rat liver tissue slice pretreated with H_2O_2 (100 μM) and then incubated with Lyso-HP (30 μM) at depths of approximately 0–150 μm with a magnification of 20×. Excitation at 780 nm. (Reprinted from ref. [22] Copyright (2016), with permission from Elsevier)

4 Notes

1. In the synthesis of the compound Lyso-HP, O_2 and H_2O may have an effect on the reaction. It is necessary to make sure that the reaction is set up in the air-free conditions and all the reagents are as dry as possible.
2. In the purification of Lyso-HP, silica may deprotect the pinacol to yield the boronic acid, which is hydrophilic and very difficult to eluate from the column. Thus, it is recommended to run the column as quickly as possible, keeping the solvent flowing throughout the purification process.
3. We find that it is best to prepare the fresh solution of Lyso-HP each time owing to potential degradation of the boronate moiety.
4. Boronate deprotection of the probe may yield the fluorescent product, which can in turn increase the background fluorescence of the experiment and decrease the sensitivity to H_2O_2. Care should be taken to provide Lyso-HP as pure as possible.

Acknowledgment

This work was financially supported by NSFC (21472067, 21502067, 21672083), the Natural Science Foundation of Shandong Province, China (ZR2014BP001), Taishan Scholar Foundation (TS 201511041), and the startup fund of University of Jinan (309-10004, 160082101).

References

1. Lippert AR, Van de Bittner GC, Chang CJ (2011) Boronate oxidation as a bioorthogonal reaction approach for studying the chemistry of hydrogen peroxide in living systems. Acc Chem Res 44:793–804
2. Stone JR, Yang S (2006) Hydrogen peroxide: a signaling messenger. Antioxid Redox Signal 8:243–270
3. Giorgio M, Trinei M, Migliaccio E et al (2007) Hydrogen peroxide: a metabolic by-product or a common mediator of ageing signals? Nat Rev Mol Cell Biol 8:722–728
4. Rhee SG (2006) H_2O_2, a necessary evil for cell signaling. Science 312:1882–1883
5. Ohshima H, Tatemichi M, Sawa T (2003) Chemical basis of inflammation-induced carcinogenesis. Arch Biochem Biophys 417:3–11
6. Dell'angelica EC, Mullins C, Caplan S et al (2000) Lysosome-related organelles. FASEB J 14:1265–1278
7. Zdolsek J, Zhang H, Roberg K et al (1993) H_2O_2-mediated damage to lysosomal membranes of J-774 cells. Free Radic Res Commun 18:71–85
8. Kim D, Kim G, Nam S-J et al (2015) Visualization of endogenous and exogenous hydrogen peroxide using a lysosome-targetable fluorescent probe. Sci Rep 5:8488
9. Song D, Lim JM, Cho S et al (2012) A fluorescence turn-on H_2O_2 probe exhibits lysosome-localized fluorescence signals. Chem Commun 48:5449–5451
10. Zhou J, Li L, Shi W et al (2015) HOCl can appear in the mitochondria of macrophages

during bacterial infection as revealed by a sensitive mitochondrial-targeting fluorescent probe. Chem Sci 6:4884–4888

11. Li Y, Wu W, Yang J et al (2016) Engineering a nanolab for the determination of lysosomal nitric oxide by the rational design of a pH-activatable fluorescent probe. Chem Sci 7:1920–1925
12. Wu X, Li Z, Yang L et al (2013) A self-referenced nanodosimeter for reaction based ratiometric imaging of hypochlorous acid in living cells. Chem Sci 4:460–467
13. Cui L, Peng Z, Ji C et al (2014) Hydrazine detection in the gas state and aqueous solution based on the Gabriel mechanism and its imaging in living cells. Chem Commun 50: 1485–1487
14. Liu C, Shao C, Wu H et al (2014) A fast-response, highly sensitive and selective fluorescent probe for the ratiometric imaging of hydrogen peroxide with a 100 nm red-shifted emission. RSC Adv 4:16055–16061
15. Liu XL, Du XJ, Dai CG et al (2014) Ratiometric two-photon fluorescent probes for mitochondrial hydrogen sulfide in living cells. J Org Chem 79:9481–9489
16. Wen Y, Liu KY, Yang HR et al (2014) A highly sensitive ratiometric fluorescent probe for the detection of cytoplasmic and nuclear hydrogen peroxide. Anal Chem 86:9970–9976
17. Tang Y, Kong X, Xu A et al (2016) Development of a two-photon fluorescent probe for imaging of endogenous formaldehyde in living tissues. Angew Chem Int Ed Engl 55:3356–3359
18. Lee MH, Jeon HM, Han JH et al (2014) Toward a chemical marker for inflammatory disease: a fluorescent probe for membrane-localized thioredoxin. J Am Chem Soc 136:8430–8843
19. Li G, Zhu D, Liu Q et al (2013) Rapid detection of hydrogen peroxide based on aggregation induced ratiometric fluorescence change. Org Lett 15:924–927
20. Jing X, Yu F, Chen L (2014) Visualization of nitroxyl (HNO) in vivo via a lysosome-targetable near-infrared fluorescent probe. Chem Commun 50:14253–14256
21. Zhu H, Fan J, Zhang S et al (2014) Ratiometric fluorescence imaging of lysosomal Zn^{2+} release under oxidative stress in neural stem cells. Biomater Sci 2:89–97
22. Ren M, Deng B, Wang J-Y et al (2016) A fast responsive two-photon fluorescent probe for imaging H_2O_2 in lysosomes with a large turn-on fluorescence signal. Biosens Bioelectron 79:237–243

Chapter 8

Lysophagy: A Method for Monitoring Lysosomal Rupture Followed by Autophagy-Dependent Recovery

Takanobu Otomo and Tamotsu Yoshimori

Abstract

Selective autophagy recognizes specific targets, including damaged mitochondria (mitophagy), aggregated proteins (aggrephagy), and invading bacteria (xenophagy) to engulf by isolation membrane, and degrades toxic materials within lysosomes. We recently revealed that a membrane-damaged lysosome itself also becomes a target of autophagy and named this process lysophagy. In this chapter, we describe methods for monitoring lysophagy; detecting lysosomal damage by staining of galectin and study the subsequent autophagic process in cultured mammalian cells.

Key words Autophagy, Galectin, Immunostaining, LLOMe, Lysophagy, Lysosome, Tandem fluorescent tagged

1 Introduction

Autophagy is an intracellularbulk degradation system conserved from yeast to mammals, which is induced by various stress conditions such as nutrient starvation [1–3]. Autophagy initiates from a cup-shaped double membrane structure called isolation membrane arising around ER-mitochondria contact site [4], which elongates and sequestrates cytoplasmic materials nonselectively. This double membrane vesicular structure, autophagosome, fuses with lysosomes, and the materials inside are degraded by the action of lysosomal acid hydrolases. In addition to nonselective autophagy, there is also a selective autophagy, which recognizes specific targets including damaged mitochondria, aggregated proteins, and invading bacteria, such processes are referred to as mitophagy, aggrephagy, and xenophagy, respectively [5]. Recently, we and other groups revealed that a damaged lysosome itself also becomes a target for autophagy [6–7]. Lysosomes (or endosomes) are often exposed to membrane-permeable dangers such as silica, monosodium urate, bacterial toxins, lipids, and β-amyloid [8–12] under physiological conditions. Indeed, autophagic-deficient mice showed

Karin Öllinger and Hanna Appelqvist (eds.), *Lysosomes: Methods and Protocols*, Methods in Molecular Biology, vol. 1594, DOI 10.1007/978-1-4939-6934-0_8, © Springer Science+Business Media LLC 2017

impaired processing of damaged lysosomes, leading to deteriorated nephropathy in a hyperuricemia model [6]. Experimentally, we use a lysosomotropic reagent, L-leucyl-L-leucine methyl ester (LLOMe), that accumulates within lysosomes and is activated into a membranolytic form by lysosomal cathepsin C [13]. For monitoring lysosomal membrane breakage and its removal by autophagy, we utilize galectin, a galactose-binding lectin, as a marker [14, 15]. Galectin normally localizes in the cytoplasm and nucleus, whereas galactose-containing carbohydrate chains are enriched at the luminal side of endosomes or lysosomes, because lysosomal membrane proteins such as Lamp1 are highly glycosylated. Endosomal or lysosomal membrane damage enables contact between galectins and galactoses leading to the recruitment of galectins at the ruptured membrane. In this chapter, we describe how to monitor the process of lysophagy, i.e., lysosomal breakage followed by autophagy-dependent recovery [16].

2 Materials

Various types of cells (e.g., HeLa, NIH3T3, mouse embryonic fibroblast (MEF), and human skin fibroblast) can be subjected to lysophagy assay. From a recent study, we describe lysophagy assay with LLOMe on HeLa-Kyoto cells.

2.1 Induction of Lysophagy in Cultured Cells

1. Culture medium: Dulbecco's Modified Eagle Medium (DMEM), 10% heat inactivated (56 °C, 45 min) fetal bovine serum (FBS), 100 units/ml penicillin, 100 μg/ml streptomycin.
2. Glass coverslips: Sterilized by autoclaving (121 °C, 30 min) (*see* **Note 1**).
3. Collagen coating solution: Cell matrix Type I-C (Nitta gelatin), dilute to 0.1 mg/ml with sterilized distilled water.
4. GFPgalectin 3 construct: Available from Addgene (pEGFP-hGal3, https://www.addgene.org/73080/).
5. Tandem fluorescent-tagged galectin3 (mRFP-GFP galectin3, tfGal3) construct: Available from Addgene (ptf-Galectin3, https://www.addgene.org/64149/).
6. LLOMe: Dissolve 1.13 g of LLOMe in 10 ml ethanol to prepare a 333 mM stock solution. Make small aliquots, seal firmly with parafilm and store at −20 °C.

2.2 Immunostaining and Detection

1. Fixation buffer: 4% paraformaldehyde (PFA) in PBS.
2. Blocking solution (*see* **Note 2**): 0.1% gelatin in PBS. Mix gelatin powder with PBS and autoclave it (121 °C, 30 min) to let gelatin completely dissolve. Store at 4 °C.

3. Permeabilization buffer: Stock solution 50 mg/ml digitonin in dimethyl sulfoxide (DMSO). Prepare a working solution of 50 μg/ml (1:1000) in blocking solution before use.
4. Primary antibodies: anti-galectin3 rat monoclonal antibody (Santa Cruz, M3/38, sc-23938, 1:1000); anti-Lamp1 mouse monoclonal antibody (SantaCruz, H4A3, 1:1000); anti-LC3 rabbit polyclonal antibody (MBL, PM036, 1:1000).
5. Secondary antibodies: Choose appropriate antibodies conjugated with fluorescent dyes for each primary antibody and microscope equipment.
6. Other fluorescent dyes: Dependent on your experimental purpose, e.g., LysoTracker (Red DNS-99) or DAPI.
7. Mounting reagent: Proper antifading mounting medium.

3 Methods

As mentioned earlier, lysosomal membrane damage can be detected by recruitment of galectin. In the previous report, we used galectin3 (Gal3) [6, 15], but reports from other groups indicate that Gal8 and Gal9 are usable as well for marking damaged endo/lysosomal membranes [17]. In addition to detecting exogenously expressed GFP-Gal3 (Fig. 1), or endogenous Gal3 (followed by immunostaining), it is also useful to design exogenous expression of RFP-GFP (tandem fluorescent)-tagged Gal3 (tfGal3) (Fig. 2a). This probe is based on the idea from the tfLC3 assay that we previously developed [18]. GFP fluorescence is attenuated and degraded by lysosomal acidic environment, whereas RFP fluorescence is comparatively stable even under acidic conditions. Thus, tfGal3 enables monitoring of the pH change of the surrounding environment (Figs. 2b and 3).

3.1 Cell Culture

1. Culture HeLa cells at 37 °C in a 5% CO_2 incubator.
2. When you use GFP-Gal3 or tfGal3, transfect the plasmid constructs in HeLa cells using standard transfection protocol prior to the experiment.
3. Place sterilized glass coverslips in culture dishes and coat them with collagen solution for at least 10 min at room temperature, then wash them twice with PBS (*see* **Note 1**).
4. Seed cells on the collagen coated coverslips placed in culture dish (*see* **Note 3**).
5. Incubate at 37 °C in a 5% CO_2 incubator for least 24 h until the cells have attached to the coverslips and cellular condition becomes stable before the lysophagy assay.

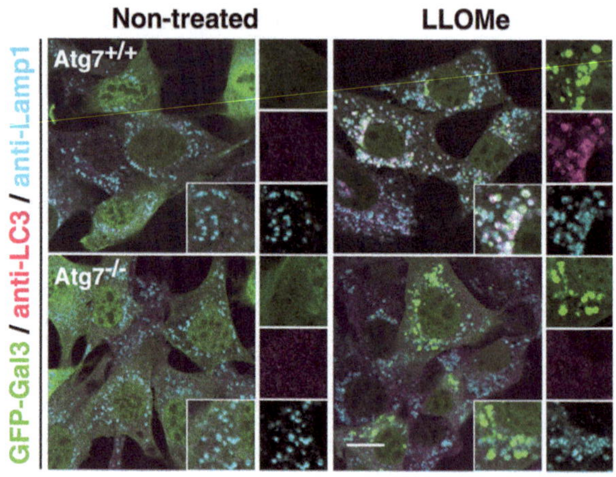

Fig. 1 The recruitment of LC3 to Gal3-positive damaged lysosomes. $Atg7^{+/+}$ and $Atg7^{-/-}$MEF cells stably expressing GFP-Gal3 are treated with LLOMe, followed by immunostaining of LC3 and Lamp1. Three hours after LLOMe treatment, punctate Gal3 recruitment appears, which colocalizes with lysosomal marker Lamp1 indicating damaged lysosomes. These structures also colocalize with the autophagosome marker LC3 in normal cells ($Atg7^{+/+}$) but not in autophagy-deficient cells ($Atg7^{-/-}$). [Adapted from ref. 6]

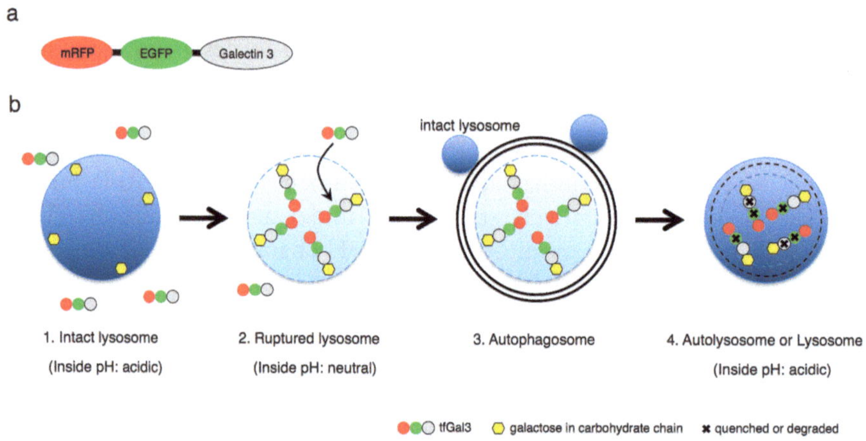

Fig. 2 The tandem fluorescent-tagged Galactin3 (tfGal3) system. (**a**) Construct of tfGal3, where galactin3 is tagged with two fluorescence reporters: GFP and RFP. (**b**) Scheme of lysophagy process: 1. Intact lysosomes include galactose containing sugar chain at the luminal side, whereas expressed tfGal3 resides in the cytosol. 2. When the lysosomal membrane rupture, the luminal pH increases and influx of tfGal3 into the lysosomal lumen occur. tfGal3 recruited to the galactose-rich inner membrane exhibits both GFP and RFP fluorescence. 3. Damaged lysosome is sequestered by autophagy-dependent process and isolated within an autophagosome, which subsequently fuses with intact lysosomes to produce an autolysosome. 4. Inside the autolysosome, GFP fluorescence is quenched in acidic environment, but RFP signal persists. This process is observed as the change of the yellow signal (green + red) into the only red signal under fluorescent microscope

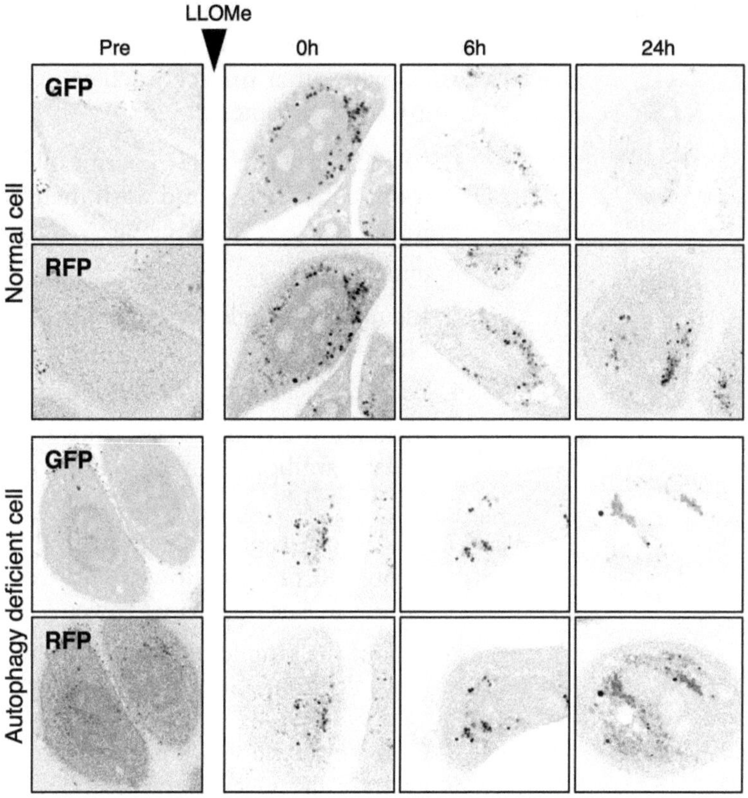

Fig. 3 Monitoring lysosomal rupture and subsequent autophagic process using the tandem fluorescent-tagged Galactin3 (tfGal3) system. LLOMe treatment induces lysosomal rupture that is detected by recruitment of Gal3 (0 h). Since the luminal pH of damaged lysosomes (i.e., pH around accumulated Gal3) is neutral at the moment, both GFP and RFP signals are detected. After several hours, GFP signal disappears in normal cells, whereas RFP signal persists during the period. This indicates pH around Gal3 changes to acidic, which is not observed in autophagy-deficient cells. [Adapted from ref. 6]

3.2 Induction of Lysophagy by LLOMe (See Notes 4 and 5)

1. Discard the culture medium and incubate the cells with 250 μM LLOMe (1:1333 dilution from stock solution) in culture medium for 15 min (*see* **Note 6**).
2. Wash out LLOMe twice with culture medium and continue incubation in culture medium at 37 °C in a 5% CO_2 incubator for several hours to allow lysophagy of the injured lysosomes (*see* **Note 7**).
3. At each selected time point, wash cells with PBS and fix with fixation buffer for 20 min at room temperature (*see* **Note 8**). After fixation, wash and keep the cells on the coverslip in PBS (*see* **Note 9**).

3.3 Immunostaining

If you have already transfected the cells with GFP-Gal3 or tfGal3, you do not need to immunostain Gal3; however, it is useful to costain with other markers such as LC3 (for autophagosomes) or Lamp1 (for lysosomes).

1. Permeabilize the cells in permeabilization buffer for 10 min at room temperature, and wash them twice with PBS.
2. Block the cells with blocking solution for 30 min at room temperature.
3. Incubate the cells on coverslips with primary antibodies (*see* **Note 10**).
4. Wash the cells with PBS.
5. Incubate the cells on coverslips with fluorescence-tagged secondary antibodies in blocking solution with/without DAPI (*see* **Note 11**).
6. Wash the cells on coverslips with PBS, rinse with distilled water to prevent precipitation of salts from the buffers, and remove extra water using a paper towel.
7. Mount the samples on objective slides with mounting medium (e.g., ProLong Gold).

3.4 Analysis

1. Observe the fluorescent signals from Gal3 and other stainings under a fluorescence microscope (Figs. 1 and 3). Take several photos from different fields.
2. Count the number of Gal3 positive dots per cell to estimate the number of damaged lysosomes (Fig. 4). Analyze the colocalization between Gal3, Lamp1, and LC3 (*see* **Note 12**).

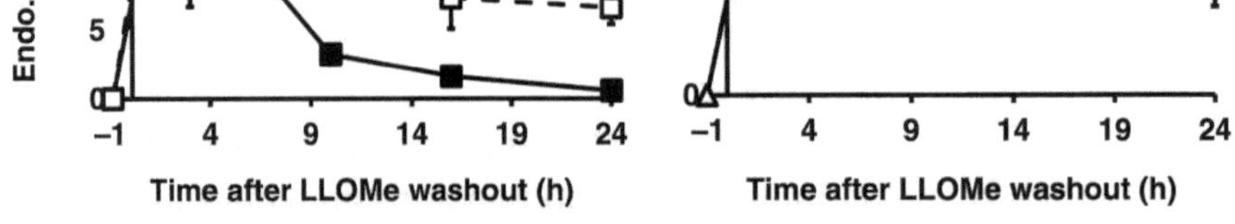

Fig. 4 Trends of endogenous Gal3 dot number during lysophagy process. Control or autophagy-deficient NIH3T3 cells (Atg4B mutant) were treated with LLOMe, washed out, and incubated with/without protease inhibitors (E64d + Pepstatin A). [Adapted from ref. 6]

4 Notes

1. We usually use round shape coverslips (12 mm diameter, 0.12–0.17 mm thickness).
2. This solution is used for diluting permeabilization buffer, primary antibodies, secondary antibodies, as well as blocking.
3. A cell density at which the cells become 60–70% confluent at the time of the lysophagy assay is recommended. With 24-well culture plate, one coverslip per well, and with 6-well plate (or 35 mm dish) maximally five coverslips side by side per well can be placed at the same time.
4. All procedures should be performed gently to prevent detachment of cells from glass coverslips.
5. Lysophagy can also be induced by silica. Different from LLOMe, which become active within lysosomes, silica may induce rupture of endosomes as well as lysosomes. Thus, the interpretation of data should be done with care. Prepare cells on coverslips as described in Subheading 3.1 and incubate the cells with 62.5–500 μg/ml silicon dioxide in culture medium for 1–3 h, depending on the cells used. A suggested stock solution is 1 mg/ml silicon dioxide dissolved in PBS and stored at −20 °C. Because silicon dioxide crystals aggregate during storage, the stock solution should be resuspended before use by passage through a 25-gauge needle. The following washing and chase steps are similar to those of LLOMe.
6. Prepare LLOMe working solution just before the experiment and warm to 37 °C, before adding to the cells. The optimal concentration of LLOMe and incubation time differs between cell types. Prior to the experiment, it is recommended to find the optimal condition to induce lysophagy with your cells, typically it ranges between 100 and 1000 μM during 15 min to 3 h. We treat NIH3T3 cells, MEFs, and human skin fibroblasts with 1000 μM for 1 h to trigger lysophagy. In case lysophagy is not triggered, try higher concentration and/or longer incubation. Another possibility might be inactivation of LLOMe. It is recommended to prepare fresh LLOMe stock solution. As experimental controls, autophagy-deficient cells such as Atg knock-out cell lines or siRNA-treated knock-down cells are possible to use for the lysophagy assay. In addition, incubating cells with membrane-permeable protease inhibitors (E64d, 10 μg/mL; Pepstatin A, 10 μg/mL) after LLOMe treatment are also usable for inhibiting lysosomal degradation during lysophagy. However, pretreatment with cysteine protease inhibitor before LLOMe treatment is not recommended because LLOMe is converted to its membranolytic form by the function of cathepsin C (Fig. 4).

7. Some cell lines show less tolerance to LLOMe treatment. In this case, do not use higher concentrations than the cell shrinkage during LLOMe treatment will be recovered a couple of hours after washout.
8. When staining living cells with fluorescent probes such as LysoTracker, add the reagent before fixation.
9. After fixation, cells on coverslips can be stored at 4 °C for several days in PBS.
10. Typically mix primary antibodies rose in different species such as anti-Gal3 (from rat), anti-LC3 (from rabbit), and anti-Lamp1 (from mouse) together in blocking solution and incubate for 1 h at room temperature. When using 12-mm-diameter coverslips, 40–50 μl antibody solution per coverslip is enough. Place the coverslips on a sheet of parafilm during incubation. Choose appropriate antibodies for each cell line according to the reactivity of antibody to the antigen of cell species (e.g., anti-Lamp1 antibody 1D4B clone for staining MEF).
11. For tracing of each cellular shape, cells are stained with HCS CellMask Deep Red stain together with the secondary antibody and DAPI solution.
12. Choose either wide field microscopy or confocal microscopy according to the specification of the equipment you have. Analyses can be done manually or automatically using software such as CellProfiler (http://www.cellprofiler.org). This is a free software with user-friendly interface and high performance.

Acknowledgment

We are grateful to all colleagues who discovered lysophagy and who developed this sophisticated assay system. This work was supported in part by the Ministry of Education, Culture, Sports, Science and Technology (MEXT) of Japan, by Japan Science and Technology Agency CREST, and by the Takeda Science Foundation.

References

1. Mizushima N, Yoshimori T, Ohsumi Y (2011) The role of Atg proteins in autophagosome formation. Annu Rev Cell Dev Biol 27S: 107–132
2. Lamb CA, Yoshimori T, Tooze SA (2013) The autophagosome: origins unknown, biogenesis complex. Nat Rev Mol Cell Biol 14:759–774
3. Kawabata T, Yoshimori T (2016) Beyond starvation: an update on the autophagic machinery and its functions. J Mol Cell Cardiol 95:2–10
4. Hamasaki M, Furuta N, Matsuda A, Nezu A, Yamamoto A, Fujita N et al (2013) Autophagosomes form at ER-mitochondria contact sites. Nature 495:389–393

5. Stolz A, Ernst A, Dikic I (2014) Cargo recognition and trafficking in selective autophagy. Nat Cell Biol 16:495–501
6. Maejima I, Takahashi A, Omori H, Kimura T, Takabatake Y, Saitoh T et al (2013) Autophagy sequesters damaged lysosomes to control lysosomal biogenesis and kidney injury. EMBO J 32:2336–2347
7. Hung YH, Chen LM, Yang JY, Yang WY (2013) Spatiotemporally controlled induction of autophagy-mediated lysosome turnover. Nat Commun 4:2111
8. Hornung V, Bauernfeind F, Halle A, Samstad EO, Kono H, Rock K et al (2008) Silica crystals and aluminum salts activate the NALP3 inflammasome through phagosomal destabilization. Nat Immunol 9:847–856
9. Emmerson BT, Cross M, Osborne JM, Axelsen RA (1990) Reaction of MDCK cells to crystals of monosodium urate monohydrate and uric acid. Kidney Int 37:36–43
10. Matsuda S, Okada N, Kodama T, Honda T, Iida T (2012) A cytotoxic type III secretion effector of *Vibrio parahaemolyticus* targets vacuolar H+-ATPase subunit c and ruptures host cell lysosomes. PLoS Pathog 8:e1002803
11. Duewell P, Kono H, Rayner KJ, Sirois CM, Vladimer G, Bauernfeind FG et al (2010) NLRP3 inflammasomes are required for atherogenesis and activated by cholesterol crystals. Nature 464:1357–1361
12. Ditaranto K, Tekirian TL, Yang AJ (2001) Lysosomal membrane damage in soluble Abeta-mediated cell death in Alzheimer's disease. Neurobiol Dis 8:19–31
13. Thiele DL, Lipsky PE (1990) Mechanism of L-leucyl-L-leucine methyl ester-mediated killing of cytotoxic lymphocytes: dependence on a lysosomal thiol protease, dipeptidyl peptidase I, that is enriched in these cells. Proc Natl Acad Sci U S A 87:83–87
14. Paz I, Sachse M, Dupont N, Mounier J, Cederfur C, Enninga J et al (2010) Galectin-3, a marker for vacuole lysis by invasive pathogens. Cell Microbiol 12:530–544
15. Fujita N, Morita E, Itoh T, Tanaka A, Nakaoka M, Osada Y et al (2013) Recruitment of the autophagic machinery to endosomes during infection is mediated by ubiquitin. J Cell Biol 203:115–128
16. Hasegawa J, Maejima I, Iwamoto R, Yoshimori T (2015) Selective autophagy: lysophagy. Methods 75:128–132
17. Thurston TL, Wandel MP, von Muhlinen N, Foeglein A, Randow F (2011) Galectin 8 targets damaged vesicles for autophagy to defend cells against bacterial invasion. Nature 482: 414–418
18. Kimura S, Noda T, Yoshimori T (2007) Dissection of the autophagosome maturation process by a novel reporter protein, tandem fluorescent-tagged LC3. Autophagy 3:452–460

Chapter 9

Delivery of Cargo to Lysosomes Using GNeosomes

Kristina M. Hamill, Ezequiel Wexselblatt, Wenyong Tong, Jeffrey D. Esko, and Yitzhak Tor

Abstract

Liposomes have been used to improve the intracellular delivery of a variety of cargos. Encapsulation of cargos in liposomes leads to improved plasma half-lives and minimized degradation. Here, we present a method for improving the selective delivery of liposomes to the lysosomes using a guanidinylated neomycin (GNeo) transporter. The method for synthesizing GNeo-lipids, incorporating them into liposomes, and the enhanced lysosomal delivery of encapsulated cargo are presented. GNeo-liposomes, termed GNeosomes, are capable of delivering a fluorescent dye to the lysosomes of Chinese hamster ovary cells as shown using confocal microscopy. GNeosomes can also be used to deliver therapeutic quantities of lysosomal enzymes to fibroblasts isolated from patients with a lysosomal storage disorder.

Key words Cellular uptake, Guanidinium-rich molecular transporters, Guanidinoglycosides, Liposomes, Lysosomal storage disorder

1 Introduction

Liposomal delivery systems have been used to improve the drug delivery of small molecule therapeutics to large biomolecules, and several liposome-based drug formulations have been clinically approved [1, 2]. Liposomes can protect the encapsulated cargo from degradation and improve their overall pharmacokinetics. Additionally, the surface of liposomes can be modified with ligands to selectively target subcellular organelles or specific tissues, thus improving efficacy and reducing toxicity.

The lysosomes are an example of a promising subcellular target for enzyme replacement therapy and cancer treatments [3, 4]. Although various methods have been developed to deliver cargo to the lysosomes, they generally rely on covalent modification [5–8]. To improve the lysosomal delivery of liposomes, we incorporated guanidinylated neomycin (GNeo), a molecular transporter previously shown to target the lysosomes [6, 9].

Karin Öllinger and Hanna Appelqvist (eds.), *Lysosomes: Methods and Protocols*, Methods in Molecular Biology, vol. 1594, DOI 10.1007/978-1-4939-6934-0_9, © Springer Science+Business Media LLC 2017

We here present how to prepare liposomes with GNeo by synthesis of two amphiphilic guanidinylated neomycin (GNeo) derivatives. The GNeo-lipids can either be preinserted into the liposomes during liposomal formation or postinserted into preformed liposomes, generating GNeosomes. The efficacy is demonstrated by encapsulating a small fluorescent dye followed by assessment of the lysosomal delivery by confocal laser scanning microscopy. To show utility, the active lysosomal enzyme α-L-iduronidase (IDUA) is encapsulated and used to restore function in fibroblasts isolated from a patient with the lysosomal storage disorder mucopolysaccharidosis type I (MPS I). The preparation techniques and methods described in the following text can potentially be applied to improve the lysosomal delivery of essentially any entrapped cargo.

2 Materials

2.1 Synthesis of GNeo-Lipids

1. Stearic acid (TCI, >98%).
2. Di-oleic acid can be obtained commercially or synthesized as reported by Zhang et al. [10].
3. 2-[2-[2-(Prop-2-ynyloxy)ethoxy]ethoxy]ethanamine (*see* **Note 1**).
4. Azide-BocGNeo (*see* **Note 2**).
5. Dichloromethane (CH_2Cl_2), *N*-(3-dimethylaminopropyl)-*N′*-ethylcarbodiimide hydrochloride (EDC), *N*,*N*-diisopropylethylamine (DIEA), tris[(1-benzyle-1*H*-1,2,3-triazol-4-yl)methyl]amine (TBTA), sodium ascorbate, $CuSO_4·5H_2O$, triisopropylsilane (TIPS), and trifluoroacetic acid (TFA) were obtained from commercial suppliers and used without further purification.
6. Sodium ascorbate: Dissolve 100 mg in 0.5 mL deionized water and purge with argon for 15 min.
7. $CuSO_4·5H_2O$: Dissolve 125 mg in 0.5 mL deionized water and purge with argon for 15 min.
8. Automated flash chromatography system.
9. Corning(R) Spin-X(R) Centrifuge Tube Filters (0.45 μm pore).
10. C18 HPLC columns to purify stearyl-GNeo: 5 μm, 10 × 250 mm (CLIPEUS) for purification and 5 μm, 4.6 × 150 mm (Higgins Eclipse, XDB-C18) for analytical analysis.
11. C4 HPLC columns to purify di-oleyl-GNeo: 5 μm, 10 × 250 mm (SepaxGP-C4) for purification and 5 μm, 4.6 × 150 mm column for analytical analysis.

2.2 Preparation of GNeosomes

1. 1,2-Dioleoyl-snglycero-3-phosphocholine (DOPC): 1 g powder dissolved in 10 mL chloroform; 1,2-dioleoyl-sn-glycero-3-phospho-ethanolamine (DOPE): 100 mg powder dissolved in 3 mL chloroform and 1 mL ethanol; Cholesterol: 1 g dissolved in 16.7 mL chloroform and 16.7 mL ethanol. Lipids were stored as solutions at −20 °C.
2. PBS: Dulbecco's phosphate buffered saline (DPBS) without calcium or magnesium.
3. Cy5 is commercially available. The structure for the variant in this protocol is seen in Fig. 1 (*see* **Note 3**).
4. α-L-Iduronidase (IDUA) manufactured by BioMarin Pharmaceuticals (Aldurazyme) and purified by FPLC on a heparin column prior to use (0–2 M NaCl in DPBS over 40 min).
5. Mini-Extruder set with holder/heating block and two 1 mL syringes (Avanti Polar Lipids).
6. Polycarbonate membranes (0.1 μm, 19 mm) and filter supports (10 mm) (Avanti Polar Lipids).
7. Allow Sephadex G-50 M to swell in water for 3 h prior to use. Fill a 30 cm glass column with stopcock about two-thirds with water then pour the well-mixed Sephadex slurry down the side wall of the column using a glass rod and avoiding introduction of air bubbles. Run three column volumes of water or DPBS through the column.
8. Sepharose 4B is supplied preswollen. The column is prepared as outlined above for Sephadex G-50.
9. Ammonium ferrothiocyanate (0.1 M): Dissolve 27.03 g ferric chloride hexahydrate ($FeCl_3{\cdot}6H_2O$) and 30.4 g ammonium thiocyanate (NH_4SCN) in deionized water to make a 1 L solution.
10. Zetasizer Nano (Malvern Instruments Inc.) for size and zeta potential measurements.

2.3 Measuring Concentration of Cy5 in GNeosomes

1. A solution of 45 mL isopropanol and 5 mL 0.075 M HCl is used to measure the concentration of Cy5 in the liposomes.
2. A fluorimeter to measure fluorescence intensity (Excite at 640 nm and measure emission intensity at 675 nm).

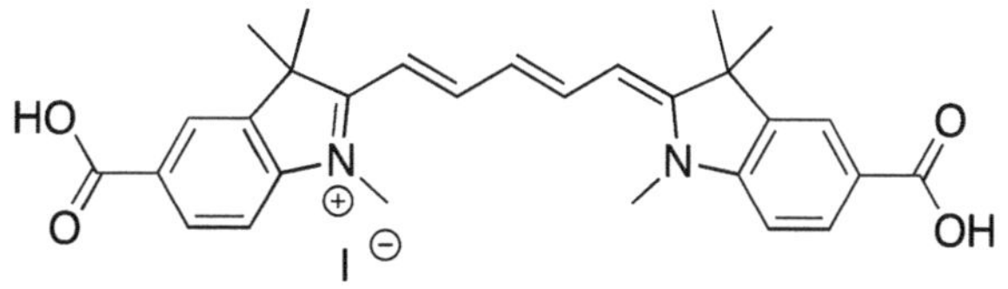

Fig. 1 Chemical structure of Cy5

2.4 Measuring Concentration and Enzyme Activity of IDUA in GNeosomes

1. Tween 20, nonionic, aqueous solution, 10% (w/v).
2. 4–12% Bis-Tris protein gel.
3. LDS sample buffer (4×).
4. 2-Mercaptoethanol (>99.0%) to reduce protein disulfide bonds before polyacrylamide gel electrophoresis.
5. MES SDS running buffer: Add 25 mL (20×) to 475 mL ultrapure water.
6. Coomassie Blue R-250 protein stain.
7. Infrared imaging system.
8. 4-Methylumbelliferone (4-MU).
9. Dilution buffer: 0.1 M Na citrate, 150 mM NaCl, pH 4.5. Weigh out 7.35 g sodium citrate and 2.19 g sodium chloride. Dissolve in 225 mL ultrapure water. Mix and adjust the pH to 4.5 with HCl. Make up to 250 mL with water. Store at 4 °C.
10. Stop buffer: 0.4 M glycine/NaOH, pH 10.4. Weigh out 7.50 g glycine and dissolve in 225 mL ultrapure water. Mix and adjust the pH to 10.4 with NaOH solution (50% in water). Make up to 250 mL with water. Store at 4 °C.
11. Dissolve 5 mg of 4-methylumbelliferyl α-L-idopyranosiduronic acid (Carbosynth) in 2.67 mL dilution buffer.

2.5 Cell Culture

1. Chinese hamster ovary (CHO) cells originate from American Type Culture Collection (CCL-61).
2. F12 medium: 500 mL supplemented with 50 mL fetal bovine serum and 5 mL penicillin/streptomycin (10,000 U/mL).
3. Metachromatic leukodystrophy (control HFF) and mucopolysaccharidosis type I (MPS I) fibroblasts were obtained from Coriell (GM00200 and GM00338).
4. Dulbecco's modified Eagle medium (DMEM): 500 mL supplemented with 50 mL fetal bovine serum, 5 mL penicillin/streptomycin (10,000 U/mL), 5 mL sodium pyruvate (100 mM), and 5 mL glutamine (200 mM).

2.6 Intracellular Localization of Cy5-Containing Liposomes Using Confocal Microscopy

1. Glass bottom dish, 35 mm, poly-D-lysine coated (MatTek).
2. Hoechst 33342, trihydrochloride, trihydrate, (100 mg).
3. LysoTracker® Green DND-26 (1 mM stock solution in anhydrous DMSO).

2.7 Intracellular Delivery of Lysosomal Enzyme

1. DMEM/F12 (Ham): 500 mL supplemented with 50 mL fetal bovine serum.
2. Sulfur-35, as sulfuric acid in 1 mL water (25 mCi).
3. DEAE Sephacel in 20% ethanol (GE Healthcare).

2. Add 28 μL of a freshly prepared 1 M solution of sodium ascorbate in water (6 mg, 28 μmol) and 28 μL of a freshly prepared 1 M solution of $CuSO_4{\cdot}5H_2O$ in water (8.75 μg, 28 μmol). Stir vigorously at room temperature overnight.
3. Dilute the reaction with CH_2Cl_2 (10 mL) and water (10 mL). Wash the organic phase twice with EDTA (0.1 M, pH 8, 12 mL), once with aqueous KCN (5%, 12 mL), and once with brine (12 mL).
4. Dry the organic phase over sodium sulfate. Filter off the sodium sulfate.
5. Concentrate the organic phase using rotary evaporation under reduced pressure.
6. Dissolve the resulting off-white residue in CH_2Cl_2 (0.7 mL). Add TIPS (50 μL) and TFA (0.5 mL). Stir for 12 h at room temperature.
7. Concentrate the reaction using a rotary evaporation under reduced pressure and coevaporate with toluene three times to remove all the trifluoroacetic acid.
8. Dissolve the residue in 5% aqueous acetonitrile and filter using centrifuge tube filters.
9. Purify the reaction using reversed phase HPLC (*see* **Note 4**). Fractions containing the desired product were lyophilized to yield the GNeo-lipid as white amorphous powders.
10. Yield of the product is 51% for stearyl-GNeo or 35% for di-oleyl-GNeo.

3.3 Preparation of Pre-inserted GNeosomes

1. Add 500 μL of chloroform, 120 μL DOPC, 67.5 μL DOPE, and 43.5 μL cholesterol (15 mg total) to a 15 mL round bottom flask.
2. Remove organic solvents by rotary evaporation under reduced pressure and further dry under high vacuum for two nights to form a lipid film.
3. Rehydrate the lipid film with 1 mL DPBS containing 0.9 mol % GNeo-lipid and the cargo to be encapsulated (0.1 mM Cy5 or 5 mg/mL α-L-iduronidase) and rotate on a rotary evaporator for 10 min in a 37 °C water bath.
4. Sonicate the liposomes for 30 s to completely remove the lipid film from the flask wall (*see* **Note 5**).
5. Freeze the liposomes in a dry ice/acetone bath in a Dewar flask for 1 min and then thaw in a 37 °C water bath for 1 min. Repeat this freezing and thawing treatment six times to obtain multilamellar liposomes.
6. Extrude the liposome suspension through a polycarbonate membrane with pore diameter of 100 nm 17 times using a mini-Extruder set to obtain unilamellar liposomes.

7. Remove the nonencapsulated cargo and GNeo-lipid by gravitational size exclusion chromatography (Sephadex G-50 for Cy5 or Sepharose 4B for IDUA) using DPBS as eluent (*see* **Note 6**).
8. To measure the concentration of lipids in the liposomes using the Stewart method, add 50 μL of liposomes (diluted to around 1 mg/mL) and 1.5 mL of chloroform to a 15 mL falcon tube. Vortex vigorously for 10 s. Add 1.5 mL ammonium ferrothiocynate (0.1 M) and vortex vigorously for 15 s. Centrifuge the biphasic system at high speed for 1 min. Measure the optical density of the organic phase at 480 nm using chloroform as a blank (*see* **Note** 7).
9. Measure the zeta potential and the size of the liposomes using Zetasizer Nano. Average diameter of liposomes is 150 nm with a polydispersity index below 0.2.

3.4 Preparation of Postinserted GNeosomes

1. Prepare the lipid film as outlined in the previous Subheading 3.3, **steps 1** and **2**.
2. Rehydrate the lipid film 1 mL PBS containing the cargo to be encapsulated (0.1 mM Cy5 or 5 mg/mL α-L-iduronidase) and rotate on a rotary evaporator for 10 min in a 37 °C water bath.
3. Prepare purified unilamellar liposomes as outlined in the previous Subheading 3.1, **steps 4–8**.
4. Dilute the liposomes to 3 mg/mL with PBS in a 2 mL glass vial. Add 0.9 mol % or 1.8 mol % GNeo-lipid and stir at room temperature for 1 h.
5. Remove the nonincorporated GNeo-lipid by size exclusion chromatography using Sephadex G-50 minicolumn [11] (*see* **Note 8**). Centrifuge at high speed for 15 min. Liposomes will pass through the column and GNeo-lipid will be retained. The liposome concentration should be unchanged and confirmed using the Stewart method.
6. Measure the zeta potential and the size of the liposomes using Zetasizer Nano. Average diameter of liposomes is 150 nm with a polydispersity index below 0.2.

3.5 Measuring Concentration of Cy5 in GNeosomes

1. Dilute liposomes to 1 mg/mL with water.
2. Dilute 100 μL of liposome solution to 1 mL with isopropanol: 0.075 N HCl (9:1).
3. Add the 1 mL solution to cuvette and measure fluorescence with excitation 640 nm and emission 670 nm.

3.6 Measuring Concentration and Enzyme Activity of IDUA in GNeosomes

1. Add 25 μL of 10% Tween 20 solution to 50 μL of 3 mg/mL liposome solution. Heat for 10 min at 55 °C to open liposomes (*see* **Note 9**).
2. Mix 14 μL of opened liposomes with 5 μL LDS sample buffer and 1 μL reducing agent. Prepare IDUA standards by mixing 14 μL of known IDUA concentrations with 5 μL sample buffer and 1 μL reducing agent. Heat all samples at 70 °C for 10 min. Load samples onto NuPAGE Novex 4–12% bis-tris protein gel and run according to manufacturer's protocol (35 min with MES buffer at 200 V constant). Stain the gel for 1 h with Thermo Scientific Imperial Protein Stain then wash overnight in water.
3. Measure protein bands on an Odyssey Infrared imaging system and quantitate by densitometry to determine the concentration of enzyme in the liposomes (*see* **Note 10**).
4. Make 0, 0.05, 0.1, 0.25, 0.5, 1.25, and 2.5 mM stocks of 4-MU standard in DMSO. Add 40 μL of 4-MU standards to 360 μL of dilution buffer and 3600 μL of stop buffer (*see* **Note 11**).
5. Mix 10 μL opened IDUA-containing liposomes with 80 μL of dilution buffer and 10 μL substrate. Incubate 1 h at 37 °C.
6. Stop reaction by adding 900 μL of stop buffer.
7. Take 200 μL of sample and 200 μL of each standard (triplicate) to 96-well black plate.
8. Measure fluorescence with excitation 340 nm and emission 483 nm. Fluorescence intensity of the liposome samples can be converted to units of enzyme using the 4-MU standard curve to confirm liposome preparation does not interfere with enzyme activity.

3.7 Using Confocal Microscopy to Determine Intracellular Localization of Cy5-Containing Liposomes

1. Seed CHO cells (250,000 cells/well) into a 35 mm dish equipped with a glass bottom coverslip coated with poly-D-lysine and culture overnight in 1.5 mL complete F-12 culture medium.
2. Wash the cells with 1 mL DPBS and add 1.5 mL plain liposomes or GNeosomes loaded with Cy5 diluted to 1 mg/mL in complete F-12 culture medium. Incubate for 1 h at 37 °C.
3. Wash cells with 1 mL PBS and add 1.5 mL of fresh F-12 culture medium and 50 μL of the nuclear stain Hoechst 33342 (1.6 μM). Incubate 10 min at 37 °C.
4. Wash cells with 1 mL PBS and add 1.5 mL fresh F-12 culture medium and 30 μL LysoTracker Green DND-26 (17 μM). Incubate for 5 min at 37 °C.

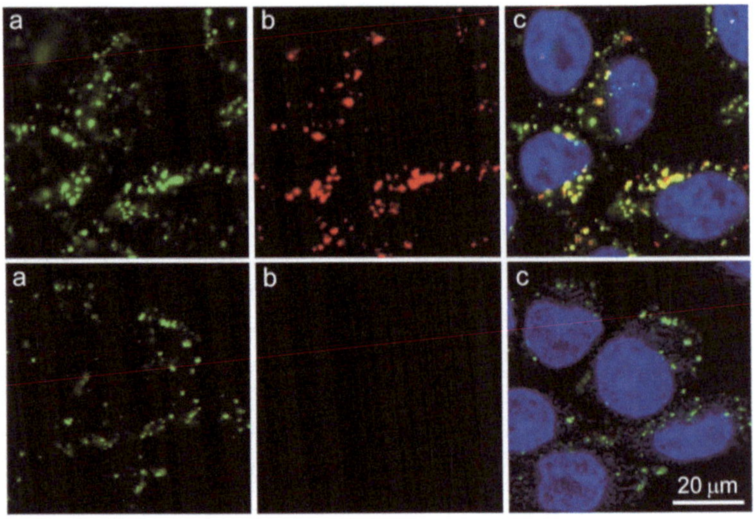

Fig. 4 Intracellular localization of lipid vesicles in CHO cells. *Upper panels*: GNeosomes. *Lower panels*: Plain liposomes. (**a**) LysoTracker Green DND-26, (**b**) vesicles loaded with Cy5, and (**c**) merged images with nuclear Hoechst dye

5. Wash cells with 1 mL PBS and add 1.5 mL complete culture medium. Image the live cells with a confocal laser scanning microscope with 405 nm, 488 nm, and 647 nm lasers for detection of Hoechst 33342, LysoTracker Green, and Cy5, respectively (Fig. 4).

3.8 Intracellular Delivery of Lysosomal Enzyme

1. Seed control HFF and MPS I fibroblasts (30,000 cells/well) in a 12-well plate in complete DMEM culture medium.
2. After 48 h, add 1 mL DMEM/F12 medium supplemented with 10% FBS containing 50 μCi $[^{35}S]O_4$ and incubate for 48 h at 37 °C.
3. Wash cells with 1 mL PBS and add 1 mL plain liposomes or GNeosomes loaded with IDUA diluted to 0.1 mg/mL in DMEM/F12 culture medium. Incubate 1 h at 37 °C.
4. Wash cells with 1 mL PBS twice and add 1 mL fresh culture medium (*see* **Note 12**).
5. After 24 h, wash cells with 1 mL PBS and add 200 μL trypsin/EDTA. Incubate for 10 min at 37 °C.
6. Pipette detached cells into 1.5 mL Eppendorf tube. Centrifuge cells for 5 min at 500 × *g*.
7. Pipette out trypsin, add 200 μL PBS to the sedimented cells, and centrifuge for 5 min at 500 × *g*.
8. Pipette out PBS and add 20 μL 0.1 M NaOH to lyse the cells and incubate 10 min at room temperature.
9. Add 1.52 μL 12 M HCl and 30 μL wash buffer.

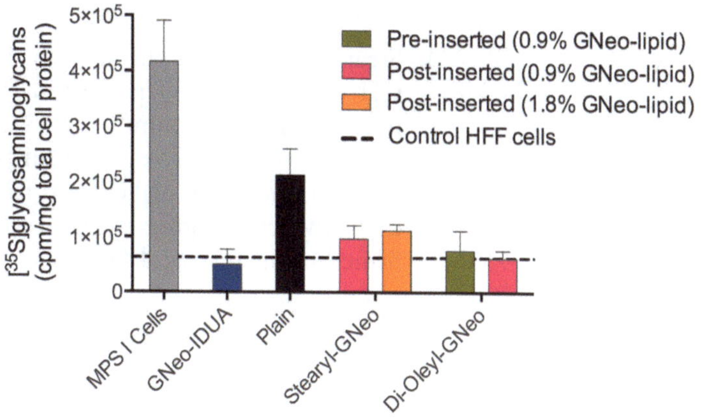

Fig. 5 Liposomal delivery of IDUA. MPS I fibroblasts were radiolabeled with $^{35}SO_4$ and chased for 24 h with 100 μg/mL plain or GNeo-decorated liposomes containing 0.2 μg/mL IDUA. The amount of [^{35}S]glycosaminoglycan remaining was measured. Cells were also treated with 0.2 μg/mL of GNeo-conjugated IDUA (GNeo-IDUA) for comparison. The *dotted line* represents the amount of [^{35}S]glycosaminoglycan remaining in control fibroblasts (HFF) radiolabeled with $^{35}SO_4$ and chased for 24 h without enzyme supplementation. The experiment was performed twice (with different batches of liposomes) in triplicate. Analysis of variance showed that the differences between plain liposomes and GNeosomes were significant ($P < 0.0005$). The difference between untreated MPS I cells and plain liposomes was also significant ($P = 0.0002$)

10. Remove 25 μL and add 75 μL wash buffer and 0.6 μL shark cartilage chondroitin sulfate. Use the other 25 μL to determine the amount of total protein using a BCA Protein Assay Kit.
11. Add 500 μL of a 50% slurry of DEAE beds in equilibration buffer to a Bio-Rad column.
12. Wash beads with 1 mL equilibration buffer.
13. Load 100 μL sample on column and allow it to completely enter the resin.
14. Wash the column with 5 mL DEAE wash buffer.
15. Elute glycosaminoglycans with 1.5 mL elution buffer.
16. Add the 1.5 mL eluent to 5 mL Scintillator Ultima Gold and count by liquid scintillation counting. Normalize the data to the total amount of cell protein (Fig. 5).

4 Notes

1. 2-[2-[2-(Prop-2-ynyloxy)ethoxy]ethoxy]ethanamine can be synthesized as reported by Natarajan et al. [12].
2. Azide-BocGNeo was synthesized as reported by Inoue et al. [13].

3. Cy5 can be synthesized as reported by Kojima et al. [14]
4. Stearyl-GNeo was purified on a C18 column with 5–95% CH_3CN (0.1% TFA) in H_2O (0.1% TFA). Di-oleyl-GNeo was purified on a C4 column with 5–95% CH_3CN (0.1% TFA) in H_2O (0.1% TFA). The oxidation of the double bond is a major side product and makes purification of the di-oleyl-GNeo derivative challenging. Purity of the product should be confirmed using analytical HPLC and ^1H NMR.
5. Liposomes containing the lysosomal enzyme were lightly vortexed instead of sonicated to avoid denaturing the enzyme.
6. To measure zeta potentials, liposomes were eluted with water instead of PBS.
7. The amount of lipids present was compared to a standard curve generated using liposomal suspensions with a known lipid concentration.
8. Protocol was adapted from Fry et al. [14]. Empty 4 mL cartridges were filled with 4 mL Sephadex (50% slurry in H_2O). Columns were washed twice with H_2O and twice with DPBS and centrifuged for 14 min at high speed to remove excess buffer prior to purifying liposomes.
9. Other detergents can be used; however, Tween 20 was used here because it did not interfere with the enzyme activity assay.
10. Protein gel was used to determine enzyme concentration because the detergents gave false positive results when using the Pierce BCA Protein Assay Kit.
11. If you use 200 μL of this standard, it corresponds to the following standard curve: 0, 0.1, 0.2, 0.5, 1, 2.5, and 5 nmol of standard.
12. MPS I cells detach very easily in 12-well plates and should be washed gently, down the sides of the well.

References

1. Pattni BS, Chupin VV, Torchilin VP (2015) New developments in liposomal drug delivery. Chem Rev 115(19):10938–10966. doi:10.1021/acs.chemrev.5b00046
2. Allen TM, Cullis PR (2013) Liposomal drug delivery systems: from concept to clinical applications. Adv Drug Deliv Rev 65(1):36–48
3. Appelqvist H, Waster P, Kagedal K, Ollinger K (2013) The lysosome: from waste bag to potential therapeutic target. J Mol Cell Biol 5(4):214–226. doi:10.1093/jmcb/mjt022
4. Fehrenbacher N, Jaattela M (2005) Lysosomes as targets for cancer therapy. Cancer Res 65(8):2993–2995. doi:10.1158/0008-5472.can-05-0476
5. Grubb JH, Vogler C, Sly WS (2010) New strategies for enzyme replacement therapy for lysosomal storage diseases. Rejuvenation Res 13(2–3):229–236. doi:10.1089/rej.2009.0920
6. Sarrazin S, Wilson B, Sly WS, Tor Y, Esko JD (2010) Guanidinylated neomycin mediates heparan sulfate-dependent transport of active enzymes to lysosomes. Mol Ther 18(7): 1268–1274
7. Maniganda S, Sankar V, Nair JB, Raghu KG, Maiti KK (2014) A lysosome-targeted drug delivery system based on sorbitol backbone towards efficient cancer therapy. Org Biomol Chem 12(34):6564–6569. doi:10.1039/c4ob01153h

8. Xiang HJ, Deng Q, An L, Guo M, Yang SP, Liu JG (2016) Tumor cell specific and lysosome-targeted delivery of nitric oxide for enhanced photodynamic therapy triggered by 808 nm near-infrared light. Chem Commun (Camb) 52(1):148–151. doi:10.1039/c5cc07006f
9. Wexselblatt E, Esko JD, Tor Y (2015) GNeosomes: highly lysosomotropic nanoassemblies for lysosomal delivery. ACS Nano 9(4):3961–3968. doi:10.1021/nn507382n
10. Zhang Z, Zhang X, Xu X, Li Y, Li Y, Zhong D, He Y, Gu Z (2015) Virus-inspired mimics based on dendritic lipopeptides for efficient tumor-specific infection and systemic drug delivery. Adv Funct Mater 25(33):5250–5260. doi:10.1002/adfm.201502049
11. Fry DW, White JC, Goldman ID (1978) Rapid separation of low molecular weight solutes from liposomes without dilution. Anal Biochem 90(2):809–815
12. Natarajan A, Du W, Xiong CY, DeNardo GL, DeNardo SJ, Gervay-Hague J (2007) Construction of di-scFv through a trivalent alkyne-azide 1,3-dipolar cycloaddition. Chem Commun (Camb) (7):695–697. doi:10.1039/b611636a
13. Inoue M, Tong W, Esko JD, Tor Y (2013) Aggregation-mediated macromolecular uptake by a molecular transporter. ACS Chem Biol 8(7):1383–1388. doi:10.1021/cb400172h
14. Kojima R, Takakura H, Ozawa T, Tada Y, Nagano T, Urano Y (2013) Rational design and development of near-infrared-emitting firefly luciferins available in vivo. Angew Chem Int Ed 52(4):1175–1179. doi:10.1002/anie.201205151

Chapter 10

Lysosomal Acidification in Cultured Astrocytes Using Nanoparticles

Camilla Lööv and Anna Erlandsson

Abstract

Following cellular engulfment, nanoparticles end up in the lysosomes, making them an ideal tool for modifying the lysosomal environment. Here, we describe how acidic nanoparticles can be used to lower the pH of lysosomes in cultured, primary astrocytes and thereby increase their degradation capacity. To guarantee that the cell culture is completely devoid of professional phagocytes, we isolate, expand, and differentiate neural stem cells from embryonic mouse cortex to achieve astrocytes for these experiments. Immunostainings with LAMP2-specific antibodies can be performed to verify the lysosomal localization of the nanoparticles, and the effect on lysosomal acidification can easily be followed with LysoTracker dye.

Key words Nanoparticles, Acidification, Astrocyte, Degradation, LysoTracker, LAMP2, pHrodo, Primary cell culture

1 Introduction

Increasing evidence indicates that insufficient lysosomal degradation is highly involved in the pathogenesis of different neurodegenerative diseases [1, 2]. Hence, treatment to enhance or restore the degradation efficiency of lysosomes is of high therapeutic and scientific interest. Acidic nanoparticles (NPs) of poly-DL-lactide-co-glycolide (PLGA) or polylactic acids (PLA) are known to be internalized and transferred to the lysosomes where they can dramatically affect the lysosomal pH [3–5]. Importantly, the NPs prepared from biodegradable polymers, such as PLGA or PLA, are nontoxic [6] and suitable for long-term lysosomal acidification [5]. The protocol described in this chapter is based on our studies of primary astrocytes [5, 7] but may also be used for other nonprofessional phagocytic cells.

Astrocytes, the most prominent glial cell in the brain, respond to all pathological conditions through a process referred to as reactive astrogliosis, in which the astrocytes convert to an inflammatory, phagocytic state [5, 7–9]. Astrocytes effectively engulf

Karin Öllinger and Hanna Appelqvist (eds.), *Lysosomes: Methods and Protocols*, Methods in Molecular Biology, vol. 1594, DOI 10.1007/978-1-4939-6934-0_10, © Springer Science+Business Media LLC 2017

dead cells, synapses, and protein aggregates of Aβ and α-synuclein [7, 10–15]. We have previously shown that compared to professional phagocytes, astrocytes store, rather than degrade, the ingested material. Using pHrodo-labeling technique, we have demonstrated that cell corpses ingested by macrophages fuse with acidic lysosomes within 5 h, while cell corpses ingested by astrocytes had still not been acidified by day 3 (Fig. 1) [7]. The poor degradation in astrocytes can be traced to several inhibitory mechanisms, which together result in the poor acidification of the lysosomes. By lowering the pH in the astrocytic lysosomes, using NPs, we could positively affect the astrocytic degradation [5]. For this kind of studies it is important that the culture is completely devoid of professional phagocytes. We have solved this problem by using a cell culture system based on embryonic, cortical stem cells that exclusively differentiate to cells within the neural lineage and do

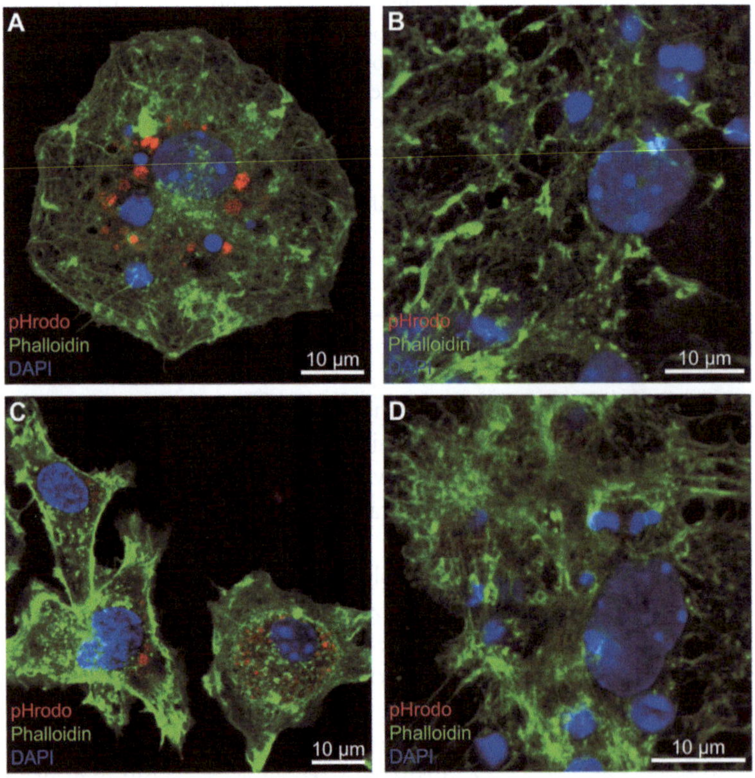

Fig. 1 Astrocyte-engulfed cells are not acidified to the same extent as cells phagocytosed by macrophages. pHrodo was used to label dead cells that were added to macrophages (**a**, **c**) and to astrocytes (**b**, **d**). The cultures were fixed after 5 h (**a**, **b**) or 3 days (**c**, **d**). Macrophages had successfully acidified and started to degrade the phagocytosed dead cells by 5 h and completed the degradation by day 3 (**a**, **c**). Astrocytes did not contain red fluorescing material at any time point, indicating that the lysosomes where neutral or alkalic (**b**, **d**). Figure has been modified from Loov et al. [7]

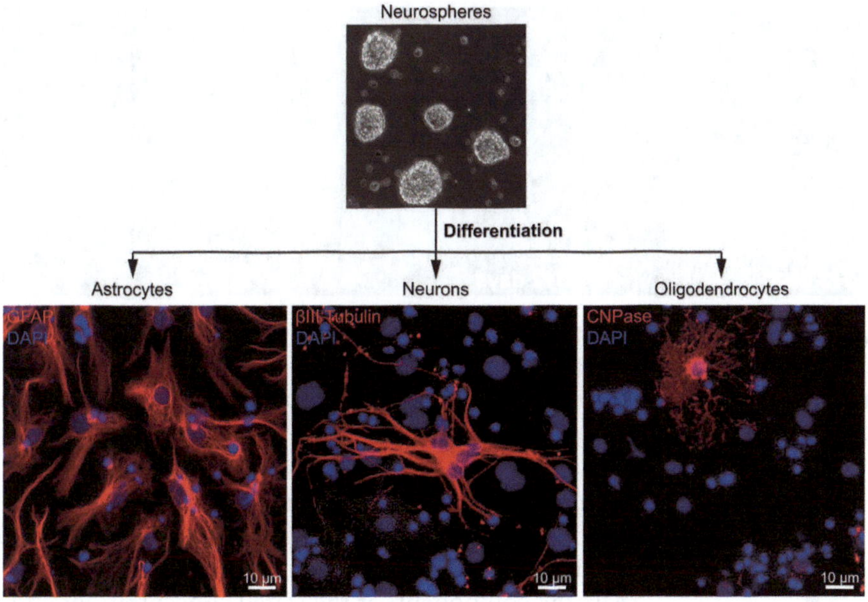

Fig. 2 Cortical stem cells differentiate mainly to astrocytes. Embryonic, cortical stem cells were expanded as neurospheres in the presence of EGF and FGF2. The stem cells exclusively differentiated within the neural lineage and did not form microglia or macrophages. Following one week of differentiation, the cultures consisted predominately of astrocytes (GFAP$^+$), but also included neurons (βIII Tubulin$^+$) and a few oligodendrocytes (CNPase$^+$)

not form microglia or macrophages [17–19]. The differentiated cultures consist predominately of astrocytes (75%), but include also neurons and a few oligodendrocytes (Fig. 2). Addition of ciliary neurotrophic factor (CNTF) during the differentiation period may be added to drive the cortical stem cells towards an astrocytic fate [18, 20]. In the following text, we describe how lysosomal degradation of dead cells can be analyzed and how NPs can be used to promote lysosomal acidification and subsequently degradation. Immunocytochemical stainings for the lysosomal marker LAMP2 can be performed to demonstrate lysosomal localization of NPs (Fig. 3) and LysoTracker can be used to study the effects of different NPs on lysosome acidification (Fig. 4).

2 Materials

Prepare all mediums and buffers in a sterile environment before starting the protocol.

2.1 Medium and Buffers

1. Neurosphere medium: High glucose Dulbecco's Modified Eagle Medium/Ham's F-12 nutrient mixture (DMEM/F12) with GlutaMAX™ (1×) (*see* **Note 1**), 50 units/ml penicillin, 50 μg/ml streptomycin, 8 mM HEPES buffer, 1× B-27

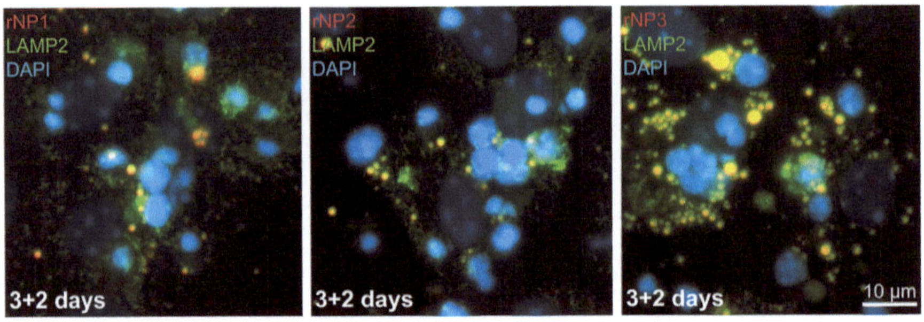

Fig. 3 The added NPs are taken up and translocated to the lysosomes. Nile Red-labeled NPs (rNPs) were added to the cultures, which allow for intracellular tracking of the engulfed particles. LAMP2-immunocytochemistry shows that the rNPs colocalize with the lysosomal marker, demonstrating successful uptake and fusion with the degradative cell compartments

serum-free supplement (ThermoFischer Scientific), 10 ng/ml fibroblast growth factor 2 (FGF2) (*see* **Note 2**), 20 ng/ml natural mouse epidermal growth factor (EGF) (*see* **Note 2**). Mix the components at the stated concentrations and warm to 37 °C prior to use.

2. Culture medium: High glucose Dulbecco's Modified Eagle Medium/Ham's F-12 nutrient mixture (DMEM/F12) with GlutaMAX™ (1×) (*see* **Note 1**), 50 units/ml penicillin, 50 μg/ml streptomycin, 8 mM HEPES buffer, 1× B-27 serum-free supplement, optional: 10 ng/ml CNTF can be added during the differentiation period to drive the cells toward an astrocytic fate (*see* **Note 2**). Mix the components at the stated concentrations and warm to 37 °C prior to use.
3. Dissociation buffer (HBSS): Hank's balanced salt solution (1×), 50 units/ml penicillin, 50 μg/ml streptomycin, 8 mM HEPES buffer.
4. Phosphate buffered saline (PBS) pH 7.2–7.4 (1 L): 8 g sodium chloride (NaCl), 0.2 g potassium chloride (KCl), 1.44 g disodium hydrogen phosphate (Na_2HPO_4), 0.24 g potassium dihydrogen phosphate (KH_2PO_4), 800 ml distilled water. Set pH between 7.2 and 7.4 and fill up to 1000 ml with distilled water.
5. Red blood cell lysis buffer (250 ml): 2.075 g ammonium chloride (NH_4Cl), 0.25 g potassium bicarbonate ($KHCO_3$), 25 ml 0.5 M EDTA solution, 225 ml distilled water.
6. DMEM medium: DMEM, high glucose, 10% fetal bovine serum (FBS), 50 units/ml penicillin, 50 μg/ml streptomycin.

2.2 Animals

Here, we describe lysosomal acidification of cultured astrocytes, using differentiated, embryonic stem cells from mouse cortex. Primary astrocytes can, however, successfully be isolated from other species and brain regions. Most common is to use astrocytes

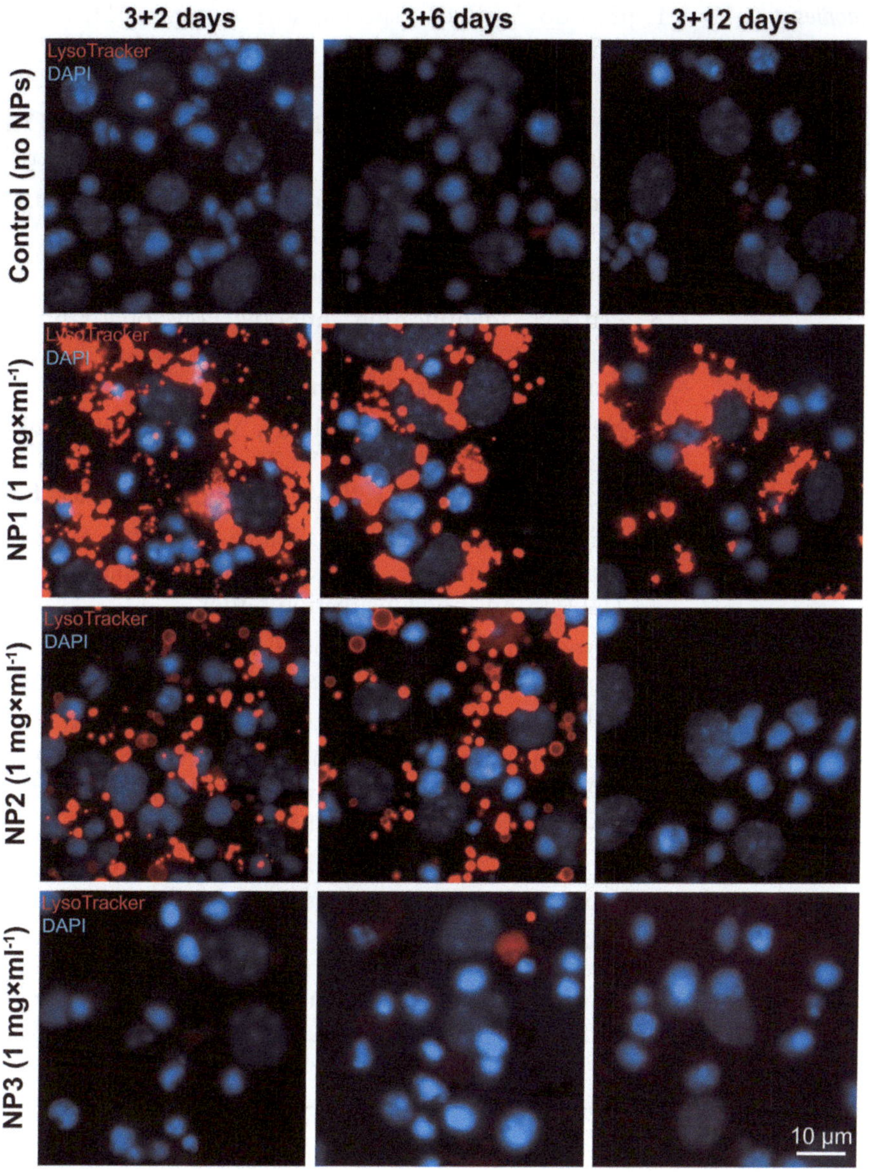

Fig. 4 Only the most acidic of the NPs are able to successfully acidify the astrocytic lysosomes for longer periods of time. Unlabeled dead cells were added to the astrocyte cultures for 3 days. Following washing, three different unlabeled NPs were added for 2 days. To study acidification and subsequent degradation, cultures were either incubated with LysoTracker and fixed (3 + 2 days) or incubated for longer time, without additional doses of NPs (a total of 6 days and 12 days) before LysoTracker labeling and fixation. The most acidic NP, NP1, was able to induce a long-term acidification of the astrocytic lysosomes, but not the less acidic NPs (NP2 and NP3). For more information on the result of degradation of dead cells in the treated cultures, *see* Ref. [5]

isolated from embryonic or postnatal rodent brains (mice or rats). Cultured astrocytes derived from postnatal animal brains will, however, always contain contaminating microglia.

2.3 Antibodies and Fluorophores

1. pHrodo™ Red, succinimidyl ester (pHrodo) 1 mg (ThermoFisher Scientific) dissolved in 150 μl DMSO to an approximate concentration of 10 mM (*see* **Note 3**).
2. 1 mM LysoTracker® Red DND-99 solution in DMSO diluted to a final concentration of 500 nM in culture medium (*see* **Note 4**).
3. Permeabilizing and blocking solution: PBS with 0.1% Triton X-100, 5% normal goat serum (NGS).
4. Antibody incubation buffer: PBS with 0.5% NGS.
5. Astrocyte identification antibody: Polyclonal rabbit anti-glial fibrillary acidic protein (GFAP, Z0334, Dako) diluted 1:400 in Antibody incubation buffer.
6. Neuron identification antibody: Mouse monoclonal neuronal class III β-tubulin clone TUJ1 (βIII Tubulin, MMS-435P, BioLegend) diluted 1:200 in antibody incubation buffer.
7. Oligodendrocyte marker: Monoclonal anti-CNPase antibody produced in mouse (CNPase, C5922, Sigma-Aldrich) diluted 1:400 in Antibody incubation buffer.
8. Lysosomal marker: Anti-LAMP2 antibody [ABL-93] (ab25339, Abcam) diluted 1:200 in Antibody incubation buffer.
9. Filamentous actin stain: Phalloidin-FITC (Sigma-Aldrich) dissolved in methanol according to manufacturer and diluted to a working concentration of 5 μM in PBS.
10. Vectashield HardSet antifade medium with DAPI (Vector Laboratories) for mounting the coverslips on slides.

2.4 Nanoparticles

1. Nanoparticles were prepared by Dr. Uday Kompella at the University of Colorado, but the base materials are commercially available from Sigma-Aldrich Inc. (*see* **Note 5**).

3 Methods

Carry out all procedures at room temperature unless otherwise specified.

3.1 Dissection of Embryonic Cortices from Mouse

Before starting the dissection, fill up a 15 ml centrifuge tube with sterile HBSS and place on ice. Pour HBSS into two 100 mm in diameter Petri dishes and one 60 mm nontissue treated dish and place on ice.

1. Euthanize a time pregnant female C57BL/6 mouse on day 14 (E14) (*see* **Note 6**). Spray abdomen with 70% ethanol and dissect out the uterus with the embryos.

2. Place the uterus in a 100 mm Petri dish with HBSS on ice. Dissect out the embryos from the uterus and meninges and transfer them to a new dish with ice cold HBSS.
3. Decapitate embryos and move the heads to a 60 mm dish with HBSS and place on ice. Make sure the volume is enough to properly cover the tissue so they do not dry out.
4. Dissect out the cortices under a dissecting microscope and collect them in the 15 ml centrifuge tube on ice.

3.2 Primary Neural Cultures

1. Centrifuge the collected cortices at 300 × *g* for 3 min.
2. Remove supernatant and dissolve the pellet in 1 ml sterile HBSS by pipetting up and down 10–20 times using a P1000 autopipette.
3. Add 9 ml HBSS and let any undissolved tissue sediment for 10 min before carefully transferring the nonsedimented cell solution to a new 15 ml centrifuge tube.
4. Centrifuge the cell suspension at 300 × *g* for 5 min.
5. Aspirate supernatant and add 1 ml Neurosphere medium to the pelleted cells.
6. Dissociate the cells carefully, using a P1000 autopipette (*see* **Note 7**).
7. Add 9 ml Neurosphere medium and transfer the cells to a nontissue treated T25 flask and incubate at 37 °C, 5% CO_2 humidified air for 2–3 days.

3.3 Passage of Neurospheres

1. Collect the neurospheres in a 15 ml centrifuge tube and centrifuge for 5 min at 300 × *g*.
2. Aspirate the supernatant and resuspend the pellet in 10 ml prewarmed, sterile HBSS by pipetting up and down 10 times with a 10 ml sterile, serological pipette. In this step the spheres should be separated, but still be intact.
3. Incubate the cells in HBSS at 37 °C on a rocking board for 15 min.
4. Centrifuge the cell suspension as aforementioned and remove the supernatant.
5. Mechanically dissociate the cells in 1 ml Neurosphere medium by pipetting up and down 50–75 times with some force using a P1000 autopipette (*see* **Note 7**).
6. Add 9 ml Neurosphere medium and transfer the cell suspension to a T75 flask with 20 ml medium and note the passage number. Repeat the procedure after 2–3 days, or seed the cells as described in the following section.

3.4 Cell Seeding and Differentiation

Make sure autoclaved 24 × 24 mm coverslips for 6-well plates, 15 mm in diameter coverslips for 12-well plates, or 12 mm in diameter for 24-well plates are available before proceeding with the following steps.

1. Dilute poly-L-ornithine solution (0.01%) 1:4 in sterile, ddH_2O (double distilled H_2O). Place the sterilized coverslips on the bottom of tissue treated plates and add the poly-L-ornithine solution to the wells (*see* **Note 8**) and incubate the plates at 37 °C for 1 h.
2. Prepare laminin (*see* **Note 9**) solution in sterile PBS at a final concentration of 1 μg/ml. Wash the plates twice in sterile ddH_2O and incubate the plates with the laminin solution for 1 h at 37 °C (*see* **Note 8**).
3. Follow the steps for "Passage of neurospheres Primary cell culture Neurospheres (Subheading 3.3) **steps 1–5**.
4. Count the number of live cells using Trypan blue solution and seed the cells in Neurosphere medium at a concentration of 1.5×10^5 cells/ml. Add 2 ml/well to 6-well plates, 1 ml/well to 12-well plates, and 0.5 ml/well to 24-well plates. Make sure to seed back cells as neurospheres for "Labeling and addition of dead cells" (*see* Subheading 3.6).
5. Incubate the cells in the Neurosphere medium over night at 37 °C, 5% CO_2 humidified air.
6. To start the differentiation, exchange the Neurosphere medium with Culture medium, without FGF2 and EGF, and incubate the cells for 7–8 days, while changing the Culture medium in full every 2–3 days.

3.5 Primary Macrophages from Spleen

Primary mouse macrophages can be used as a positive control for effective lysosomal degradation.

1. Coat coverslips in a 6-well plate with poly-L-ornithine solution (stock 0.01%, diluted 1:4 in sterile, ddH_2O) for 1 h at 37 °C as described earlier (*see* **Note 8**), but leave out laminin. Wash the plates twice in PBS just before seeding the cells.
2. Dissect out the spleen from an adult C57BL/6 mouse and place in HBSS on ice.
3. Use a syringe plunger to mash the spleen in 1–3 ml HBSS before collecting the cells in a 15 ml centrifuge tube.
4. Centrifuge the cells at 300 × *g* for 5 min and resuspend the cells in 1 ml sterile Red blood cell lysis buffer.
5. Following incubation at room temperature for 5 min, centrifuge the cells as aforementioned and carefully remove the supernatant.
6. Resuspend the cells in DMEM medium and seed the cells on the coated coverslips at a density of one 6-well plate per spleen.

7. Incubate the plate at 37 °C, 5% CO_2 in humidified air for 2 h, wash twice to remove nonadhered cells and administer new DMEM medium.
8. Culture the cells for 2 days prior to addition of the pHrodo-labeled dead cells (*see* Subheading 3.6).

3.6 Addition of pHrodo-Labeled Dead Cells to Study Acidification of the Engulfed Cells

The following protocol has been modified from Miksa et al. [21].

1. Follow the steps for "Passage of neurospheres" (Subheading 3.3) **steps 1–4**, but dissociate in Culture medium instead of Neurosphere medium.
2. Add approximately 5 ml medium to the cells (total volume 6 ml) and transfer the cell suspension to a nontissue treated 100 mm in diameter Petri dish and swirl the cell suspension to evenly distribute the cells in a thin layer.
3. Place the cells in a UV chamber (e.g., GS Gene linker, BioRad) and remove the lid and irradiate the cells with 240 mJ/cm^2 of UV light.
4. Put the lid back on, twirl the dish in an eight shape to mix the cells and put the dish back into the chamber and remove the lid.
5. Treat with a second burst of 240 mJ/cm^2 UV light before collecting the dying cells in a 50 ml tube and count the number of cells (*see* **Note 10**).
6. Add HBSS to 1×10^6 cells to a total volume of 50 ml and add 1 μl pHrodo of the stock solution (*see* Subheading 2.2).
7. Incubate at room temperature in the dark on a rocker for 45 min.
8. Pellet the cells by centrifugation at room temperature at $300 \times g$ for 5 min, wash once in HBSS, and redissolve in Culture medium.
9. Add the pHrodo-labeled dead cells to differentiated, astrocyte cultures and control macrophage cultures at a concentration of 1.5×10^5 cells/ml.
10. At, for example 5 h and 3 days (*see* **Note 11**), either fix the cells or wash off any free floating cells and continue to grow the cells for optional time. Fix the cells in the dark in 4% paraformaldehyde in PBS for 15 min. Wash the fixed cells twice in PBS and store the cells in PBS in the dark at 4 °C until performing the immunostaining.

3.7 Addition of Acidic Nanoparticles to Increase Degradation in Astrocytes

The low lysosomal acidity of astrocytic can be enhanced by the addition of acidic nanoparticles (NPs) that are taken up into the lysosomes where they decrease the pH [3].

1. Incubate the astrocyte cell culture with the dead cells for 3 days (24 h might be enough for degradation studies of protein aggregates).

2. Prepare prewarmed Culture medium with 1 mg/ml PLGA Resomer® H RG 502 H (NP1), PLGA Resomer® H RG 503 H (NP2), or PLA Resomer® R 203S (NP3).
3. Wash away free-floating cells from the astrocyte cultures and add the Culture medium with the NPs to the cells.
4. Incubate the cells for 2 days with the particles, control cells are left untreated. Prior to fixing the cells, incubate a portion of the cells with 500 nM of LysoTracker in Culture medium for 2 h at 37 °C, 5% CO_2. After 2 h of incubation, fix the cells in 4% paraformaldehyde in PBS for 15 min at room temperature followed by two washes in PBS.
5. Quickly mount the LysoTracker-stained and fixed cells using HardSet Vectashield with DAPI (*see* **Note 12**) to reduce leakage of the dye and place in the dark at 4 °C until analysis.
6. To study degradation over time, incubate the remainder of the cells for additional days in Culture medium, without adding more NPs.
7. Repeat the incubation with LysoTracker (**step 4**) before fixing and mounting cells at the later time points.
8. Wait for the mounting medium to harden in the dark before micrographing the cells.

3.8 Image Analysis and Statistics

1. Micrograph the cells with a fluorescence microscope, using the same magnification, same exposure time, and linear exposure for all the slides.
2. Use the free and open-source software CellProfiler (www.cellprofiler.org) [22] (*see* **Note 13**) and automatic threshold methods to measure dead and live cells (blue channel). The smaller the area of highly condensed DAPI stain, the more degraded are the dead cells.
3. Since the difference between the various NP-treatments causes a dramatic shift in the intensity of the LysoTracker dye, thresholds have to be set manually in order to include the entire area of interest. This setting will then be used to analyze the total area of the dye and integrated fluorescence intensity.
4. Export the data as a comma separated value file (.csv), which can be opened in Excel prior to analysis in a statistical program. The file can also be imported directly into the statistical program if the program can read comma separated value files.
5. Test for normal distribution before choosing the appropriate statistical parameters (i.e., parametric or nonparametric).

4 Notes

1. GlutaMAX™ is a trademark of ThermoFisher Scientific and is an alternative supplement to L-glutamine. The use of GlutaMAX™ compared to L-glutamine has shown to decrease the amount of ammonia buildup, which improves cell health, especially in sensitive cells. Some medium are sold already supplemented with GlutaMAX™, but the additive can also be found as a 100× concentrate.
2. Add the growth factors after any sterile filtering as they may bind to the filters and reduce the final concentration.
3. pHrodo™ Red, succinimidyl ester (pHrodo™ Red, SE, ThermoFisher Scientific): The dye reacts with amine groups on the cell surface and is nonfluorescent at neutral pH, but emits red light at an increasing intensity as the pH is lowered.
4. LysoTracker cannot be used to measure relative pH changes as pHrodo. However, the dye requires protonation and will not be visible at neutral or alkalic pH.
5. More information about the preparation procedures of the PLGA Resomer® H RG 502 H (NP1), PLGA Resomer® H RG 503 H (NP2), and PLA Resomer® R 203S (NP3) can be found in Baltazar et al. [3].
6. The procedures described herein has been conducted using neuroprogenitor cells from mouse embryos at 14 days of gestation (E14) but may be performed with older or younger embryos (e.g., E12-E17).
7. When dissociating the cells using a P1000 autopipette, make sure that the pipette tip does not press directly against the bottom of the tube or the cells will tear by the force applied while pipetting up and down when passed between the pipette tip and tube.
8. Air bubbles may form underneath the coverslips, which can lift the glass and prevent proper coating. A serological pipette can be used to gently tap the coverslips to dissociate the air and submerge the glass. As the tapping may remove the coating, this should be done early and with great care, both not to damage the glass and to remove any of the coating.
9. Laminin should be stored at −80 °C in aliquots to prevent freeze-thaw. To prevent the laminin from forming a gel, the frozen aliquot should be defrosted slowly at 4 °C before use.
10. Most cells will still be alive directly after UV treatment, i.e., by Trypan blue negative, but will start to die through apoptosis over time.
11. Macrophages will have completely degraded any phagocytosed cells by 3 days, which means that longer incubation times will be futile, compared to astrocyte-engulfed cells (*see* Fig. 1).

12. DAPI can be used to distinguish between dead cells and live cells. The nuclei of dead cells appear highly condensed and smaller in size than live cells. DAPI is highly stable in the pH range of 3–8 [23], which means it can be used to measure degradation over time.
13. Fiji (ImageJ), which is also available for free download, can be used similarly to CellProfiler to measure both fluorescence intensities and area.

Acknowledgment

The authors would like to thank Claire H. Mitchell at the University of Pennsylvania and Uday Kompella at the University of Colorado for supplying the nanoparticles and Martin Simonsson at Luleå University of Technology for help with image analysis. This work was supported by the Swedish Research Council, the Alzheimer Foundation, the Parkinson Foundation, Hedlund Foundation, Åhlén Foundation, O.E. and Edla Johansson Foundation, the Uppsala Berzelii Centre, and the Swedish Brain Foundation.

References

1. Appelqvist H, Wäster P, Kågedal K, Öllinger K (2013) The lysosome: from waste bag to potential therapeutic target. J Mol Cell Biol 5:214–226
2. Nixon RA, Yang DS, Lee JH (2008) Neurodegenerative lysosomal disorders: a continuum from development to late age. Autophagy 4:590–599
3. Baltazar GC, Guha S, Lu W, Lim J, Boesze-Battaglia K et al (2012) Acidic nanoparticles are trafficked to lysosomes and restore an acidic lysosomal pH and degradative function to compromised ARPE-19 cells. PLoS One 7:e49635
4. Bourdenx M, Daniel J, Genin E, Soria FN, Blanchard-Desce M et al (2016) Nanoparticles restore lysosomal acidification defects: implications for Parkinson and other lysosomal-related diseases. Autophagy 12:472–483
5. Loov C, Mitchell CH, Simonsson M, Erlandsson A (2015) Slow degradation in phagocytic astrocytes can be enhanced by lysosomal acidification Glia 63:1997–2009. 10.1002/glia.
6. Sahay G, Alakhova DY, Kabanov AV (2010) Endocytosis of nanomedicines. J Control Release 145:182–195
7. Loov C, Hillered L, Ebendal T, Erlandsson A (2012) Engulfing astrocytes protect neurons from contact-induced apoptosis following injury. PLoS One 7:e33090
8. Pekny M, Wilhelmsson U, Pekna M (2014) The dual role of astrocyte activation and reactive gliosis. Neurosci Lett 565:30–38
9. Buffo A, Rolando C, Ceruti S (2010) Astrocytes in the damaged brain: molecular and cellular insights into their reactive response and healing potential. Biochem Pharmacol 79:77–89
10. Jones RS, Minogue AM, Connor TJ, Lynch MA (2013) Amyloid-beta-induced astrocytic phagocytosis is mediumted by CD36, CD47 and RAGE. J Neuroimmune Pharmacol 8:301–311
11. Chung WS, Clarke LE, Wang GX, Stafford BK, Sher A et al (2013) Astrocytes mediumte synapse elimination through MEGF10 and MERTK pathways. Nature 504:394–400
12. Chang GH, Barbaro NM, Pieper RO (2000) Phosphatidylserine-dependent phagocytosis of apoptotic glioma cells by normal human microglia, astrocytes, and glioma cells. Neuro Oncol 2:174–183

13. Magnus T, Chan A, Linker RA, Toyka KV, Gold R (2002) Astrocytes are less efficient in the removal of apoptotic lymphocytes than microglia cells: implications for the role of glial cells in the inflamed central nervous system. J Neuropathol Exp Neurol 61:760–766
14. Sokolowski JD, Nobles SL, Heffron DS, Park D, Ravichandran KS et al (2011) Brain-specific angiogenesis inhibitor-1 expression in astrocytes and neurons: implications for its dual function as an apoptotic engulfment receptor. Brain Behav Immun 25:915–921
15. Fellner L, Irschick R, Schanda K, Reindl M, Klimaschewski L et al (2013) Toll-like receptor 4 is required for alpha-synuclein dependent activation of microglia and astroglia. Glia 61:349–360
16. Sollvander S, Nikitidou E, Brolin R, Soderberg L, Sehlin D, et al (2016) Accumulation of amyloid-beta by astrocytes result in enlarged endosomes and microvesicle-induced apoptosis of neurons. Mol Neurodegener 11:38
17. Davis AA, Temple S (1994) A self-renewing multipotential stem cell in embryonic rat cerebral cortex. Nature 372:263–266
18. Johe KK, Hazel TG, Muller T, Dugich-Djordjevic MM, RD MK (1996) Single factors direct the differentiation of stem cells from the fetal and adult central nervous system. Genes Dev 10:3129–3140
19. Ravin R, Hoeppner DJ, Munno DM, Carmel L, Sullivan J et al (2008) Potency and fate specification in CNS stem cell populations in vitro. Cell Stem Cell 3:670–680
20. Bonni A, Sun Y, Nadal-Vicens M, Bhatt A, Frank DA et al (1997) Regulation of gliogenesis in the central nervous system by the JAK-STAT signaling pathway. Science 278:477–483
21. Miksa M, Komura H, Wu R, Shah KG, Wang P (2009) A novel method to determine the engulfment of apoptotic cells by macrophages using pHrodo succinimidyl ester. J Immunol Methods 342:71–77
22. Szabo AG, Krajcarski DT, Cavatorta P, Masotti L, Barcellona ML (1986) Excited state pKa behaviour of DAPI. A rationalization of the fluorescence enhancement of DAPI in DAPI-nucleic acid complexes. Photochem Photobiol 44:143–150
23. Carpenter AE, Jones TR, Lamprecht MR, Clarke C, Kang IH et al (2006) CellProfiler: image analysis software for identifying and quantifying cell phenotypes. Genome Biol 7:R100

Chapter 11

Analysis of Lysosomal pH by Flow Cytometry Using FITC-Dextran Loaded Cells

Ida Eriksson, Karin Öllinger, and Hanna Appelqvist

Abstract

The acidic environment of the lysosomal lumen provides an optimal milieu for the acid hydrolases and is also essential for fusion/fission of endo-lysosomal compartments and sorting of cargo. Evidence suggests that maintaining lysosomal acidity is essential to avoid disease. In this chapter, we describe a protocol for analyzing the lysosomal pH in cultured cells using the fluorescent probe fluorescein isothiocyanate (FITC)-dextran together with a dual-emission ratiometric technique suitable for flow cytometry. Fluorescence-labeled dextran is endocytosed and accumulated in the lysosomal compartment. FITC shows a pH-dependent variation in fluorescence when analyzed at maximum emission wavelength and no variation when analyzing at the isosbestic point, thereby the ratio can be used to determine the lysosomal pH. A standard curve is obtained by equilibrating intralysosomal pH with extracellular pH using the ionophore nigericin. The protocol also includes information regarding procedures to induce lysosomal alkalinization and lysosomal membrane permeabilization.

Key words Lysosomes, pH, Flow cytometry, Dual emission ratiometry, Lysosomal membrane permeabilization, Bafilomycin A

1 Introduction

1.1 The Central Role of Lysosomal pH

Lysosomes are characterized by acidic intraluminal pH (pH ≤ 5), which is a result of the action of a transmembrane multimeric protein complex, the vacuolar H^+-ATPase (V-ATPase) [1]. The V-ATPase is an ATP-dependent proton pump composed of a peripheral ATPase domain (V_1) and a membrane-integral proton-translocating domain (V_0) [2]. As the terminal organelles in the endocytic pathway, lysosomes digest macromolecules, and the acidic environment aids the degradation process by loosening the molecular structures, which facilitates the access of lysosomal hydrolases [1].

The reversible assembly of the V_1 and V_0 domains regulates the activity of the V-ATPase. Upon amino acid starvation, V-ATPase assembly increases, and this effect is reversed upon readdition of amino acids [2]. The antibiotic bafilomycin A is a potent and

Karin Öllinger and Hanna Appelqvist (eds.), *Lysosomes: Methods and Protocols*, Methods in Molecular Biology, vol. 1594, DOI 10.1007/978-1-4939-6934-0_11, © Springer Science+Business Media LLC 2017

relatively specific inhibitor of V-ATPases [3]. In addition, ammonium chloride (NH_4Cl) can be used to effectively decrease the acidity of lysosomes [4].

In addition to facilitating lysosomal degradation, the low pH is also important for lysosomal fusion with autophagosomes and the release of lysosomal calcium into the cytoplasm [5]. Therefore, defects in lysosomal acidification may cause pathological alterations. In retinal pigmented epithelial cells, disruption of lysosomal acidity by the use of chloroquine triggers a retinopathy characterized by the accumulation of nondegradable lipofuscin-like material in in vivo and in vitro systems [6, 7]. Activation of microglia and subsequent lysosomal acidification potentiate degradation of Aß amyloid fibrils [8], suggesting that an increase in lysosomal pH is a contributing factor in Alzheimer's disease. Moreover, alterations in intralysosomal pH have been detected in cancer cells [9], and the efficacy of the cytostatic drug cisplatin has been correlated to a low lysosomal pH [10]. Likewise, resistance to cisplatin was found to be associated with defects in the lysosomal compartment resulting in reduced drug accumulation [10–12]. The absolute rise in pH is often small under pathological conditions, but elevations of only a few tenths of a pH unit can have a major impact on both lysosomal function and the accumulation of waste, seen over decades. Reacidification of defective lysosomes following poly(DL-lactide-co-glycolide) (PLGA) acidic nanoparticles treatment restores lysosomal function in different pathological contexts [13]. Thus, reacidifying treatments to restore the degradative function is a future promising approach to treat pathologies associated with lysosomal defects.

1.2 Lysosomal Membrane Permeabilization

Due to their high content of hydrolytic enzymes, lysosomes are potentially harmful to cells. Already in 1959, Christiane de Duve introduced the name suicide bags to describe lysosomes, as damage to the lysosomal membrane results in leakage of lysosomal content to the cytosol and subsequent cell death [14]. Limited or partial permeabilization of the membrane induces apoptosis, while massive membrane damage induces necrosis [15]. The process by which lysosomal content is released to the cytosol is called lysosomal membrane permeabilization (LMP). LMP is evident in cell death signaling induced by classical apoptosis inducers such as UV irradiation, p53 activation, oxidative stress, growth factor deprivation, and ligation of death receptors [16–18]. In addition, there are cell death–inducing compounds that directly target lysosomes, such as lysosomotropic detergents, e.g., *O*-methyl-serine dodecylamide hydrochloride (MSDH) and Leu-Leu-OMe [19, 20]. As previously reported, apoptosis-associated LMP is accompanied by lysosomal alkalinization [21, 22]. The mechanism how lysosomal content is released to the cytosol during LMP is, however, still not clear, but several mechanisms have been presented [16, 23].

1.3 pH Measurements Using Fluorescent Probes

Fluorescent pH probes have been proven effective tools to investigate the role of intracellular pH in diverse physiological and pathological processes. The lysosome targeting probes are often weak bases selectively concentrated in acidic organelles as a result of protonation. The probes are covalently attached to large molecules such as dextrans, to ensure retention in the lysosome when pH is changing, [4, 24]. Dextran conjugates are actively taken up by endocytosis, and once loaded, they are retained in viable cells for long periods. Several different variants of pH-sensitive probes conjugated to dextran exist on the market, including Oregon Green, FITC, and LysoSensor probes.

The FITC-molecule can be used as a dual-emission ratiometric pH indicator, entailing that after excitation, the emission is analyzed by two different wavelengths. As presented in Fig. 1, FITC shows a pH-dependent variation in fluorescence when analyzed at the maximum emission wavelength and no variation when analyzed at the isosbestic point. Thus the ratio of the emission intensities detected at these two wavelengths will vary with pH but will not be affected by the amount of probe available. pH-sensitive probes can be analyzed by fluorescence microscopy [22]; however,

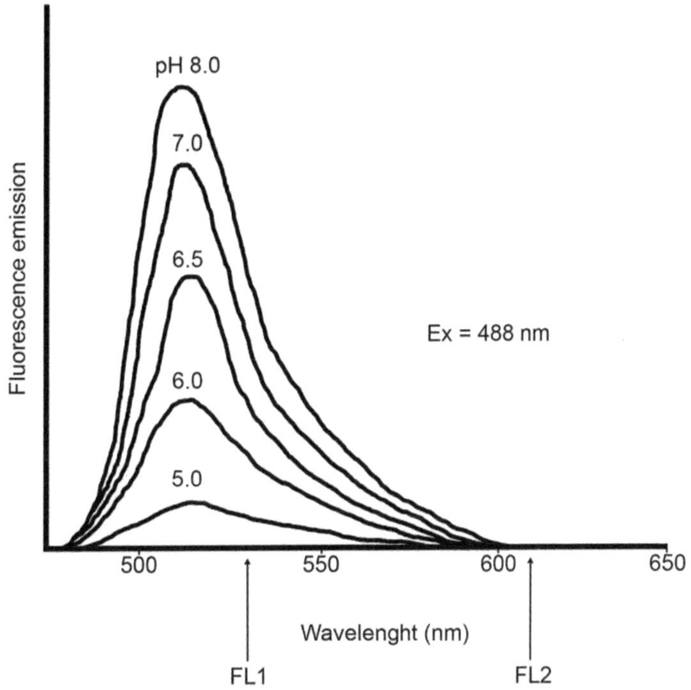

Fig. 1 Fluorescence emission spectra of FITC excited at 488 nm. Note the changes in emission intensities with changes in pH. The emission isosbestic point is near 600 nm. FITC emissions are optimally analyzed at 515 (FL1) and 610 nm (FL2). (Modified from Nilsson et al. (2003) Analysis of cytosolic and lysosomal pH in apoptotic cells by flow cytometry. Meth Cell Sci 25:185–194)

the use of flow cytometry has several advantages, including that large number of cells are rapidly measured and analyzed, subpopulations of cells can be distinguished and analyzed, and photobleaching does not impair the analysis.

In the protocol described in this chapter, lysosomal pH is determined by creating a standard curve obtained from cells loaded with FITC-dextran. Standard curves used for pH measurements are often sigmoidal or exponential [25], which usually is explained by a reduction in the buffer capacity of the wide range of pH used, and the lack of correct osmolarity, which should be 300 mmol/kg. In the method presented in this chapter, we have modified the incubation buffer in several ways to obtain a linear standard curve between pH 4.0 and 6.0. Previous work has shown that ΔpH across the lysosomal membrane is dependent on the potassium ion concentration and that the $[K^+]$ must be above 125 mM to maintain stable ΔpH across the membrane [26]. In order to equilibrate the buffer pH and the lysosomal pH in standard curve samples before measurement, the ionophore nigericin is used. Moreover, to avoid counteraction of the V-ATPase, its activity is inhibited by the addition of 2-deoxyglucose, and ATP production is suppressed by the addition of sodium azide. These measures are necessary to obtain a linear standard curve [27].

2 Materials

1. Normal human foreskin fibroblasts.
2. Cell culture medium: Eagle's minimum essential medium GlutaMAX, 10% fetal calf serum, 50 IU/m penicillin-G, and 50 mg/m streptomycin.
3. FITC-dextran of a molar weight of 40,000 Da.
4. PBS (phosphate buffered saline) without calcium and magnesium.
5. Britton-Robinson buffers with modifications.
   (a) Prepare a solution containing citric acid ($C_6H_8O_7{\cdot}H_2O$), KH_2PO_4, diethyl-barbituric acid, and boric acid (H_2BO_3), all at the concentration 0.02857 M.
   (b) Titrate with 0.2 M KOH to reach pH values of 4.0, 4.5, 5.0, 5.5, and 6.0 and note the amount of KOH added.
   (c) For an efficient equilibration of the buffer across the lysosomal membrane, the concentration of potassium ions should be around 125 mM. Therefore, calculate the potassium ion concentration in the buffers made and add KCl to a final concentration of 125 mM potassium ions (Table 1).
   (d) Measure the osmolarity of the buffers. If necessary, add NaCl to increase the osmolarity to approximately 300 mmol/kg (*see* **Note 1**).

Table 1
Guide to the addition of KCl in the Britton-Robinson buffer to obtain correct potassium concentration

pH	Approximate volume (ml) of 0.2 M KOH added to 100 ml	Total K^+ after addition of KOH[a] (mmol)	Needed amount of K^+ to reach 125 mM (mM)	Mass KCl to add to 100 ml buffer (g)[b]
4.0	16	6.06	73	0.544
4.5	25	7.86	61	0.455
5.0	30	8.86	55	0.410
5.5	40	10.86	47.5	0.354
6.0	45	11.86	43	0.320

[a]The amount of K^+ is obtained by summing up the amount of potassium ions in KOH and in the Britton Robinson buffer (2.857 mmol/100 ml)
[b]Note that these masses are calculated to correspond to 100 ml buffer

6. Sodium azide: 2.5 M.
7. 2-Deoxy-D-glucose: 2.5 M.
8. Nigericin: 1 mM in ethanol.

3 Methods

3.1 Culturing of Cells and Loading with FITC-Dextran (See Note 2)

1. Human fibroblasts cultured in cell culture medium are incubated in humidified air with 5% CO_2 at 37 °C and subcultured once a week (*see* **Note 3**).
2. Trypsinize, count and seed cells at a density of 9000 cells/cm^2 (*see* **Note 4**) in a ∅ 35 mm cell culture dish. We recommend using at least triplicates for samples and a standard curve consisting of five pH-values (e.g., pH 4.05, 4.5, 5.0, 5.5, and 6) (*see* **Note 5**). Also include an unstained sample.
3. Make a solution of 1 mg/ml FITC-dextran in cell culture medium. Filter-sterilize the solution through a hydrophilic polyethersulfone membrane by using a syringe-driven filter unit with 0.22 μM pore size.
4. Prepare cell culture medium containing FITC-dextran at a final concentration of 0.1 mg/ml (*see* **Note 6**).
5. Aspire the cell culture media from the cells and add 1 ml of FITC-dextran containing medium and incubate for 3 days (*see* **Note 7**) at 37 °C with 5% of CO_2 in air.

3.2 Preparations for Measurement of Lysosomal pH (See Note 7)

1. Chase the FITC-dextran to lysosomes by aspirating the media and add fresh cell culture medium, incubate cells for 2 h (*see* **Note 8**). If desired, cells can be examined in a fluorescence microscope to verify FITC-dextran accumulation in a punctate pattern corresponding to lysosomes (Fig. 2). During this period, also expose cells to LMP-inducers or inhibitors, if desired (*see* **Note 9**).

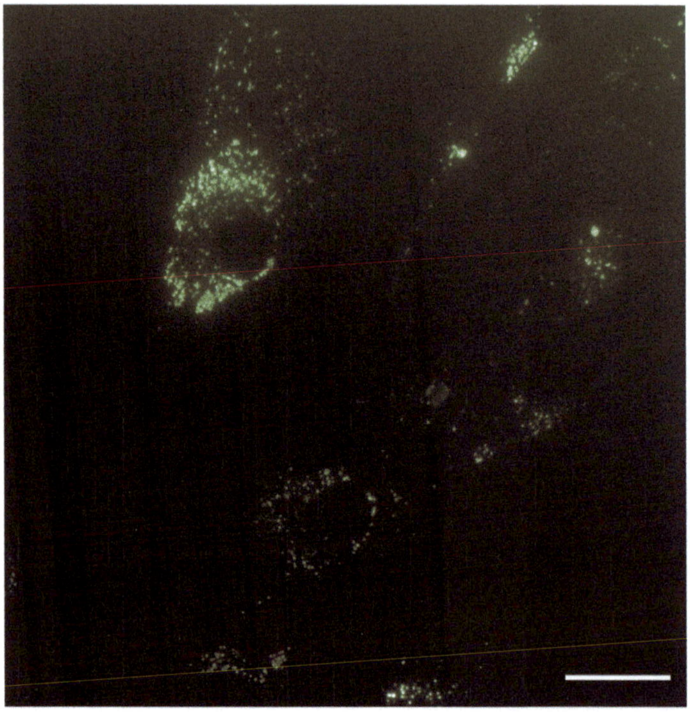

Fig. 2 Cellular uptake of FITC-dextran. Fibroblasts incubated with 0.1 mg/ml FITC-dextran in complete cell culture medium for 72 h, followed by a 4 h chase. Fluorescence image taken with Zeiss fluorescence microscope, λ_{exc} 488 nm. Scalebar 10 μM

2. After the chase period, detach cells by trypsination and transfer to tubes.
3. Centrifuge at 300 × *g* for 5 min and remove the medium by aspiration.
4. Wash the cells in 1 ml of PBS (room temperature). Recentrifuge the tubes at 300 × *g* for 5 min.
5. Pour off the PBS and place the samples on ice.
6. Prepare an appropriate volume Britton-Robinson buffer of each pH (0.5 ml/standard sample). Add sodium azide and 2-deoxyglucose to a final concentration of 50 mM and nigericin to a final concentration of 10 μM (*see* **Note 10**). Keep the buffers on ice.

3.3 Measurement of Lysosomal pH on a Flow Cytometer

1. Analyze lysosomal pH using a flow cytometer equipped with a 488 nm argon laser. For the FL1 channel, use a SP 550 nm filter and a BP 525 ± 40 nm filter, and for FL2, a 595 SP filter and a BP 575 nm filter (*see* **Note 11**).

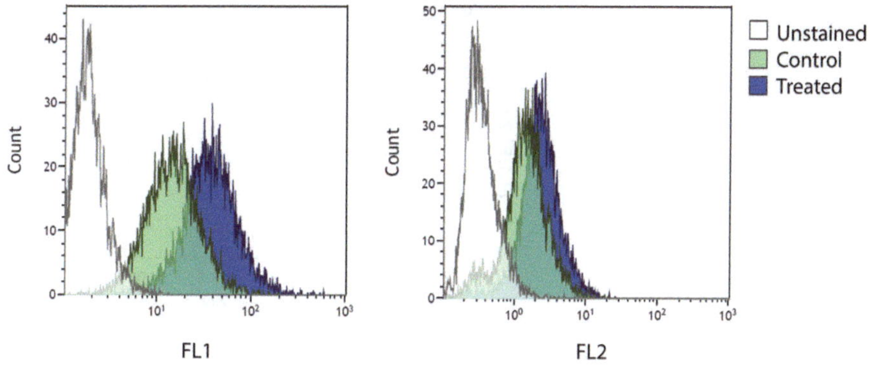

Fig. 3 Examples of histograms from flow cytometric analyses showing FL1 and FL2 emission in fibroblasts from controls and samples in which lysosomal proton gradient is destroyed. Cells were loaded with FITC-dextran for 72 h, followed by a 4 h chase period

2. Analyze an unstained sample and adjust the sensitivity of the SSC- and FSC-channel to ensure a proper collection of cell data (*see* **Note 12**).
3. For samples belonging to the standard curve, add 0.5 ml of the appropriate buffer to the cell pellet and resuspend. Filtrate the cell suspension to remove large particles, which may disturb the flow cytometric measurements, using a filter with 30 μm pores, and allow the sample to equilibrate for 10 min on ice before measurement. Start by measuring the samples containing buffers with the lowest and the highest pH and adjust voltages on the FL1- and FL2-channel to ensure correct fluorescence measurements. Acquire and store data from the pH series using the adjusted settings. Collect data from at least 5000 cells in each sample.
4. For experimental samples, resuspend the cell pellet in 0.5 ml PBS, filtrate and measure immediately using the adjusted settings (Fig. 3).

3.4 Analysis of FITC-Dextran Flow Cytometry Data

1. Create a standard curve by calculating FL1/FL2 ratios, using the median values, obtained from the pH series (*see* Fig. 4).
2. For calculation of lysosomal pH in cells: Calculate FL1/FL2 ratios using the median values and use the linear equation obtained from the standard curve (*see* Fig. 4). As seen in Fig. 5, the lysosomal pH in cells is acidic under control conditions. However, treatment with bafilomycin A or NH_4Cl both increases lysosomal pH significantly (*see* **Note 9**) and can be used as inhibitors of lysosomal function. Lysosomal membrane permeabilization induced by the lysosomotropic detergents MSDH and Leu-Leu-OMe likewise result in lysosomal alkalinization, which previously has been associated with acidification of the cytosol [21, 28].

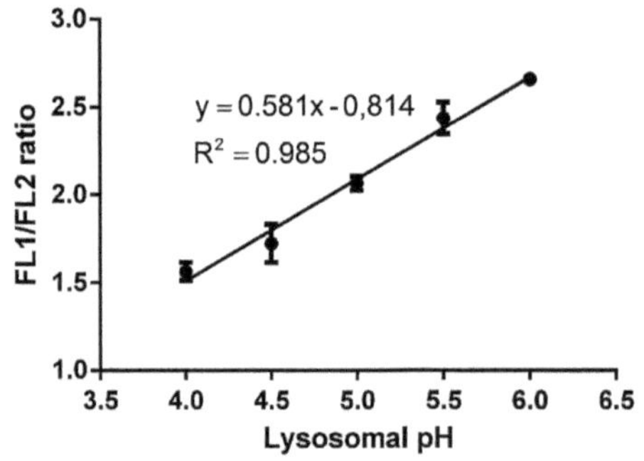

Fig. 4 Example of a standard curve obtained for lysosomal pH using the fluorescent probe FITC-dextran and a dual-emission ratiometric technique, where FITC is excited at 488 nm and emission optimally analyzed at 530 (FL1) and 610 nm (FL2). The FL1/FL2 ratio is plotted on the y-axis against the lysosomal pH on x-axis. The standard curve was obtained by the addition of nigericin to a modified Britton-Robinson buffer. The values shown are mean ± standard deviation, n = 4. The equation of a straight line is applied and R^2 represents the regression coefficient

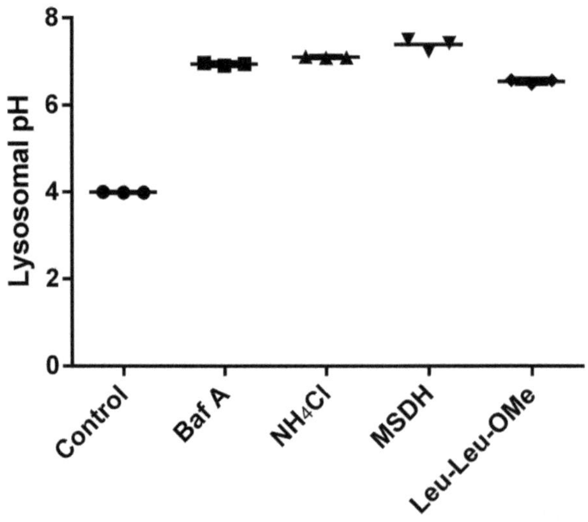

Fig. 5 Results from measurement of lysosomal pH in normal human skin fibroblast. Control cells have a pH of 4.0 while cultures treated with Baf A (bafilomycin A; 20 nM, 20 min); NH_4Cl (ammonium chloride; 10 mM, 30 min), MSDH (15 μM, 2 h) or Leu-Leu-OMe (0.5 mM, 2 h) show a higher pH, n = 3

4 Notes

1. For the buffers with a pH of 4.0, 4.5, and 5.0, the NaCl addition should be approximately 20 mM and for the buffer of pH 5.5 and 6.0 approximately 30 and 40 mM, respectively.

2. Alternative pH-sensitive probes are commercially available. Oregon green has, however, in our hands not given linear standard curves. LysoSensor yellow/blue should be excited at 335 nm, thus a flow cytometer with UV-laser is needed.
3. The protocol is described for human fibroblasts but has successfully been modified to fit a number of different cell lines, both adherent and suspension cells [21, 29]. The cells should be cultured in their regular cell culture medium.
4. The seeding density needs to be optimized to cell type used and is dependent on, e.g., proliferation rate and plating efficiency. The cells should be at approximately 70% confluence at the day of pH measurement.
5. We recommend that the standard curve initially is prepared using at least five standard points (pH 4.0, 4.5, 5.0, 5.5, and 6.0). During repeated measurements, a standard curve consisting of pH 4, 5, and 6 might be sufficient.
6. The concentration of FITC-dextran and incubation time needed for the cells to take up enough FITC-labeled dextran may vary depending on the endocytic capacity of the cells used. Optimization for each cell type should be performed.
7. Cells are fluorescently labeled. Keep protected from light as much as possible.
8. If the lysosomal pH of the control cells is higher than expected (around 5), the chase time can be prolonged in order to get all FITC-dextran to the final and most acidic destination in the endocytic compartment, the lysosomes. Importantly, keep in mind that, e.g. cancerous cells might have defect lysosomal pH (higher or lower) [9, 10].
9. Lysosomal pH can be increased by treating cells with bafilomycin A (20 nM, 30 min) or ammonium chloride (NH_4Cl; 10 mM, 30 min). The effect of NH_4Cl is reversible; therefore, it needs to be present throughout all preparation steps. Lysosomal membrane permeabilization can be induced using the lysosomotropic detergent MSDH (15 μM, 2 h) or Leu-Leu-OMe (0.5 mM, 2 h). The concentrations and incubations times might need optimization depending on cell type and cell density.
10. Sodium azide and nigericin are toxic substances that should be handled according to the regulations of safe handling.
11. In this protocol, no gating of different cell populations is described. If different population of cells are to be measured, gating can be performed using a FSC/SSC plot or cell population differentiated using separate markers not interfering with the FITC fluorescence. In this case, create FL1 and FL2 histograms over gated regions and acquire histogram statistics. Example of gating of apoptotic cells can be found in Nilsson et al. [21].

12. As the standard curve is obtained for the pH range 4–6, the lysosomal pH in treated cells can, in this case, only be estimated to be above pH 6.

References

1. Mindell JA (2012) Lysosomal acidification mechanisms. Annu Rev Physiol 74:69–86. doi:10.1146/annurev-physiol-012110-142317
2. Stransky LA, Forgac M (2015) Amino acid availability modulates vacuolar H+-ATPase assembly. J Biol Chem 290(45):27360–27369. doi:10.1074/jbc.M115.659128
3. Bowman EJ, Siebers A, Altendorf K (1988) Bafilomycins: a class of inhibitors of membrane ATPases from microorganisms, animal cells, and plant cells. Proc Natl Acad Sci U S A 85(21):7972–7976
4. Ohkuma S, Poole B (1978) Fluorescence probe measurement of the intralysosomal pH in living cells and the perturbation of pH by various agents. Proc Natl Acad Sci U S A 75(7):3327–3331
5. Mauvezin C, Nagy P, Juhasz G, Neufeld TP (2015) Autophagosome-lysosome fusion is independent of V-ATPase-mediated acidification. Nat Commun 6:7007. doi:10.1038/ncomms8007
6. Ramsey MS, Fine BS (1972) Chloroquine toxicity in the human eye. Histopathologic observations by electron microscopy. Am J Ophthalmol 73:229–235
7. Sundelin SP, Terman A (2002) Different effects of chloroquine and hydroxychloroquine on lysosomal function in cultured retinal pigment epithelial cells. APMIS 110(6):481–489
8. Majumdar A, Cruz D, Asamoah N, Buxbaum A, Sohar I, Lobel P, Maxfield FR (2007) Activation of microglia acidifies lysosomes and leads to degradation of Alzheimer amyloid fibrils. Mol Biol Cell 18(4):1490–1496. doi:10.1091/mbc.E06-10-0975
9. Kokkonen N, Rivinoja A, Kauppila A, Suokas M, Kellokumpu I, Kellokumpu S (2004) Defective acidification of intracellular organelles results in aberrant secretion of cathepsin D in cancer cells. J Biol Chem 279(38):39982–39988. doi:10.1074/jbc.M406698200
10. Nilsson C, Roberg K, Grafström RC, Öllinger K (2010) Intrinsic differences in cisplatin sensitivity of head and neck cancer cell lines: correlation to lysosomal pH. Head Neck 32(9):1185–1194
11. Chauhan SS, Liang XJ, Su AW, Pai-Panandiker A, Shen DW, Hanover JA, Gottesman MM (2003) Reduced endocytosis and altered lysosome function in cisplatin-resistant cell lines. Br J Cancer 88(8):1327–1334. doi:10.1038/sj.bjc.6600861
12. Safaei R, Larson BJ, Cheng TC, Gibson MA, Otani S, Naerdemann W, Howell SB (2005) Abnormal lysosomal trafficking and enhanced exosomal export of cisplatin in drug-resistant human ovarian carcinoma cells. Mol Cancer Ther 4(10):1595–1604. doi:10.1158/1535-7163.MCT-05-0102
13. Bourdenx M, Daniel J, Genin E, Soria FN, Blanchard-Desce M, Bezard E, Dehay B (2016) Nanoparticles restore lysosomal acidification defects: implications for Parkinson and other lysosomal-related diseases. Autophagy 12(3):472–483. doi:10.1080/15548627.2015.1136769
14. Turk B, Turk V (2009) Lysosomes as "suicide bags" in cell death: myth or reality? J Biol Chem 284(33):21783–21787. doi:10.1074/jbc.R109.023820
15. de Duve C (1959) Lysosomes, a new group of cytoplasmic particles. In: Hayashi T (ed) Subcellular particles. The Ronald Press Co., New York, pp 128–159
16. Li W, Yuan X, Nordgren G, Dalen H, Dubowchik GM, Firestone RA, Brunk UT (2000) Induction of cell death by the lysosomotropic detergent MSDH. FEBS Lett 470(1):35–39
17. Thiele DL, Lipsky PE (1992) Apoptosis is induced in cells with cytolytic potential by L-leucyl-L-leucine methyl ester. J Immunol 148(12):3950–3957
18. Johansson AC, Appelqvist H, Nilsson C, Kågedal K, Roberg K, Öllinger K (2010) Regulation of apoptosis-associated lysosomal membrane permeabilization. Apoptosis 15(5):527–540
19. Guicciardi ME, Deussing J, Miyoshi H, Bronk SF, Svingen PA, Peters C, Kaufmann SH, Gores GJ (2000) Cathepsin B contributes to TNF-alpha-mediated hepatocyte apoptosis by promoting mitochondrial release of cytochrome c. J Clin Invest 106(9):1127–1137
20. Bivik C, Larsson P, Kågedal K, Rosdahl I, Öllinger K (2006) UVA/B-induced apoptosis in human melanocytes involves translocation of cathepsins and Bcl-2 family members. J Invest Dermatol 126(5):1119–1127

21. Nilsson C, Johansson U, Johansson AC, Kågedal K, Öllinger K (2006) Cytosolic acidification and lysosomal alkalinization during TNF-alpha induced apoptosis in U937 cells. Apoptosis 11(7):1149–1159
22. Stroikin Y, Mild H, Johansson U, Roberg K, Ollinger K (2008) Lysosome-targeted stress reveals increased stability of lipofuscin-containing lysosomes. Age (Dordr) 30(1):31–42. doi:10.1007/s11357-007-9045-9
23. Guicciardi ME, Leist M, Gores GJ (2004) Lysosomes in cell death. Oncogene 23(16): 2881–2890
24. Geisow MJ (1984) Fluorescein conjugates as indicators of subcellular pH. A critical evaluation. Exp Cell Res 150(1):29–35
25. Vergne I, Constant P, Lanéelle G (1998) Phagosomal pH determination by dual fluorescence flow cytometry. Anal Biolchem 255(1):127–132
26. Reijngoud D, Tager JM (1975) Effect of ionophores and temperature on intralysosomal pH. FEBS Lett 54(1):76–79
27. Myers BM, Tietz PS, Tarara JE, LaRusso NF (1995) Dynamic measurements of the acute and chronic effects of lysosomotropic agents on hepatocyte lysosomal pH using flow cytometry. Hepatology 22(5):1519–1526
28. Appelqvist H, Johansson AC, Linderoth E, Johansson U, Antonsson B, Steinfeld R, Kågedal K, Öllinger K (2012) Lysosome-mediated apoptosis is associated with cathepsin D-specific processing of Bid at Phe24, Trp48, and Phe183. Ann Clin Lab Sci 42(3):231–242
29. Eriksson I, Joosten M, Roberg K, Ollinger K (2013) The histone deacetylase inhibitor trichostatin A reduces lysosomal pH and enhances cisplatin-induced apoptosis. Exp Cell Res 319(1):12–20. doi:10.1016/j.yexcr.2012.10.004

Chapter 12

Detection of Lysosomal Exocytosis in Platelets by Flow Cytometry

Anna L. Södergren and Sofia Ramström

Abstract

Flow cytometry is a method that allows high throughput analysis of individual cells in suspension. By inclusion of fluorescently labelled antibodies, it is possible to analyze the abundance of one or more surface antigens, such as LAMP-1, without prior lysis of cells. Here we describe the special considerations required for the investigation of lysosomal exocytosis from platelets analyzed with flow cytometry.

Key words Flow cytometry, Granule, Lysosomal-associated membrane protein-1 (LAMP-1), Lysosomal exocytosis, Lysosome, Platelet, Platelet activation, Platelet secretion

1 Introduction

Platelets are small discoid cell fragments (2–5 μm in diameter) formed from megakaryocytes in the bone marrow. They circulate in the blood and function to detect damages to the vessel wall and to prevent bleeding from these, thus maintaining hemostasis. When a platelet comes across a vessel damage, it will adhere to the damaged surface and become activated. Activated platelets will secrete the content of their granules, aggregate into a platelet plug and facilitate coagulation, which will produce fibrin fibers and a blood clot (a thrombus).

Platelets have three types of granules that may be released upon platelet activation: dense granules, α-granules, and lysosomes [1, 2]. The dense granules are released first upon platelet activation followed by the release of α-granules and lysosomes [2, 3]. The α-granules (50–80/platelet) contain different proteins involved in hemostasis, inflammation, angiogenesis, etc. The dense granules (3–8/platelet) contain small molecules such as ADP, ATP, calcium, and serotonin, which may activate other platelets and recruit them to the growing thrombus or reinforce the platelet activation state through autoactivation [1, 2].

Karin Öllinger and Hanna Appelqvist (eds.), *Lysosomes: Methods and Protocols*, Methods in Molecular Biology, vol. 1594,
DOI 10.1007/978-1-4939-6934-0_12, © Springer Science+Business Media LLC 2017

Platelets also contain 0–3 lysosomes/platelet [1, 2] that contain degrading proteins. Lysosomes are released from platelets in situations resulting in platelet activation both in vivo [4, 5], and in vitro [6, 7]. Lysosomal exocytosis does, however, not occur in all platelets, and extensive exocytosis requires strong platelet activation levels [2, 6]. The function of lysosomal exocytosis in platelets is currently unknown, but it has been speculated that it may play a role in clearance of the thrombi, remodeling of the extracellular matrix, cleavage of cell surface receptors, and facilitating adhesion of cells [2, 5].

Upon lysosomal exocytosis the lysosomal membrane fuses with the plasma membrane, leading to release of the lysosomal contents [8]. The lysosome will then expose the luminal surface of its membrane to the extracellular side of the plasma membrane and proteins therein can thus be detected on the platelet surface without lysis of the cells. Dense granules are lysosome-related organelles [1], and therefore several proteins commonly thought of as lysosomal can also be found on the membrane of dense granules, e.g., lysosomal-associated membrane protein (LAMP)-2 [9] and CD63 (LIMP-1) [10], whereas LAMP-1 seems to be specific for lysosomes.

LAMP-1 has been suggested as a potential marker for platelet activation, which might be interesting for highly activated platelets, where other markers, such as CD62P, (P-selectin), might not be as useful [6, 11]. In stored platelet concentrates used for transfusion, LAMP-1 exposure in response to platelet activation might also be a possible marker of platelet reactivity and hence quality of the concentrate [11].

Flow cytometry can be used for cells in suspension. In a high throughput manner the cells are passed through one or more laser beams and individually evaluated for a number of characteristics. Size and granularity are evaluated with the forward and side scatter parameters (FSC and SSC, respectively). Which parameter is the best to use for size determinations varies with the model of the used flow cytometer [12]. It is also possible to evaluate a fluorescent signal from the cells. This is most commonly done by evaluating the binding of fluorescently labeled antibodies, directed toward the antigen of interest. The light from the laser excites the fluorophore(s) on the bound antibodies, which then emit light. The emitted light is passed through bandpass filters to separate light of different wavelengths and converted into an electrical signal by a photomultiplier tube (PMT).

The output signal can be reported either as fluorescence intensity or as percent of cells binding the antibody [13, 14]. A higher fluorescence intensity indicates more binding of the fluorescently labeled antibody. It is, however, important to remember that flow cytometry only gives relative units of fluorescence intensity.

Further the exact values obtained are dependent on the settings on the individual flow cytometer, thus the fluorescence intensity values cannot be compared between instruments (*see* **Note 1**). Mean fluorescence intensity results are often reported but should preferably be avoided as this parameter is sensitive to outliers. Median or geometric mean fluorescence intensity gives more reliable results if the fluorescence is high or the antibody has bound most cells. However, if the antibody binds less than half of the cells, it is important to recognize that the median fluorescence intensity parameter will report the fluorescence intensity of a negative cell.

In cases where the antigen exposure is relatively low, as is the case with lysosomal markers on activated platelets, results can be better presented as percent of cells expressing the antigen. Such results are also easier to compare between instruments. In this case, results are obtained by comparison of samples to a carefully prepared negative control. If the fluorescence intensity of a cell is above the level in the negative control, the antibody has bound, indicating that the antigen is present.

Using flow cytometry, it is possible to determine the number of platelets in which lysosomes have been exocytosed and the relative extent of this. Flow cytometry does, however, not provide information on the amount of lysosomal content that has been released. The release of lysosomal contents may instead be quantified by measuring of the release of lysosomal enzymes such as β-hexosaminidase [15].

2 Materials

1. HEPES-buffer: 137 mM NaCl, 2.7 mM KCl, 1 mM $MgCl_2$, 5.6 mM glucose, 1 g/l bovine serum albumin (BSA), 20 mM 4-(2-hydroxyethyl)piperazine-1-ethanesulfonic acid, *N*-(2-hydroxyethyl)piperazine-*N'*-(2-ethanesulfonic acid) (HEPES), set the pH to 7.4.
2. Agonist: solution of human α-thrombin 6 U/ml in HEPES-buffer.
3. GPRP (a peptide with sequence glycyl-L-prolyl-L-arginyl-L-proline) inhibits fibrin polymerization [16].
4. Anti-CD41a-R-phycoerythrin (PE) antibody (clone: HIP8 facilitates identification of platelets in the flow cytometer).
5. Anti-LAMP-1 (CD107a; clone: H4A3) conjugated to fluorescein isothiocyanate (FITC) or PE-Cyanine 5 (Cy5), (*see* **Notes 2** and **3**).
6. Isotype control antibody (IgG_1 κ; clone: MOPC-21) conjugated to the corresponding fluorophore FITC or PE-Cy5 (*see* **Note 4**).

3 Methods

The procedure described in the following text contains information on how to investigate lysosomal exocytosis on platelets in citrated whole blood upon platelet activation with thrombin. Information on how the protocol can be varied to fit other research questions can be found in the notes section. Unless otherwise indicated, all procedures are carried out at room temperature.

3.1 Obtaining a Platelet Suspension

1. Draw blood through standard venipuncture procedures into evacuated blood collection tubes containing sodium citrate (3.2% or 3.8%). Take care not to activate the platelets during blood collection (*see* **Notes 5–8**).
2. Let the blood rest for 30 min to 2 h before activating your platelets. To avoid spontaneous platelet activation the platelets should be used as soon as possible after the initial 30 min of rest.
3. Flow cytometry allows for detection of platelets in whole blood [17]. Analysis in whole blood is beneficial as it avoids artefactual platelet activation from handling. If, however, whole blood is not appropriate for the specific research question, prepare platelet-rich plasma (PRP) or a suspension of isolated platelet, as required (*see* **Notes 9** and **10**).
4. If required for the specific research question, pretreat your platelets as appropriate (e.g., with a drug or inhibitor) before the experiment.

3.2 Preparation of Tubes for Platelet Activation and Labeling

The following procedure describes the preparation of tubes for simultaneous platelet activation and labeling with antibodies (*see* **Note 11**). Each experiment should include at least an isotype control sample tube (negative control), a control sample tube, and the number of sample tubes required. We highly recommend to always include a sample with resting platelets to get the baseline for the results although this might not always be needed for the specific research question. Unless otherwise indicated, all concentrations given are final concentrations and based on a total sample volume of 36 μl.

1. Prepare a mastermix of the following constituents and add 27 μl/sample tube:
   (a) GPRP (2 mM), to avoid fibrin polymerization in the sample tube (*see* **Note 12**).
   (b) Anti-LAMP-1, conjugated to FITC (4.2 μg/ml) or PE-Cy5 (0.5 μg/ml) (*see* **Notes 2** and **3**).

(c) Anti-CD41a-PE antibody (25 ng/ml), simplifies identification of platelets in the flow cytometer (*see* **Note 13**).

(d) HEPES-buffer, to fill up the volume to 27 μl/tube.

If the specific research question requires additional compounds to be added, reduce the volume of HEPES-buffer added (*see* **Note 14**).

2. Add 6 μl of thrombin solution or HEPES buffer (for resting control samples) to each sample tube (making the final concentration of thrombin 1 U/ml). Often, it is desired to run several samples with different agonists, thus these are added separately. If this is not the case, they may be added as part of the mastermix.
3. Prepare an isotype control solution with the following constituents and add 33 μl to a separate isotype control sample tube.

(a) GPRP (2 mM).

(b) Anti-CD41a-PE antibody (25 ng/ml).

(c) Isotype control antibody conjugated to FITC (2.5 μg/ml) or PE-Cy5 (0.5 μg/ml) (*see* **Notes 4** and **15**).

(d) HEPES-buffer, to fill up the volume to 33 μl/tube.

In the next step, 3 μl of blood will be added to each tube making the total sample volume 36 μl/tube (*see* **Note 16**).

3.3 Platelet Activation and Labeling

1. Make sure the blood is well mixed before the start of the experiment. If needed, gently mix the blood by inverting the blood tube approximately ten times.
2. Add 3 μl of blood to each tube, and mix gently by pipetting up and down 2–3 times, gently shaking the tube, or tapping the tube with your finger. Vigorous shaking should be avoided as this might activate the platelets. Add the blood in intervals of, e.g., 15 s (*see* **Note 17**).
3. Incubate the platelets for exactly 10 min (*see* **Note 18**). Incubated in the dark to avoid bleaching of the fluorophores.
4. "Terminate" the platelet activation and labeling by adding 690 μl of HEPES-buffer to each tube (in 15 s intervals, corresponds to a 1:20 dilution of the sample) (*see* **Note 19**). Mix gently with a table vortex, at a low speed (e.g., 1000 rpm).
5. Analyze the samples on the flow cytometer as soon as possible, within 1 h (*see* **Note 20**).

3.4 Setting Up an Acquisition Protocol on the Flow Cytometer

Setting up an acquisition protocol on the flow cytometer requires some time and should therefore be done prior to the analysis of samples.

Detection of platelets with flow cytometry is different from detection of other cells [13]. To allow detection of the small platelets the flow cytometer scatter parameters should be set to logarithmic scale and the threshold used for acquisition should be set as low as possible. To simplify detection of platelets, the threshold is set on the FL2-detector (corresponding the platelet marker CD41a-PE) (*see* **Note 13**) (Fig. 1). The optimal voltage and gain settings vary for each flow cytometer and have to be determined for each new acquisition protocol. In general, when setting up a protocol, one should aim at getting the largest separation possible between the negative and positive peak for the fluorophores included. Both the negative and the positive peaks should fit in the plot. If the negative peak lies below the minimum fluorescence shown, the voltage should be turned up, and if the positive peak disappears to the right in the plot the voltage should be turned down.

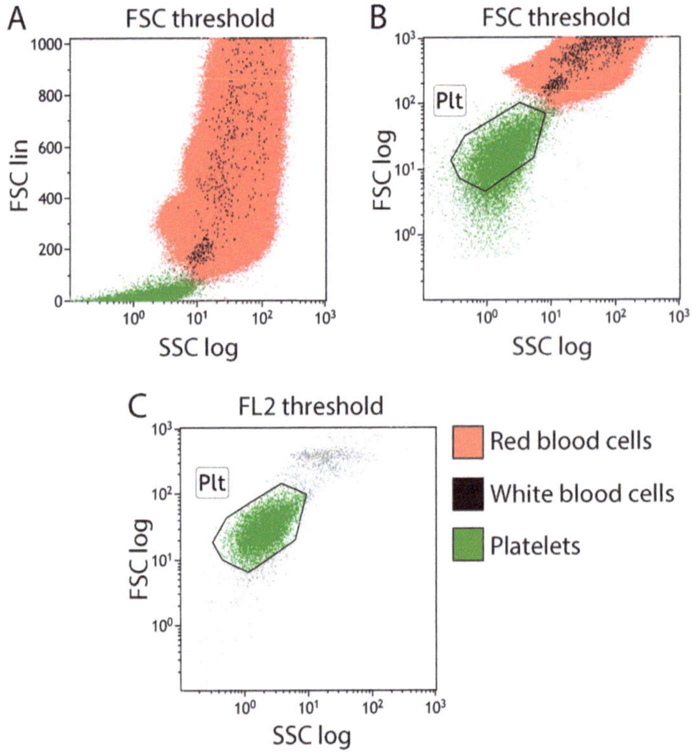

Fig. 1 The appearance of cells in whole blood in a flow cytometer using different acquisition settings. (**a**) Forward scatter (FSC) in linear scale and side scatter (SSC) in logarithmic scale, acquisition performed using an FSC threshold. (**b**) FSC and SSC in logarithmic scale, acquisition with FSC threshold. (**c**) FSC and SSC in logarithmic scale. Acquisition performed using an FL2 threshold corresponding to the platelet marker CD41a-PE. Dots outside of the platelet (Plt)-gate (in **c**) are coincidence events of platelets and red/white blood cells

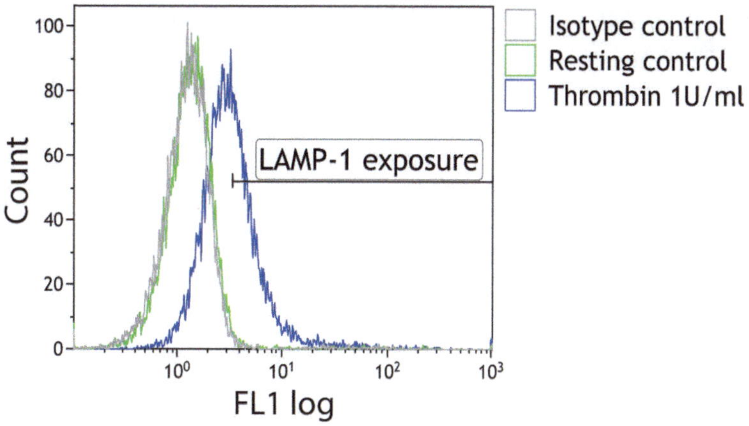

Fig. 2 Histogram of anti-LAMP-1-FITC and FITC-isotype control antibody binding (seen in the FL1 channel) using flow cytometry. Note the position of isotype control peak (negative control) and the peak for the resting control sample where lysosomal exocytosis is low. The gate "LAMP-1 exposure" includes 1–2% of cells in the isotype control sample and delimits the boundary between positive and negative cells. The right shift of the peak seen in the thrombin stimulated sample indicates increased antibody binding due to lysosomal exocytosis

For LAMP-1 exposure, however, the shift is gradual; therefore the separation between negative and positive peaks is not obvious. In this case, it is wise to aim at locating the negative peak in the first decade in a histogram plot, *see* Fig. 2. When multiple fluorophores are used in the same sample, spectral overlap also has to be taken into consideration. This is handled by selecting fluorophores with as little spectral overlap as possible. The remaining spectral overlap is handled by creating compensation matrices [18]. Often the flow cytometers have protocols for setting the compensation, and these can be used to get a rough estimate of the compensation needed. However, as the numbers in compensation matrices seen in platelet research may differ from those normally seen for other cells, the readymade protocols should be checked for accuracy. In many newer flow cytometers the compensation is digital and can thus be adjusted after acquisition if needed. If this is not the case the accuracy of the compensation settings should be checked from time to time.

3.5 Platelet Acquisition

1. To simplify detection of platelets the threshold is set on the FL2-detector (corresponding the platelet marker CD41a-PE) (*see* **Note 13**) (Fig. 1).
2. Exclude coincidence events by placing a gate delimiting the platelet population, *see* Fig. 1 (*see* **Note 21**). The position of the platelet population may vary somewhat between individuals. Adjust the location of the platelet gate if needed.
3. Continue acquisition until 5000–10,000 platelets from each sample have been analyzed.

3.6 Analyzing the Results

1. Make a histogram plot of the appropriate fluorescence parameter.
2. Place a gate including 1–2% of the cells in the isotype control sample (negative control; the 1–2% with the highest fluorescence intensity), *see* Fig. 2.
3. Apply these settings to the results of the samples you wish to analyze.
4. Determine the fraction of cells having bound the antibody in each sample (i.e., percent positive cells)
5. If the percent positive cells is high (>80%), use median or geometric mean fluorescence intensity instead. This value is determined from all platelets (both positive and negative platelets).

3.7 Examples of Results

Lysosomal exocytosis occurs in platelets upon activation, in vivo (upon infliction of a wound) and in vitro upon agonist stimulation. The extent of lysosomal exocytosis varies with the level of activation but does not occur in all platelets. However, in resting platelets, lysosomal exocytosis is absent or very low (Fig. 3).

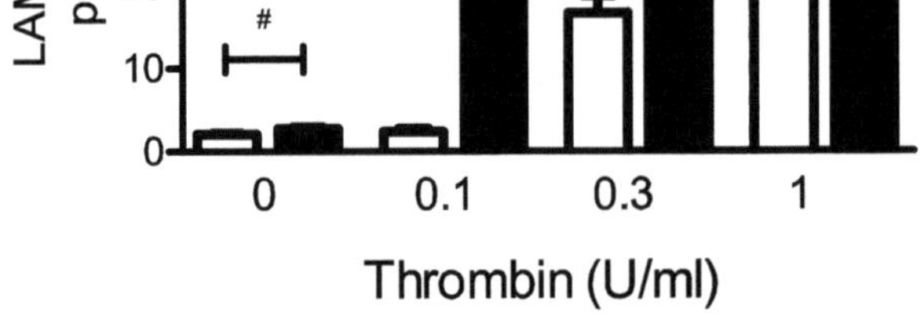

Fig. 3 Lysosomal exocytosis in platelets upon stimulation with thrombin for 1 or 10 min. Lysosomal exocytosis determined with flow cytometry using the protocol described. Note that the anti-LAMP-1 antibody is available conjugated to several different fluorophores, here it is conjugated to FITC (*see* **Note 3**). Figure from [6] Thrombin-induced lysosomal exocytosis in human platelets is dependent on secondary activation by ADP and regulated by endothelial-derived substances by Sodergren et al. (2016) Platelets 27 (1):86-92, reprinted by permission of the publisher (Taylor & Francis Ltd., http://www.tandfonline.com)

4 Notes

1. It is possible to utilize the Molecules of equivalent soluble fluorochrome (MESF) procedure to get an estimation on the absolute number of antigens present [13, 19].
2. The optimal concentration of antibodies may vary from lot to lot of the antibody and between manufacturers. Thus the optimal concentration to use should be titrated before starting the experiment. The concentration should be chosen so that it is the lowest antibody concentration that is able to fully saturate the antigen in maximally activated platelet samples. For platelet experiments, the required antibody concentration is often lower than that stated by the manufacturer.
3. The LAMP-1 antibody is available conjugated to several different fluorophores. When choosing a fluorophore, choose fluorophores with the least spectral overlap possible. Further, try to reserve bright fluorophores for less abundant antigens [20].
4. The concentration of the isotype control should be determined based on the concentration of the anti-LAMP-1 antibody, to which it is supposed to be a negative control. Consideration should also be taken to the F/P (fluorescence/protein) ratio of the two antibodies, as this may vary between different antibodies and lots. The F/P ratio is mainly an issue for smaller fluorophores, such as FITC. For larger fluorophores the F/P ratio is almost always 1. If the antibody of interest has F/P ratio = 2 and the isotype control has F/P ratio = 4, the final concentration of the antibody in the sample tubes should be twice that of the isotype control antibody in the isotype control sample tube. Check that the concentration is appropriate by comparing an isotype control sample to a resting sample where the LAMP-1 antigen is not exposed. These should essentially overlap, *see* Fig. 2.
5. When drawing blood through venipuncture for platelet research, care should be taken not to activate the platelets during the procedure [21]. Use minimal stasis and a large enough needle (21 gauge or larger) when drawing blood. Discard the first portion of blood to avoid contamination from vessel wall material and platelet activation. Mix the blood with the anticoagulant immediately after filling each tube, by gently making 5–10 full inversions of the blood tube.
6. Platelets are commonly obtained from venous blood, but if available, platelets from arterial blood can also be used (if both venous and arterial blood will be used and compared in the study, it is advisable to validate that the results are comparable in the specific research setting). Capillary blood should not be used as platelets in this are activated by the collection procedure.

7. If using other agonists than thrombin, blood can also be collected in hirudin (or if necessary heparin) [13]. If the platelets are to be isolated the blood is preferably collected in tubes containing acid citrate dextrose (ACD). If this is not available, heparin or hirudin should be used, as citrate is not optimal for isolation of platelets.
8. If platelets from other species than human are used, follow the common procedures for blood collection from these. It is also needed to ensure appropriate antibody specificity.
9. Preparation of platelet rich plasma (PRP)
   (a) To prepare PRP, centrifuge the citrate tube at 150 × *g* for 15 min. The centrifuge should be set at a slow deceleration.
   (b) Gently transfer the top layer containing plasma and platelets to another tube. Leave approximately 1 cm above the red cell layer.
10. Preparation of isolated platelets. The procedures differ somewhat between labs, following is the protocol used in our lab.
    (a) Draw blood in blood collection tubes (9 ml) with ACD as anticoagulant. If unavailable, use heparin tubes and mix blood with 1:6 ACD. ACD will reduce platelet activation from centrifugation and handling.
    ACD buffer: sodium citrate (85 mM), citric acid (71 mM), glucose (20 g/l), set the pH to 4.5.
    (b) Prepare PRP as outlined earlier.
    (c) Transfer PRP (approximately 5 ml/tube) to tubes with a cone-shaped bottom. If needed, apyrase (1 U/ml, grade 3 or higher) can be added to the PRP to further prevent platelet activation during centrifugation.
    (d) Centrifuge the PRP once more at 480 × *g* for 20 min. The centrifuge should be set at a slow deceleration. The platelets will form a pellet in the bottom of the tube.
    (e) Remove most of the plasma, leaving approximately 1 ml above the pellet. Wash the pellet by gently adding Krebs-Ringer glucose (KRG)-buffer (up to 2 ml above the pellet). Be careful not to disturb the platelet pellet. Remove the plasma/buffer, again leaving approximately 1 ml buffer above the pellet. Repeat twice.
    (f) After the third wash remove all buffer.
    (g) Resuspend the platelet pellet in 1 ml KRG-buffer supplemented with prostacyclin (PGI_2; 1 U/ml), by very gently pipetting up and down. Do not resuspend the red blood cells in the pellet.
    (h) Count your platelets and dilute the platelet suspension to the desired concentration, often 200–300 × 10^6/ml.

(i) Recalcify the platelet suspension by adding 1.8 mM calcium (final concentration) and incubate for 30 min.

11. Simultaneous platelet activation and labelling are appropriate when platelet activation is to be carried out for 10 min. If platelet activation is to be carried out for much shorter periods of time, activation should be stopped by addition of, e.g., hirudin (*see* **Note 19**). Labelling with antibodies should then be carried out in a separate step. If platelet activation is to be extended to much longer times, antibodies should be added during the last 10 min of stimulation or be added in a separate subsequent step.

12. If platelet activation is carried out in situations where clotting is prevented, GPRP can be omitted. This is the case if other agonists than thrombin is used or if washed platelets are used, as most of the fibrinogen has then been removed.

13. As platelets are smaller than other blood cells, it is possible to distinguish them also by their forward and side scatter properties. This makes it possible to omit the platelet marker CD41a if needed and use a forward or side scatter threshold for detection. Platelet detection via CD41a using a fluorescence threshold is however more reliable and reduces the interference from noise compared to a forward or side scatter threshold. If desired, it is possible to use the CD41a-antibody conjugated to other fluorophores. Alternatively, other platelet markers such as CD61 can be used.

14. In certain situations, it may be of interest to add other compounds to the mastermix than those listed, e.g., an inhibitor of some sort. In some experiments, it may also be beneficial to supplement the mastermix (and dilution buffer) with calcium. This is not necessary for lysosomal exocytosis but may still influence the results somewhat [13]. In our lab, we use a final concentration of 1.5 mM calcium, when regarded as beneficial.

15. If additional antibodies are included to evaluate several parameters simultaneously, isotype control antibodies for these may be added to the same isotype control sample tube.

16. To avoid platelet aggregation in the sample, the blood should be quite diluted when added to the sample tubes. Here the dilution is 1:12. A too high concentration of platelets or stirring of the sample during activation may result in platelet aggregation in the samples, leading to "loss" of platelets and potentially clogging of the flow cytometer sample probe.

17. As whole blood is rather viscous, it is wise to carefully wipe the outside of the pipette tip removing additional blood before adding it to the sample tube.

18. Incubation of platelets at other temperatures than RT is easily accomplished by incubation of the sample tubes in a water bath set at the desired temperature.

19. If required, hirudin can be added to the buffer to ensure that further platelet activation by thrombin is prevented once the reaction is "terminated" through dilution. The concentration of hirudin is given in antithrombin units and should at least match the concentration of thrombin, or preferably be slightly higher.
20. If the time for sample analysis has to be extended, a light fixative, such as 0.5% formaldehyde can be added to the dilution-buffer when "terminating the reaction". In such cases the stability of the samples need to be tested and the affinity of any antibodies verified. Fluorescence intensity of the antibody may also change. Therefore the median (or mean) fluorescence intensity results from unfixed and fixed samples should not be directly compared.
21. When running whole blood on the flow cytometer, there will inevitably be a few coincidences of platelets and red or white blood cells passing the laser simultaneously [22]. These events should not be analyzed, as it is impossible to determine if other characteristics detected are from the platelet or the other cell. If platelets and other cells frequently pass the laser simultaneously, this can often be decreased by lowering the acquisition speed on the flow cytometer or by increasing the sample dilution when "terminating" the platelet activation and labeling.

References

1. Flaumenhaft R (2013) Platelet secretion. In: Michelson AD (ed) Platelets, 3rd edn. Academic Press, Amsterdam, p 343
2. Rendu F, Brohard-Bohn B (2001) The platelet release reaction: granules' constituents, secretion and functions. Platelets 12(5):261–273. doi:10.1080/09537100120068170
3. Jonnalagadda D, Izu LT, Whiteheart SW (2012) Platelet secretion is kinetically heterogeneous in an agonist-responsive manner. Blood 120(26):5209–5216. doi:10.1182/blood-2012-07-445080
4. Ciferri S, Emiliani C, Guglielmini G, Orlacchio A, Nenci GG, Gresele P (2000) Platelets release their lysosomal content in vivo in humans upon activation. Thromb Haemost 83(1):157–164
5. Emiliani C, Ciferri S, Mencarelli S, Mezzasoma AM, Momi S, Orlacchio A, Gresele P (2006) Defective platelet beta-N-acetyl hexosaminidase content and release in chronic myeloproliferative disorders. Platelets 17(1):20–29. doi:10.1080/09537100500235958
6. Sodergren AL, Svensson Holm AC, Ramstrom S, Lindstrom EG, Grenegard M, Ollinger K (2016) Thrombin-induced lysosomal exocytosis in human platelets is dependent on secondary activation by ADP and regulated by endothelial-derived substances. Platelets 27(1):86–92. doi:10.3109/09537104.2015.1042446
7. Febbraio M, Silverstein RL (1990) Identification and characterization of LAMP-1 as an activation-dependent platelet surface glycoprotein. J Biol Chem 265(30): 18531–18537
8. Reddy A, Caler EV, Andrews NW (2001) Plasma membrane repair is mediated by Ca(2+)-regulated exocytosis of lysosomes. Cell 106(2):157–169
9. Israels SJ, McMillan EM, Robertson C, Singhory S, McNicol A (1996) The lysosomal granule membrane protein, LAMP-2, is also present in platelet dense granule membranes. Thromb Haemost 75(4):623–629
10. Nishibori M, Cham B, McNicol A, Shalev A, Jain N, Gerrard JM (1993) The protein CD63 is in platelet dense granules, is deficient in a patient with Hermansky-Pudlak syndrome, and appears identical to granulophysin. J Clin

Invest 91(4):1775–1782. doi:10.1172/JCI116388

11. Sodergren AL, Tynngard N, Berlin G, Ramstrom S (2016) Responsiveness of platelets during storage studied with flow cytometry - formation of platelet subpopulations and LAMP-1 as new markers for the platelet storage lesion. Vox Sang 110(2):116–125. doi:10.1111/vox.12324
12. Poncelet P, Robert S, Bouriche T, Bez J, Lacroix R, Dignat-George F (2016) Standardized counting of circulating platelet microparticles using currently available flow cytometers and scatter-based triggering: Forward or side scatter? Cytometry A 89(2):148–158. doi:10.1002/cyto.a.22685
13. Ramstrom S, Sodergren AL, Tynngard N, Lindahl TL (2016) Platelet function determined by flow cytometry: new perspectives? Semin Thromb Hemost 42(3):268–281. doi:10.1055/s-0035-1570082
14. Goodall AH, Appleby J (2004) Flow-cytometric analysis of platelet-membrane glycoprotein expression and platelet activation. Methods Mol Biol 272:225–253. doi:10.1385/1-59259-782-3:225
15. Holmsen H, Dangelmaier CA (1989) Measurement of secretion of lysosomal acid glycosidases. Methods Enzymol 169:336–342
16. Michelson AD (1994) Platelet activation by thrombin can be directly measured in whole blood through the use of the peptide GPRP and flow cytometry: methods and clinical applications. Blood Coagul Fibrinolysis 5(1):121–131
17. Shattil SJ, Cunningham M, Hoxie JA (1987) Detection of activated platelets in whole blood using activation-dependent monoclonal antibodies and flow cytometry. Blood 70(1):307–315
18. Maecker HT, Trotter J (2006) Flow cytometry controls, instrument setup, and the determination of positivity. Cytometry A 69(9):1037–1042. doi:10.1002/cyto.a.20333
19. Schwartz A, Gaigalas AK, Wang L, Marti GE, Vogt RF, Fernandez-Repollet E (2004) Formalization of the MESF unit of fluorescence intensity. Cytometry B Clin Cytom 57(1):1–6. doi:10.1002/cyto.b.10066
20. Maecker HT, Frey T, Nomura LE, Trotter J (2004) Selecting fluorochrome conjugates for maximum sensitivity. Cytometry A 62(2):169–173. doi:10.1002/cyto.a.20092
21. Harrison P, Mackie I, Mumford A, Briggs C, Liesner R, Winter M, Machin S, British Committee for Standards in H (2011) Guidelines for the laboratory investigation of heritable disorders of platelet function. Br J Haematol 155(1):30–44. doi:10.1111/j.1365-2141.2011.08793.x
22. Harrison P, Ault KA, Chapman S, Charie L, Davis B, Fujimoto K, Houwen B, Kunicka J, Lacombe F, Machin S, Raynor R, van Hove L, van Assendelft OW, International Society of Laboratory Hematology Task Force for the Reference Platelet C (2001) An interlaboratory study of a candidate reference method for platelet counting. Am J Clin Pathol 115(3):448–459. doi:10.1309/91PR-E4G6-XBAF-N8DY

Chapter 13

Detection of Lysosomal Exocytosis by Surface Exposure of Lamp1 Luminal Epitopes

Norma W. Andrews

Abstract

Elevation in the cytosolic Ca^{2+} concentration triggers exocytosis of lysosomes in many cell types. This chapter describes a method to detect lysosomal exocytosis in mammalian cells, which takes advantage of the presence of an abundant glycoprotein, Lamp1, on the membrane of lysosomes. Lamp1 is a transmembrane protein with a large, heavily glycosylated region that faces the lumen of lysosomes. When lysosomes fuse with the plasma membrane, epitopes present on the luminal domain of Lamp1 are exposed on the cell surface. The Lamp1 luminal epitopes can then be detected on the surface of live, unfixed cells using highly specific monoclonal antibodies and fluorescence microscopy. The main advantage of this method is its sensitivity, and the fact that it provides spatial information on lysosomal exocytosis at the single cell level.

Key words Lysosome, Exocytosis, Secretion, Calcium, Lamp1, Lysosomal glycoprotein, Monoclonal antibodies, Plasma membrane, Wounding, Repair

1 Introduction

"Lysosome" is the Greek word for "digestive body," reflecting how lysosomes have been viewed since they were first described by Christian de Duve in 1955 [1]. Each eukaryotic cell has hundreds of these membrane-bound organelles, which have an acidic luminal pH and contain about 50 different hydrolases [2]. In addition to their role as digestive compartments where molecules taken up by endocytosis or autophagy are degraded, lysosomes also behave as regulated exocytic vesicles, fusing with the plasma membrane and releasing their contents extracellularly when the cytosolic Ca^{2+} concentration rises above 1 μM [3, 4]. Prior to this discovery, the ability to respond to Ca^{2+} influx with exocytosis was thought to be an exclusive property of the "secretory" lysosomes of specialized cell types such as cytotoxic T cells, neutrophils, and mast cells [5]. It is now clear that intracellular Ca^{2+} transients trigger exocytosis of conventional lysosomes in many cell types [3, 4]. This exocytic response is required for the mechanism by which eukaryotic cells

Karin Öllinger and Hanna Appelqvist (eds.), *Lysosomes: Methods and Protocols*, Methods in Molecular Biology, vol. 1594,
DOI 10.1007/978-1-4939-6934-0_13, © Springer Science+Business Media LLC 2017

repair wounds on their plasma membrane [6, 7] and appears to involve mainly a peripheral population of lysosomes [8]. Although most lysosomes move along microtubules toward the cell center, recent studies are beginning to uncover the molecular machinery responsible for retaining part of the lysosomal population at the cell periphery (9–12), where they are well positioned for exocytosis.

Ca^{2+}-regulated exocytosis of lysosomes can be detected using several methods including: (a) measurement of lysosomal enzyme activity released into the supernatant; (b) extracellular detection of fluorescent fluid phase tracers previously loaded into lysosomes; (c) extracellular detection of the lysosomally processed, mature form of cathepsin D; (d) surface exposure of luminal epitopes of the lysosomal glycoprotein Lamp1 [3, 9]. In this chapter, we describe in detail the latter method, which takes advantage of the abundance of Lamp1 on lysosomal membranes and the availability of highly specific monoclonal antibodies against luminal epitopes of these glycoproteins. The major advantages of this approach are its sensitivity and the fact that it allows lysosomal exocytosis to be detected locally on the plasma membrane, at the single cell level.

2 Materials

1. Mammalian cells: this method was developed for adherent cells, but it is possible to also analyze cells that grow in suspension (*see* **Note 1**).
2. Sterile culture medium: Dulbecco's Modified Eagle's Medium (DMEM) or other, depending on cell type supplemented with sterile fetal bovine serum (10 or 20%).
3. PBS^{++}: phosphate buffered saline with Ca^{2+} and Mg^{2+}.
4. Sterile PBS^{++}/dextrose: PBS^{++} containing 4.5% dextrose.
5. Agent to trigger intracellular Ca^{2+} transients*:* ionomycin, receptor agonists, pore-forming toxins, microneedle, cell scraper, etc. [3, 6].
6. PBS^{++} with 10% bovine serum albumin (BSA).
7. Anti-Lamp1 monoclonal antibodies (mAb)*:* must be specific for the cells used, such as 1D4B anti-mouse Lamp1 or H4A3 anti-human Lamp1 (species-specific anti-Lamp1 mAbs are available from the Developmental Studies Hybridoma Bank).
8. Secondary antibodies coupled to a fluorophore (such as Alexa 488): must be specific for the animal species where the anti Lamp1 mAb was generated (generally mouse, but can also be rat such as in case of the 1D4B anti-mouse Lamp1).
9. PBS^{++} with 4% paraformaldehyde (PFA): to fix cells after treatment with anti-Lamp1.

3 Methods

3.1 Cell Culturing

1. Plate adherent mammalian cells on plastic dishes containing coverslips in the bottom, in the appropriate culture medium for the cell used (*see* **Notes 1** and **2**).
2. Place dishes with cells (carefully, so glass coverslips don't slide on top of each other) in a 37 °C 5% CO_2 humidified incubator for 24 h. After this period, observe the cells using an inverted phase-contrast microscope to make sure they are not too confluent.

3.2 Induction of Lysosomal Exocytosis and Staining of Lamp1

1. Prepare the following solutions (1 ml of each will be required for each set of 3 coverslips):
   (a) PBS^{++} 10% BSA containing the appropriate anti-Lamp1 mAb. If using hybridoma supernatants obtained from the Developmental Studies Hybridoma Bank dilutions of 1:50 or 1:100 are usually appropriate (*see* **Note 3** on how to determine the correct mAb concentration).
   (b) PBS^{++} 10% BSA containing the appropriate fluorescent secondary antibodies (*see* **Note 3** on how to determine the correct dilution). Keep both solutions and the remainder PBS^{++} 10% BSA on ice.
2. Fill a shallow container (such as a Styrofoam box) with crushed ice, and place a metal plate (solid metal) on top of the ice bed.
3. Leave it inside a cold room for at least 20 min before starting the experiment.
4. Place ~100 ml of PBS^{++}/dextrose on ice and a similar volume of the same buffer in a 37 °C water bath, for at least one 1 h before starting the experiment.
5. Use the 37 °C PBS^{++}/dextrose to wash the cell-containing dishes six times (if using 35 mm dishes, use ~3 ml for each wash). Add the PBS^{++}/dextrose gently, on the sides of the dish, to avoid lifting the cells.
6. After the last wash add 37 °C PBS^{++} to the dishes (2 ml for 35 mm dishes) and perform the procedure to trigger intracellular Ca^{2+} elevation and lysosome exocytosis for 30 s to 5 min maximum (*see* **Note 4**).
7. Following this treatment, immediately aspirate the 37 °C PBS^{++}/dextrose and replace it by ice-cold PBS^{++}/dextrose.
8. Transfer the dishes to the metal plate on top of the ice bed. This rapid cooling will inhibit endocytosis, keeping the luminal epitopes of Lamp1 on the cell surface.
9. Aspirate all the liquid in the dishes, and add on top of each 2 cm^2 coverslip 50 μl of the ice-cold anti-Lamp1 mAb solu-

tion, or just PBS^{++} 10% BSA (important no mAb control). If using larger coverslips, increase the volume accordingly. Work quickly and treat coverslips one by one, so the cells don't dry.

10. Cover the dishes with the lid and leave on the cold metal plate for 30 min. This live treatment of the cells with the mAb on ice will ensure that only surface-exposed Lamp1 is recognized (*see* **Note 5**).
11. Wash the coverslips with ice-cold PBS^{++}/dextrose five times and add the fixative (PBS^{++} 4% PFA). This can be done on single 2 cm^2 coverslips with just 50 µl, but if possible, it is better to add a larger volume covering all coverslips (such as 2 ml for a 35 mm dish). Remove the dishes/coverslips from the cold metal plate and leave for 1 h at room temperature.
12. Wash the fixed coverslips six times with PBS$^{+/+}$ and add on top of each coverslip 50 µl of the secondary fluorescent antibody solution. If using larger coverslips, increase the volume accordingly. Cover the dishes with the lid and leave at room temperature for 30 min–1 h (*see* **Note 6**).
13. Prepare glass microscopy slides, one for each condition of the experiment.
14. Wash the coverslips six times with PBS^{++}, aspirate all liquid in the last wash.
15. Pick up the coverslip with sharp-pointed tweezers, touch it lightly to a paper towel to remove excess liquid, and invert it on top of a drop (~10 µl) of the anti-fade mounting medium placed on the glass microscopy slide. Work quickly and treat coverslips one by one, so the cells don't dry. Make sure you place the cells face down on the drop of mounting medium (cells should be located between the glass of the coverslip and the glass of the microscopy slide).
16. Let the mounting medium harden overnight in the dark (to avoid bleaching the fluorophore conjugated to the secondary antibodies).

3.3 Microscopy

1. Observe the cells in a epifluorescence or laser confocal fluorescence microscope equipped with a high numerical aperture (NA = 1.2) 63× or 100× oil objective, filters appropriate for visualizing the secondary antibody fluorophore and a digital camera. Start with the controls that were not treated with the anti-Lamp1 mAbs but were incubated with secondary antibodies alone. This should give you an idea of whether there is unspecific fluorescence (this should be minimum if washes are done correctly, and if a good quality secondary antibody is used).
2. Examine the coverslips treated with both antibody solutions. The Lamp1 fluorescence should appear punctate over the

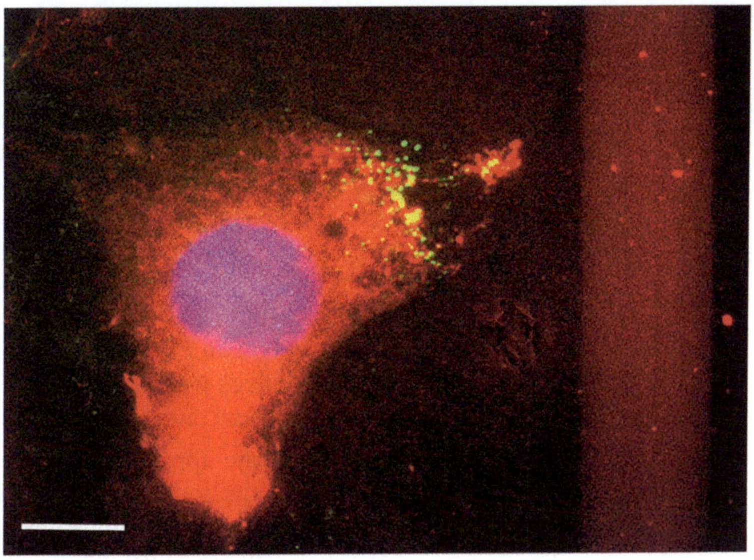

Fig. 1 A culture of NRK rat fibroblasts was scratched with a razor blade in the presence of Texas red-dextran (*red*) and subjected to the method described in this chapter to detect luminal epitopes of the Lamp1 lysosomal glycoprotein (*green*). The punctate Lamp1 fluorescence was detected close to the injury site, where Ca^{2+} influx occurred and triggered lysosomal exocytosis. *Blue*, nuclear DNA staining with DAPI. *Bar* = 10 μm

whole cell surface [3, 6] or localized on a specific area of the plasma membrane, if the Ca^{2+} influx was localized (*see* Fig. 1). The fluorescence pattern should be very different from what is observed with the same antibodies in the same cells after saponin permeabilization (*see* **Note 3**).

4 Notes

1. The best results are obtained by using 35 mm diameter dishes with three round glass coverslips (2 cm^2 each) on the bottom. This allows more uniform spread of the cells when compared to a 24 well dish, and more room to remove each coverslip using sharp tweezers. For the average cultured cell line the ideal number of cells to plate on a 35 mm dish is 2.5×10^5, in a culture medium volume of 2 ml, to reach ~70% confluency 24 h later. Prepare extra dishes containing higher and lower numbers of cells, to be able to select dishes containing subconfluent cell monolayers.
2. The methods section describes a procedure designed for adherent mammalian cells. The method can, however, be adapted for the use of cells in suspension. The main changes are to place the cells in small, capped test tubes and then perform the washes

and antibody incubations using a tube rotator (kept in a cold room for the cold steps) and centrifugation (500–800 × *g* for 5–10 min, in a refrigerated centrifuge). At the end of the procedure the cells can be attached to polylysine-coated coverslips (for coating, coverslips are covered with 1 mg/ml >300 K poly-l-lysine in water for 30 min at room temperature or overnight at 4 °C, followed by several washes with water) and processed for fluorescence microscopy as described in the Methods section. Alternatively, the fluorescence associated with the cells can be quantified by flow cytometry [10], and the percentage of surface-exposed Lamp1 can be compared to the total Lamp1 in cells permeabilized with 0.05% saponin before incubation with the primary and secondary antibodies (*see* **Note 3**).

3. To determine the appropriate concentrations of the primary and secondary antibodies to detect Lamp1 luminal epitopes on the cell surface, it is important to first perform immunofluorescence assays using permeabilized cells, to determine if all lysosomes in the cell type being used can be visualized. To do so, use cells plated on coverslips for 24 h as described earlier, washed 5× with PBS^{++} and fixed with PBS^{++} 4% PFA for 30 min–1 h at room temperature. There is no need to incubate with the primary anti-Lamp1 mAbs on ice in this case, since the goal is to detect all lysosomes. Important: Triton X-100 extracts Lamp1 from lysosomal membranes, so this detergent should not be used. Instead, permeabilization of the fixed cells should be performed for 15 min at room temperature with 0.05% saponin diluted in PBS^{++}. Another important point is to be aware that saponin permeabilization is reversible, so the detergent has also to be present in all washing solutions to be used after cell permeabilization (PBS^{++} 0.05% saponin) and also in the antibody-containing solutions (PBS^{++} 10% BSA 0.05% saponin, plus antibodies). Based on the manufacturer guidelines for dilution of the anti-Lamp1 mAb (which may be in the form of a hybridoma culture supernatant) and of the fluorescent secondary antibodies, test 3 or 4 different dilutions of each. It is again important to have a secondary antibody alone control, because fixed cells tend to show more unspecific background (the background can be reduced by incubating the coverslips in 15 mM NH_4Cl for 15 min after fixation with PBS^{++} 4% PFA, before permeabilization and treatment with the antibodies). After mounting the cells and observing in a fluorescence microscope, determine the highest antibody concentration that detects 100% of the cell-associated lysosomes (i.e., the concentration that shows no increase in fluorescence over the immediately lower dilution). This should be the condition to use in subsequent experiments to detect surface Lamp1.

4. In most cell types tested, exocytosis of lysosomes occurs within 15–20 s after the cytosolic Ca^{2+} concentration is elevated above 1 μM [6]. Thus, to detect maximum amounts of Lamp1 luminal epitopes on the cell surface the cells should be transferred to ice and incubated with anti-Lamp1 mAbs as soon as possible after stimulation of exocytosis, to avoid Lamp1 internalization by endocytosis.
5. It is widely assumed that cell fixation with PFA without detergent permeabilization does not allow antibodies to enter cells. However, fixation alone can create plasma membrane openings that allow some antibodies to penetrate. Avoiding this is particularly important for Lamp1, since this glycoprotein is extremely abundant and even a small level of antibody binding intracellularly can give the erroneous impression that luminal epitopes were translocated to the plasma membrane. The only safe way to avoid this problem is to perform the incubation with the anti-Lamp1 antibodies on live cells before fixation, and keep the cells at ice-cold temperatures to shut down endocytosis until the cells are washed and fixed. After this primary antibody is washed out the cells can be fixed, facilitating incubation with the secondary antibodies and subsequent steps.
6. In addition to determining the best antibody concentration as discussed in **Note 2**, it may also be useful to test different periods of incubation of the cells with the antibodies. Although 30 min is usually sufficient to achieve maximum binding, this sometimes can be improved by increasing the incubation time.

References

1. de Duve C (2005) The lysosome turns fifty. Nat Cell Biol 7(9):847–849. doi:10.1038/ncb0905-847
2. Saftig P, Klumperman J (2009) Lysosome biogenesis and lysosomal membrane proteins: trafficking meets function. Nat Rev Mol Cell Biol 10(9):623–635. doi:10.1038/nrm2745
3. Rodriguez A, Webster P, Ortego J, Andrews NW (1997) Lysosomes behave as Ca2 + −regulated exocytic vesicles in fibroblasts and epithelial cells. J Cell Biol 137:93–104
4. Andrews NW (2000) Regulated secretion of conventional lysosomes. Trends Cell Biol 10(8):316–321
5. Griffiths GM (1996) Secretory lysosomes - a special mechanism of regulated secretion in haemopoietic cells. Trends Cell Biol 6:329–332
6. Reddy A, Caler E, Andrews N (2001) Plasma membrane repair is mediated by Ca2+−regulated exocytosis of lysosomes. Cell 106:157–169
7. Andrews NW, Almeida PE, Corrotte M (2014) Damage control: cellular mechanisms of plasma membrane repair. Trends Cell Biol 24(12):734–742. doi:10.1016/j.tcb.2014.07.008
8. Jaiswal JK, Andrews NW, Simon SM (2002) Membrane proximal lysosomes are the major vesicles responsible for calcium-dependent exocytosis in nonsecretory cells. J Cell Biol 159(4):625–635. doi:10.1083/jcb.200208154
9. Castro-Gomes T, Corrotte M, Tam C, Andrews NW (2016) Plasma membrane repair is regulated extracellularly by proteases released from lysosomes. PLoS One 11(3):e0152583. doi:10.1371/journal.pone.0152583
10. Miller H, Castro-Gomes T, Corrotte M, Tam C, Maugel TK, Andrews NW, Song W (2015) Lipid raft-dependent plasma membrane repair interferes with the activation of B lymphocytes. J Cell Biol 211(6):1193–1205. doi:10.1083/jcb.201505030

Chapter 14

Using the *MEROPS* Database for Investigation of Lysosomal Peptidases, Their Inhibitors, and Substrates

Neil D. Rawlings

Abstract

This chapter describes how to retrieve data on lysosomal peptidases from the *MEROPS* database for proteolytic enzymes, their substrates and inhibitors (http://merops.sanger.ac.uk). Features described in this chapter include the summary page, pages for structure, interactions with inhibitors, substrates, literature and involvement in physiological pathways, and how to download data from the *MEROPS* FTP site. The lysosomal peptidase legumain is used as an example.

Key words Peptidase, Legumain, MEROPS database, Substrate specificity, Cathepsin

1 Introduction

The *MEROPS* database is an online resource that provides information on proteolytic enzymes, their substrates, and inhibitors [1]. The database began in 1996 and initially provided a classification and nomenclature of peptidases, derived from their classification by sequence and structural similarities [2]. The database was expanded to include protein inhibitors of peptidases, following a classification along similar means to that of peptidases [3], and small molecule inhibitors, both naturally occurring and synthetic, were added to the database in 2008 [4]. Identified cleavage positions in peptidase substrates, including peptides, proteins, and synthetic compounds, were also added to the database [5].

Human lysosomal peptidases include cathepsins B, D, F, H, K, L, O, S, V, W, X, legumain, gamma-glutamyl hydrolase, serine carboxypeptidase A, lysosomal Pro-Xaa carboxypeptidase, dipeptidyl-peptidase I, and tripeptidyl-peptidase I (*see* **Note 1**). These enzymes are required to reduce lysosomal proteins to small peptides and amino acid. Protein substrates enter the lysosome via various routes: phagocytosis, endocytosis of cytoplasm via autophagosomes, chaperone-mediated autophagocytosis, and some proteins are directed to the lysosome for degradation by the

Karin Öllinger and Hanna Appelqvist (eds.), *Lysosomes: Methods and Protocols*, Methods in Molecular Biology, vol. 1594, DOI 10.1007/978-1-4939-6934-0_14,

process of ubiquitinization. Autophagins -1, -2, -3, and -4 are important for processing the Atg8 protein at its C-terminus, allowing the exposed cysteine to become lipidated, attach to a membrane and initiate formation of the autophagosome [6]. The ubiquitin hydrolase USP6 is known to rescue proteins from lysosomal degradation by removing ubiquitin tags [7].

This paper describes how the *MEROPS* database can be used to retrieve information about lysosomal peptidases, their inhibitors, and substrates. A more complete description of how to use all the features on the website was published in 2014 [8].

2 Materials

The *MEROPS* website (http://merops.sanger.ac.uk) can be accessed using an Internet browser (such as Mozilla Firefox, Google Chrome, or Microsoft Internet Explorer). There are no restrictions on access from smart phones or tablets, but the user should be aware that some aspects of the displays may not have been formulated with small screens in mind. The website contains hypertext documents and CGI scripts which interact with an underlying MySQL database. There are some JavaScript elements, and some views of protein structures use Java applets which require the pre-installation of Java and appropriate security exceptions to be set.

The home page (Fig. 1) has a menu on the left-hand side. Clicking on any of the options in this menu will change the display on the rest of the screen. The new page displayed may either be a page of text and images, or another menu or index. By default, the home page shows data relating to peptidases, but by clicking the "Switch to Inhibitors" option in the upper left corner, pages relating to inhibitors are shown.

3 Methods

3.1 Peptidase Summary, Nomenclature, and Classification

There is a set of pages for each peptidase on the website. To access pages specific for animal legumain, from the home page do the following.

Click on "Name" from the left-hand menu. The first page of the Index of Peptidase Names is shown. Click on the "L" in the series of letters shown underneath the header. This shows the alphabetical list of peptidase names beginning with the letter "L". Scroll down to "Legumain, animal-type" and click the *MEROPS* identifier (C13.004: there is a unique identifier for each different peptidase). The summary page for the animal legumain will be displayed (*see* Fig. 2). This peptidase summary has extra information provided by an invited specialist in the field, but the description below applies to all peptidase summaries.

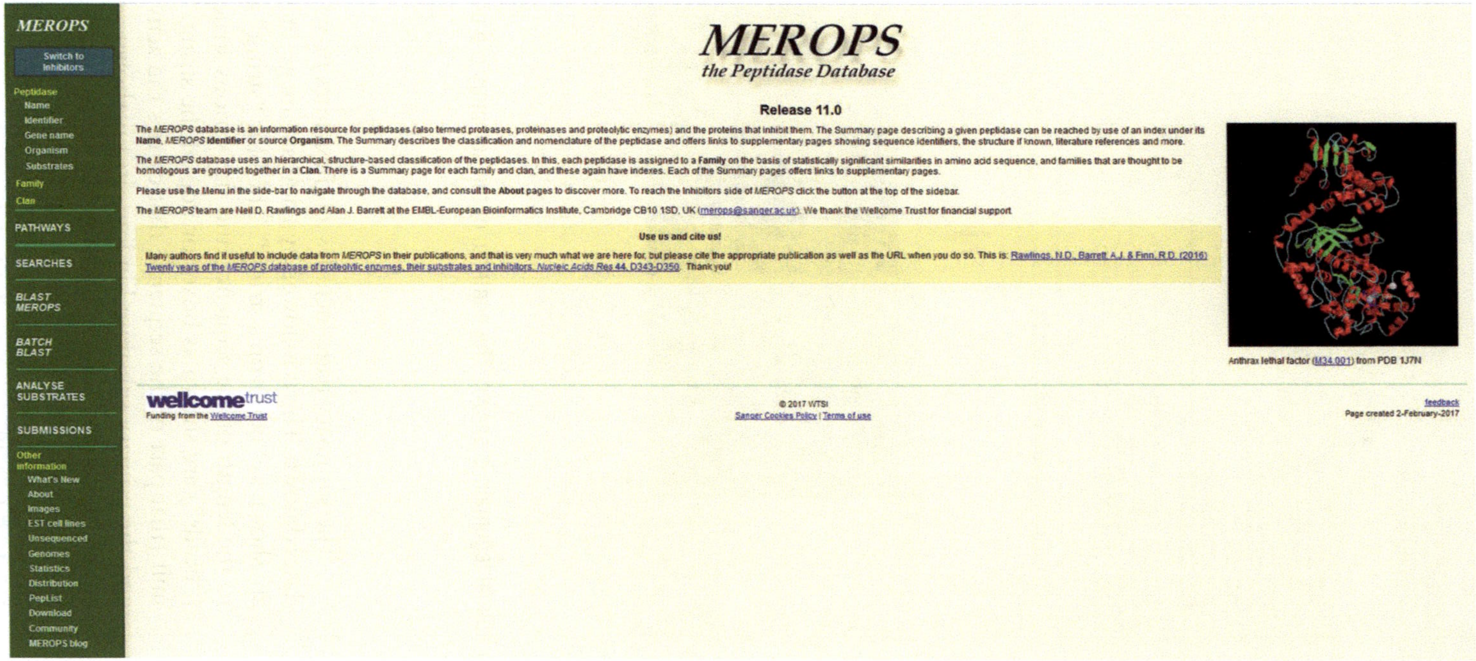

Fig. 1 The *MEROPS* website homepage. A menu of options is shown on the left-hand side with a green background. Click any of these options to see either an index or to initiate a search or analysis. The molecular structure shown is from a selection and will change each time the page is refreshed

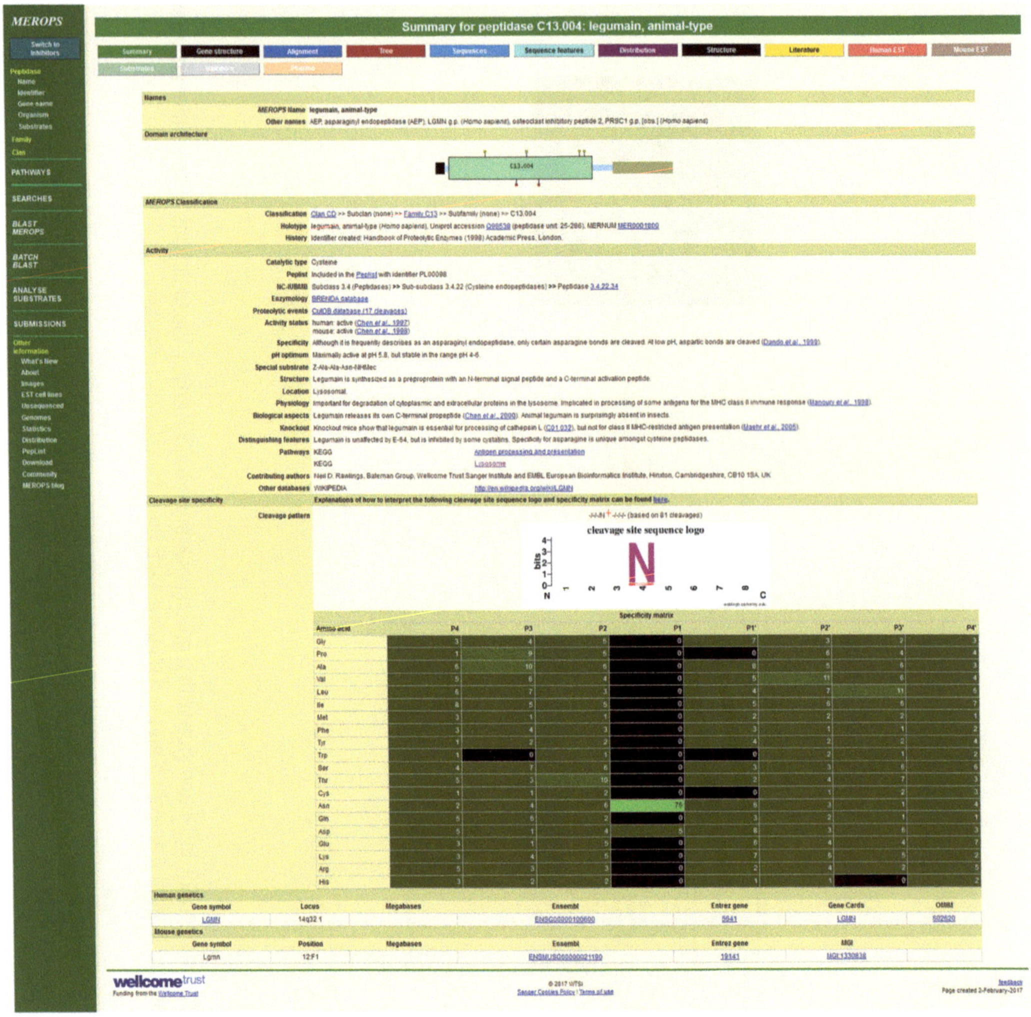

Fig. 2 The summary for animal legumain is shown. The contents of the summary are described in more detail in the text

The summary page includes all the names by which the peptidase is known. This is followed by a domain image, in which domains are shown as "beads on a string". The blue "string" is proportional to the length of the sequence and the various structural domains ("beads") are shown as boxes on top of the string. The peptidase unit (that part of the sequence that carries the active site residues and the primary substrate binding site) is shown as a green rectangle. "Lollipops" on the bottom edge of the peptidase unit rectangle indicate the active site residues, the details of which are shown in a tool tip revealed by placing the mouse cursor over the head of the lollipop. A small black box indicates a signal peptide or a transmembrane region, and a small grey box indicates a propeptide. Lollipops on the top edge show carbohydrate attachment sites and known disulfide bridges are also shown.

The next section in the table shows the *MEROPS* classification. Sequences are grouped into a "protein species", which is given a unique *MEROPS* identifier and is thought to represent the same activity in different organisms. Sequences that have homologous peptidase units are grouped into a family. Families that have peptidase units with similar tertiary structures are grouped into a clan [2]. Animal legumain is assigned the identifier C13.004, is a member of family C13 and clan CD (*see* **Note 2**).

The next section shows the activity of the peptidase. The catalytic activity is described, and there are links to other enzyme databases such as Enzyme Nomenclature [9] and Brenda [10]. References are shown where activity was first identified, for gene knockouts, known physiological roles and relevance for disease. Participation in known pathways is also shown [11].

This is followed by displays to indicate the specificity of the peptidase, provided there are at least ten known substrate cleavage sites. These displays are derived from the substrate cleavage collection, and show preferences for up to four residues either side of each known cleavage site. Cleavage preferences are shown as a Weblogo [12], in which the height of the letter indicates the greater preference for the relevant amino acid, and as a matrix in which the number of times each amino acid occurs in the substrate binding pockets S4 to S4' is shown (*see* **Note 3**), and cells are highlighted to indicate preferences. The brighter the shade of green, the greater the preference for the amino acid that is in that substrate-binding pocket. An amino acid that has never been observed in a substrate binding pocket is highlighted with a black background. The logo and matrix show that animal legumain predominantly cleaves asparaginyl bonds, but can cleave aspartyl bonds as well. Preferences in other substrate binding sites are limited, but the matrix shows that tryptophan has never been observed in the P3 position, cysteine, proline and tryptophan are unacceptable in P1′, and histidine has never been observed in P3′.

The final table in the summary shows human and mouse genetics for legumain, including the official gene symbols, and chromosomal locations.

There are additional pages which can be accessed by clicking the tabs at the top of the page. The Gene Structure page shows known gene structures from selected species, and shows the intron/exon distribution and phase of the intron–exon junctions. On the Alignment page an alignment of all sequences assigned to the same identifier, in this case C13.004, is dynamically generated. A neighbor-joining phylogenetic tree derived from this alignment can be accessed by clicking the Tree button. Details of the sequence accessions from the primary protein (UniProtKB [13]) and nucleic acid sequence databases (European Nucleotide Archive [14], GenBank [15]) of all homologs assigned to the same identifier, arranged in alphabetical order of species scientific name, can be accessed from the Sequences button. The Sequence Features button shows the extent of the peptidase unit, identifies the active site residues, and the source protein sequence used in the *MEROPS* sequence collection for each homolog assigned to the same identifier. The Distribution button shows the distribution of sequences assigned to the same identifier amongst species with a homolog in the same family (C13). The Human and Mouse EST pages show alignments of human and mouse expressed sequence tags. The Pharma page provides cross-references to websites and databases with pharmacological information on the peptidase in question, such as BindingDB [16] and ChEMBL [17]. The Structure, Literature, Substrates, and Inhibitor buttons are described below.

3.2 Structure

Click on the Structure tab from a summary page to see the structure page. The page includes a table of all known structures of the peptidase or inhibitor in question deposited in the Protein Data Bank [18]. Structures of proteins are grouped by organism. Each row in the table shows a description of the Protein Data Bank deposition; the resolution of the structure; links to the entry at the Protein Data Bank and PDBe [19], links to the structure classification databases SCOP [20] and CATH [21], links to PDBSum [22] and Proteopedia [23] (*see* **Note 4**); and a link to the reference describing the structure.

Above this table, for selected entries, a representation of the structure is shown generated in MolScript [24] and rendered in Raster3D [25]. The image is orientated so that active site residues are visible, and structures with similar folds are shown in the same orientation. Strands are shown as green arrows, helices as red coils, and turns and unstructured regions are shown as a turquoise string. Active site residues and metal ligands are shown in ball-and-stick format and colored (Arg, His, and Lys in purple; Asp, Asn, and Gln in pink; Cys in yellow; Glu in blue; Leu, Met, and Tyr in green; Ser and Thr in orange), metals as Corey-Pauling-Koltun (CPK) spheres (zinc in light grey; calcium and cobalt in light yellow; magnesium in green), carbohydrates as yellow CPK spheres and small molecule inhibitors in grey in ball-and-stick format. This image is also presented in the AstexViewer [26], which by default rotates allowing the structure to be viewed at different angles. It is also possible to superimpose a surface on the rotating structure by clicking the "Show Surface" button. The full suite of commands is available in the AstexViewer, which enables the user to display the image in a variety of ways (*see* **Note 5**). The structure page for animal legumain is shown in Fig. 3.

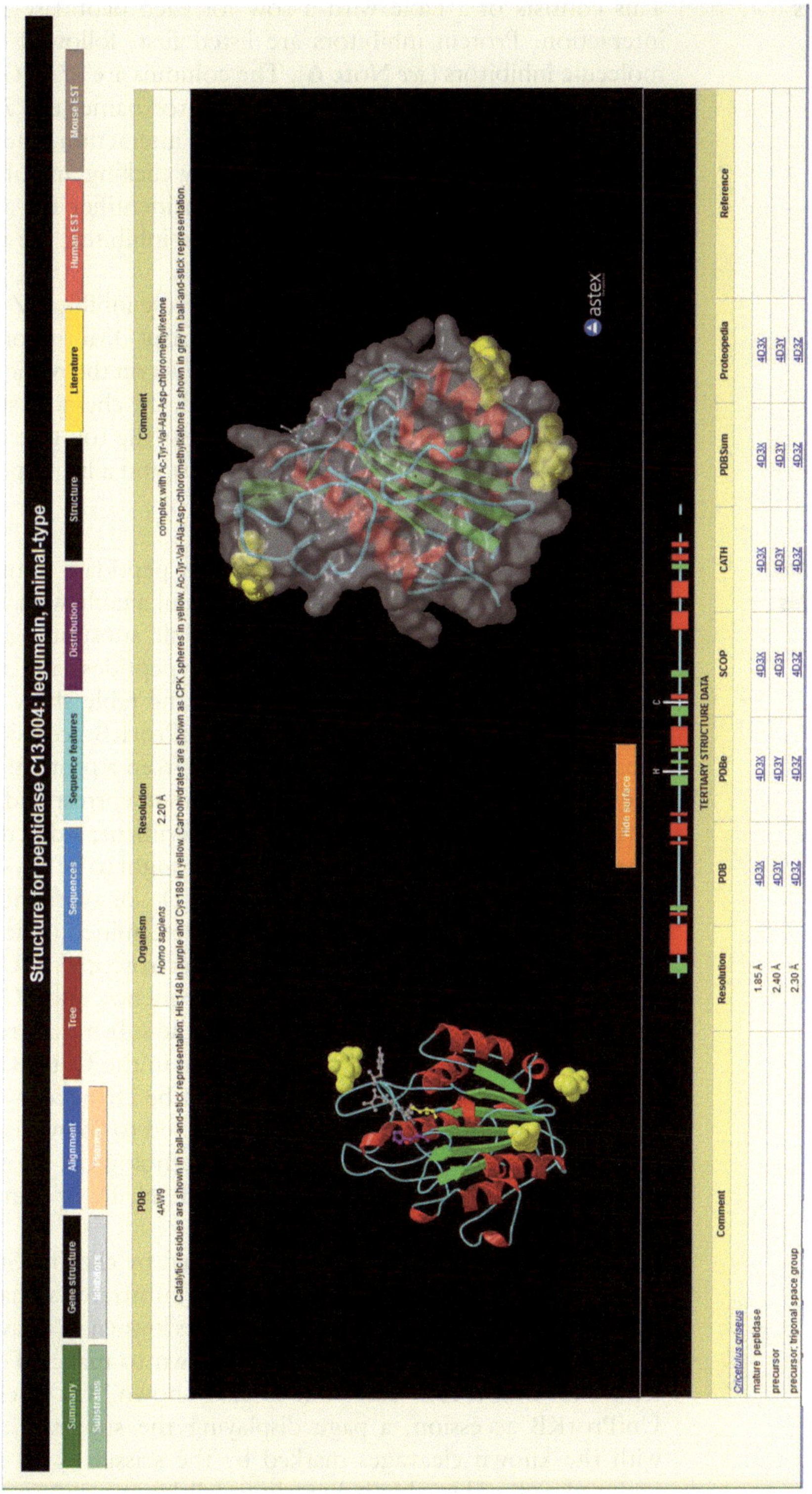

Fig. 3 The structure page for animal legumain is shown. The contents of the summary are described in more detail in the text

3.3 Interactions with Inhibitors

On clicking the Inhibitors button the Inhibitors page is presented. This consists of a table with a row for each peptidase–inhibitor interaction. Protein inhibitors are listed first, followed by small molecule inhibitors (*see* **Note 6**). The columns are *MEROPS* identifier (for protein inhibitors only), inhibitor name, the K_i of the interaction (if known), conditions of the interaction, and a reference. It is possible to reorder the table by clicking any of the column headers. On clicking the *MEROPS* identifier for a protein inhibitor or the name of a small molecule inhibitor, the inhibitor summary page is displayed.

The summary page for the small molecule inhibitor Z-DEVD-FMK (Z-Asp(OMe)-Glu(OMe)-Val-Asp(OMe)-fluoromethyl ketone) is shown in Fig. 4. There are sections on the name, mechanism of inhibition and peptidases inhibited, the chemical structure, and some general notes. The buttons across the top give access to the supplementary pages for a reference list and a list of peptidases inhibited by this compound.

3.4 Interactions with Substrates

On clicking the Substrates button from the peptidase summary the Substrates page is presented. This consists of a table with a row for each known substrate cleavage. By default substrates are listed alphabetically, and cleavages in proteins, peptides, and synthetic substrates are included. The columns in the table show the substrate name; a cross-reference to the UniProtKB accession if the substrate is a protein or a peptide derived from a protein; the residue range from the UniProtKB entry that corresponds to the sequence of the substrate used in the experiment; a description of the cleavage site; whether the substrate is thought to be physiological (P), non-physiological (N), pathological (D), or synthetic (S); the evidence by which the cleavage site was determined (where NT is N-terminal sequencing, MS is mass spectroscopy, MU is site-directed mutagenesis, CS is a consensus sequence, and LC is liquid chromatography); the P4-P4′ residues of the substrate; a reference; and a cross-reference to the same cleavage in the CutDB database [27]. The order of items in the table can be changed by clicking the column headers. The table can be filtered to show cleavages in only physiological, non-physiological, pathological, or synthetic substrates by clicking the corresponding P, N, D, or S in the text above the table.

There is an index of substrates accessed by clicking Substrates from the left-hand menu. The index lists substrates alphabetically and for each substrate shows the peptidases known to cleave it (and how many cleavages the peptidase is known to make). For a protein, the UniProtKB accession is also shown. On clicking the UniProtKB accession, a page displaying the substrate sequence with the known cleavages marked by the scissile bond symbol ł (colored red). The table below shows all known cleavages of this substrate ordered by residue number after which cleavage occurs.

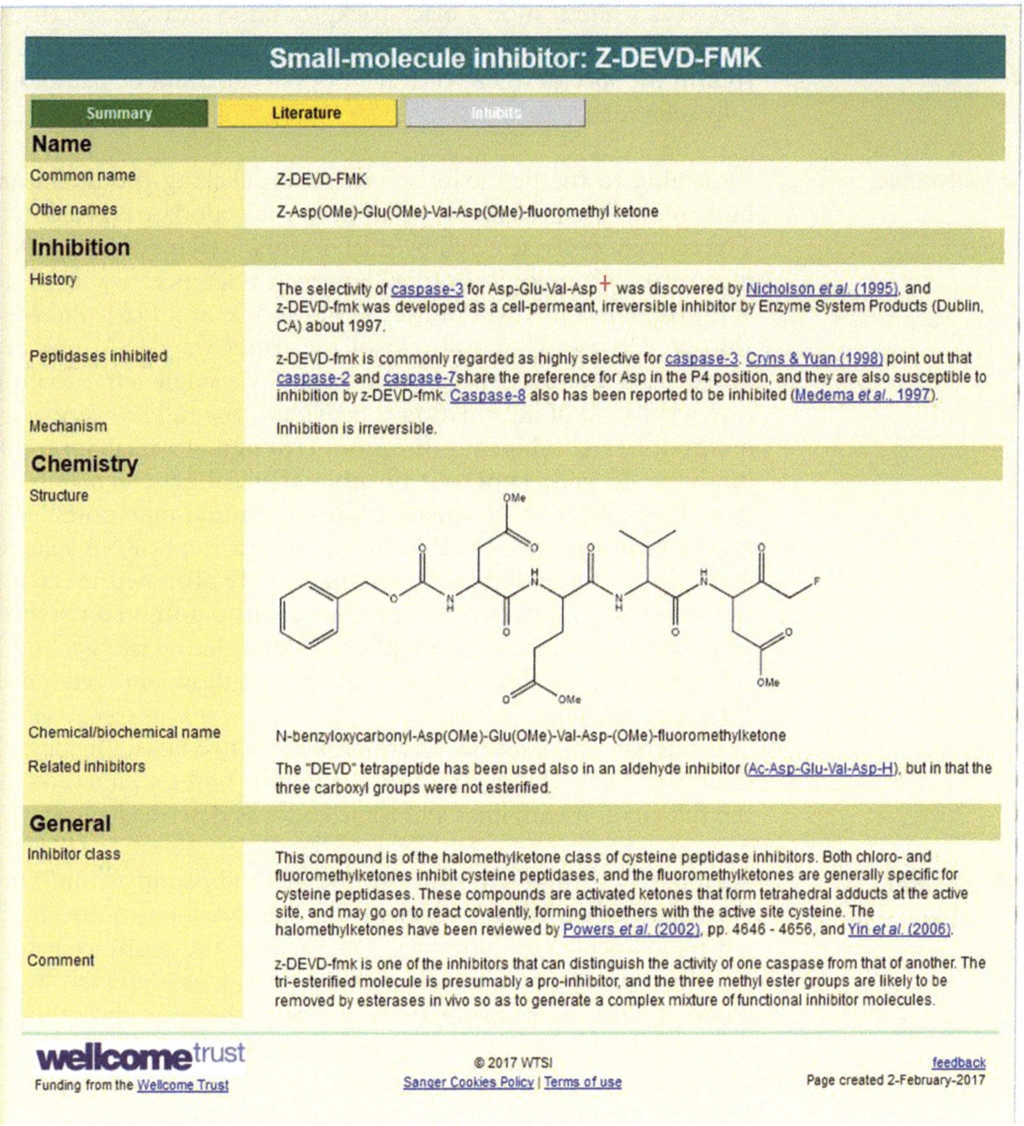

Small-molecule inhibitor: Z-DEVD-FMK

Summary | Literature | Inhibits

Name

Common name	Z-DEVD-FMK
Other names	Z-Asp(OMe)-Glu(OMe)-Val-Asp(OMe)-fluoromethyl ketone

Inhibition

History	The selectivity of caspase-3 for Asp-Glu-Val-Asp+ was discovered by Nicholson *et al.* (1995), and z-DEVD-fmk was developed as a cell-permeant, irreversible inhibitor by Enzyme System Products (Dublin, CA) about 1997.
Peptidases inhibited	z-DEVD-fmk is commonly regarded as highly selective for caspase-3. Cryns & Yuan (1998) point out that caspase-2 and caspase-7share the preference for Asp in the P4 position, and they are also susceptible to inhibition by z-DEVD-fmk. Caspase-8 also has been reported to be inhibited (Medema *et al.*, 1997).
Mechanism	Inhibition is irreversible.

Chemistry

Structure	
Chemical/biochemical name	N-benzyloxycarbonyl-Asp(OMe)-Glu(OMe)-Val-Asp-(OMe)-fluoromethylketone
Related inhibitors	The "DEVD" tetrapeptide has been used also in an aldehyde inhibitor (Ac-Asp-Glu-Val-Asp-H), but in that the three carboxyl groups were not esterified.

General

Inhibitor class	This compound is of the halomethylketone class of cysteine peptidase inhibitors. Both chloro- and fluoromethylketones inhibit cysteine peptidases, and the fluoromethylketones are generally specific for cysteine peptidases. These compounds are activated ketones that form tetrahedral adducts at the active site, and may go on to react covalently, forming thioethers with the active site cysteine. The halomethylketones have been reviewed by Powers *et al.* (2002), pp. 4646 - 4656, and Yin *et al.* (2006).
Comment	z-DEVD-fmk is one of the inhibitors that can distinguish the activity of one caspase from that of another. The tri-esterified molecule is presumably a pro-inhibitor, and the three methyl ester groups are likely to be removed by esterases in vivo so as to generate a complex mixture of functional inhibitor molecules.

wellcometrust
Funding from the Wellcome Trust

© 2017 WTSI
Sanger Cookies Policy | Terms of use

feedback
Page created 2-February-2017

Fig. 4 Example of a small molecule inhibitor summary. The summary for Z-DEVD-FMK is shown

Columns in the table are similar to those described above for the Substrate page from a peptidase summary. There is an option to display an alignment of sequences of orthologs of the substrate with the cleavages highlighted to indicate how well the cleavage sites are conserved. The sequence of the known substrate is highlighted in green. Residues corresponding to P4-P4′ are highlighted for each cleavage site. Residues identical to those in the sequence of the known substrate are highlighted in pink, replacements known to occur in the same position but in an unrelated substrate for the same peptidase are highlighted in orange, and a replacement not known to occur in any substrate for this peptidase is shown as white text on a black background. A cleavage site that is poorly conserved is presumed not to be physiologically important (but in the species where cleavage does occur could be pathologically relevant).

3.5 Literature

Returning to the peptidase summary, on clicking the Literature button the list of references relating to the peptidase is presented. The references are selected by the curators. References are shown in reverse chronological order. For each reference (where available) there is a link to PubMed, PubMedCentral [15], PubMed Europe [28], and the full text of the paper via a DOI link (*see* **Note 7**). A reference may be annotated by a single letter to indicate what kind of information is provided (the full key is A: assay method; E: recombinant expression; I: design of a small molecule inhibitor; K: gene knockout or other artificial genetic manipulation; L: localization or visualization; M: natural mutation, allelic variant or polymorphism; P: substrate specificity; R: RNA splicing variation; S: three-dimensional structure; T: proposed therapeutic target; U: suggested to have therapeutic potential itself; or V: review). By clicking the highlighted letter in the text above the reference list, the display can be filtered to show only references for a specific topic.

References shown on other pages are active links. On clicking the reference, a window opens in the middle of the screen showing the full citation with links and topic codes as described above.

3.6 Pathways

An index of pathways in which peptidases and peptidase inhibitors participate is available by clicking the Pathways button from the left-hand menu (*see* **Note 8**). On clicking the pathway named "Antigen processing and presentation" a page is presented which shows a clickable cross-reference to the pathway in the external, source database, and a table listing the peptidases involved in the pathway. Steps in the pathway that involve peptidases are numbered to give an indication of progress through the pathway. The column "Displayed As" shows how the peptidase appears in the external pathway. If substrate cleavages are involved in the pathway, then these are also shown. In this example, cleavages of the HLA class II histocompatibility antigen gamma chain by animal legumain are shown [29].

3.7 Downloading Data

Data may also be downloaded from the *MEROPS* FTP site. Click on the "Download" option from the left-hand menu to display the list of files that can be downloaded (*see* **Note 9**). The following can be downloaded.

1. A dump of the *MEROPS* MySQL database for local installation.
2. The file dnld_list.txt contains a non-redundant list of accession numbers and allows the user to create a library of peptidase and peptidase inhibitor sequences.

3. The library pepunit.lib contains sequences of all the peptidase and inhibitor domains in FastA format.
4. The library protease.lib contains complete sequences of all the peptidase and inhibitor domains in FastA format.
5. The library merops_scan.lib contains a selection of the sequences of the peptidase and inhibitor domains in FastA format. The selection contains one example of each different peptidase and peptidase inhibitor; the example that corresponds to the sequence of the best-characterized ortholog.
6. The compressed file seqlib.zip contains a sequence library for each family in *MEROPS*. When expanded, each file contains all the sequences of the peptidase or peptidase inhibitor domains for members of the family in question in FastA format.

4 Notes

1. Peptidases that occur in the mammalian lysosome are frequently referred to as "cathepsins" (from the Greek meaning "to boil down"). However, not all cathepsins are lysosomal peptidases, for example cathepsin G is stored in neutrophil granules [30]. There is also a misapprehension that most lysosomal peptidases are cysteine peptidases; however, one of the most potent and widespread is the aspartic peptidase cathepsin D [31]. Lysosomal peptidases are not necessary endopeptidases, for example cathepsin X is a carboxypeptidase [32], and dipeptidyl-peptidase I releases a dipeptide from the N-terminus of a protein [33].
2. Proteolytic enzymes are represented by seven different catalytic types, 61 clans and 257 families. A member of one clan will not be homologous to a member of a different clan. The seven different catalytic types are as follows: aspartic, cysteine, glutamic, metallo, serine, threonine, and asparagine lyases [34]. In addition, there are peptidases of mixed catalytic type, and peptidases where the catalytic type is unknown. In naming clans, families, and identifiers, the first character represents the catalytic type, namely A (aspartic), C (cysteine), G (glutamic), M (metallo), N (asparagine lyase), P (mixed), S (serine), T (threonine), and U (unknown). Each clan is given a two-letter identifier (e.g., CD), each family a letter and a number (e.g., C13), and each identifier consists of the family name, a dot, and then a consecutive number (e.g., C13.004).
3. A peptidase cleaves a substrate at the scissile bond. Usually, at the site of cleavage the peptide chain is linear, and residues are numbered either side of the scissile bond. Residues from the scissile bond toward the N-terminus are numbered P1, P2, P3,

etc., and residues from the scissile bound toward the C-terminus are numbered P1′, P2′, P3′, etc. Each residue occupies a substrate binding site, which is a pocket on the enzyme's surface. The P1 residue occupies the S1 pocket, the P1′ residue occupies the S1′ pocket and so on [35].

4. Links to external services are generated automatically and it is possible that a link might be presented which returns no data because the external service has not been updated for some time. Because there are millions of links, it is too time-consuming to check that every link to an external service retrieves the expected data. If a user discovers a link that is not retrieving the data he or she might expect, then please submit a description of the problem by clicking the "feedback" button at the footer of any page. This ensures that the problem enters our Request Tracker ticketing system.
5. To be able to display a rotating protein structure with the AstexViewer, Java must be installed. The instructions to run the AstexViewer are stored in the underlying MySQL database and are derived from the MolScript input file used to generate the fixed image. This ensures that the two displays are consistent.
6. There are two different kinds of peptidase inhibitors. Some are proteins that are gene products, which can be classified by comparing structures and sequences in much the same way as peptidases. For each of these a unique *MEROPS* identifier is created. For inhibitors that are synthetic constructs or naturally occurring compounds that are not products of genes, a collection of data is maintained but not a classification. These latter are known as "small molecule inhibitors".
7. The full text of a publication accessed by the DOI link may only be available to subscribers of the journal.
8. At present pathways information is taken only from the KEGG database [11]. Links between *MEROPS* and KEGG are automatically generated by searching for occurrences in a pathway of UniProtKB accessions common to both databases.
9. Downloading data from the *MEROPS* FTP site is free, but a user is required to obtain a username and password.

Acknowledgement

I would like to thank Dr. Alan J. Barrett for maintaining the *MEROPS* collections of references, identifiers, and small molecule inhibitors; Dr. Robert D. Finn for management and advice; and Matthew Waller and Paul Bevan from the Wellcome Trust Sanger Institute web team for maintaining the website.

References

1. Rawlings ND, Barrett AJ, Finn RD (2016) Twenty years of the *MEROPS* database of proteolytic enzymes, their substrates and inhibitors. Nucleic Acids Res 44:D343–D350
2. Rawlings ND, Barrett AJ (1993) Evolutionary families of peptidases. Biochem J 290:205–218
3. Rawlings ND, Tolle DP, Barrett AJ (2004) Evolutionary families of peptidase inhibitors. Biochem J 378:705–716
4. Rawlings ND, Morton FR, Kok CY, Kong J, Barrett AJ (2008) *MEROPS*: the peptidase database. Nucleic Acids Res 36:D320–D325
5. Rawlings ND (2009) A large and accurate collection of peptidase cleavages in the *MEROPS* database. Database (Oxford) 2009:bap015
6. Tanida I, Sou YS, Ezaki J, Minematsu-Ikeguchi N, Ueno T, Kominami E (2004) HsAtg4B/HsApg4B/autophagin-1 cleaves the carboxyl termini of three human Atg8 homologues and delipidates microtubule-associated protein light chain 3- and GABAA receptor-associated protein-phospholipid conjugates. J Biol Chem 279:36268–36276
7. Funakoshi Y, Chou MM, Kanaho Y, Donaldson JG (2014) TRE17/USP6 regulates ubiquitylation and trafficking of cargo proteins that enter cells by clathrin-independent endocytosis. J Cell Sci 127:4750–4761
8. Rawlings ND, Barrett AJ, Bateman A (2014) Using the *MEROPS* database for proteolytic enzymes and their inhibitors and substrates. Curr Protoc Bioinformatics 48:1.25.1–1.2533. doi:10.1002/0471250953.bi0125s48
9. Nomenclature Committee of the International Union of Biochemistry and Molecular Biology (1992) Enzyme Nomenclature 1992. Academic Press, London
10. Chang A, Schomburg I, Placzek S, Jeske L, Ulbrich M, Xiao M, Sensen CW, Schomburg D (2015) BRENDA in 2015: exciting developments in its 25th year of existence. Nucleic Acids Res 43:D439–D446
11. Kanehisa M, Sato Y, Kawashima M, Furumichi M, Tanabe M (2016) KEGG as a reference resource for gene and protein annotation. Nucleic Acids Res 44:D457–D462
12. Crooks GE, Hon G, Chandonia JM, Brenner SE (2004) WebLogo: a sequence logo generator. Genome Res 14:1188–1190
13. UniProt Consortium (2015) UniProt: a hub for protein information. Nucleic Acids Res 43:D204–D212. doi: 10.1093/nar/gku989
14. Gibson R, Alako B, Amid C, Cerdeño-Tárraga A, Cleland I, Goodgame N, Ten Hoopen P, Jayathilaka S, Kay S, Leinonen R, Liu X, Pallreddy S, Pakseresht N, Rajan J, Rosselló M, Silvester N, Smirnov D, Toribio AL, Vaughan D, Zalunin V, Cochrane G (2016) Biocuration of functional annotation at the European nucleotide archive. Nucleic Acids Res 44:D58–D66. doi:10.1093/nar/gkv1311
15. Resource Coordinators NCBI (2016) Database resources of the National Center for Biotechnology Information. Nucleic Acids Res 44:D7–D19. doi:10.1093/nar/gkv1290
16. Gilson MK, Liu T, Baitaluk M, Nicola G, Hwang L, Chong J (2016) BindingDB in 2015: A public database for medicinal chemistry, computational chemistry and systems pharmacology. Nucleic Acids Res 44:D1045–D1053. doi:10.1093/nar/gkv1072
17. Bento AP, Gaulton A, Hersey A, Bellis LJ, Chambers J, Davies M, Krüger FA, Light Y, Mak L, McGlinchey S, Nowotka M, Papadatos G, Santos R, Overington JP (2014) The ChEMBL bioactivity database: an update. Nucleic Acids Res 42:D1083–D1090. doi:10.1093/nar/gkt1031
18. Rose PW, Prlić A, Bi C, Bluhm WF, Christie CH, Dutta S, Green RK, Goodsell DS, Westbrook JD, Woo J, Young J, Zardecki C, Berman HM, Bourne PE, Burley SK (2015) The RCSB protein data bank: views of structural biology for basic and applied research and education. Nucleic Acids Res 43:D345–D356. doi:10.1093/nar/gku1214
19. Velankar S, van Ginkel G, Alhroub Y, Battle GM, Berrisford JM, Conroy MJ, Dana JM, Gore SP, Gutmanas A, Haslam P, Hendrickx PM, Lagerstedt I, Mir S, Fernandez Montecelo MA, Mukhopadhyay A, Oldfield TJ, Patwardhan A, Sanz-García E, Sen S, Slowley RA, Wainwright ME, Deshpande MS, Iudin A, Sahni G, Salavert Torres J, Hirshberg M, Mak L, Nadzirin N, Armstrong DR, Clark AR, Smart OS, Korir PK, Kleywegt GJ (2016) PDBe: improved accessibility of macromolecular structure data from PDB and EMDB. Nucleic Acids Res 44:D385–D395. doi:10.1093/nar/gkv1047
20. Andreeva A, Howorth D, Chandonia JM, Brenner SE, Hubbard TJ, Chothia C, Murzin AG (2008) Data growth and its impact on the SCOP database: new developments. Nucleic Acids Res 36:D419–D425
21. Sillitoe I, Lewis TE, Cuff A, Das S, Ashford P, Dawson NL, Furnham N, Laskowski RA, Lee D, Lees JG, Lehtinen S, Studer RA, Thornton J, Orengo CA (2015) CATH: comprehensive structural and functional annotations for genome sequences. Nucleic Acids Res 43:D376–D381. doi:10.1093/nar/gku947

22. de Beer TA, Berka K, Thornton JM, Laskowski RA (2014) PDBsum additions. Nucleic Acids Res 42:D292–D296. doi:10.1093/nar/gkt940
23. Hodis E, Prilusky J, Sussman JL (2010) Proteopedia: a collaborative, virtual 3D web-resource for protein and biomolecule structure and function. Biochem Mol Biol Educ 38:341–342. doi:10.1002/bmb.20431
24. Esnouf RM (1999) Further additions to MolScript version 1.4, including reading and contouring of electron-density maps. Acta Crystallogr D Biol Crystallogr 55:938–940
25. Merritt EA, Murphy ME (1994) Raster3D Version 2.0. A program for photorealistic molecular graphics. Acta Crystallogr D Biol Crystallogr 50:869–873
26. Hartshorn MJ (2002) AstexViewer: a visualisation aid for structure-based drug design. J Comput Aided Mol Des 16:871–881
27. Igarashi Y, Eroshkin A, Gramatikova S, Gramatikoff K, Zhang Y, Smith JW, Osterman AL, Godzik A (2007) CutDB: a proteolytic event database. Nucleic Acids Res 35:D546–D549
28. Europe PMC Consortium (2015) Europe PMC: a full-text literature database for the life sciences and platform for innovation. Nucleic Acids Res 43:D1042–D1048. doi:10.1093/nar/gku1061
29. Manoury B, Mazzeo D, Li DN, Billson J, Loak K, Benaroch P, Watts C (2003) Asparagine endopeptidase can initiate the removal of the MHC class II invariant chain chaperone. Immunity 18:489–498
30. Starkey PM, Barrett AJ (1976) Neutral proteinases of human spleen. Purification and criteria for homogeneity of elastase and cathepsin G. Biochem J 155:255–263
31. Barrett AJ (1967) Lysosomal acid proteinase of rabbit liver. Biochem J 104:601–608
32. Nägler DK, Zhang R, Tam W, Sulea T, Purisima EO, Ménard R (1999) Human cathepsin X: A cysteine protease with unique carboxypeptidase activity. Biochemistry 38: 12648–12654
33. McDonald JK, Zeitman BB, Ellis S (1972) Detection of a lysosomal carboxypeptidase and a lysosomal dipeptidase in highly-purified dipeptidyl aminopeptidase I (cathepsin C) and the elimination of their activities from preparations used to sequence peptides. Biochem Biophys Res Commun 46:62–70
34. Rawlings ND, Barrett AJ, Bateman A (2011) Asparagine peptide lyases: a seventh catalytic type of proteolytic enzymes. J Biol Chem 286:38321–38328. doi:10.1074/jbc.M111.260026
35. Schechter I, Berger A (1968) On the active site of proteases. 3. Mapping the active site of papain; specific peptide inhibitors of papain. Biochem Biophys Res Commun 32:898–902

Chapter 15

Next-Generation Sequencing Approaches to Define the Role of the Autophagy Lysosomal Pathway in Human Disease: The Example of LysoPlex

Giuseppina Di Fruscio, Sandro Banfi, Vincenzo Nigro, and Andrea Ballabio

Abstract

Next-Generation Sequencing (NGS) technologies have deeply changed the throughput of genetic testing allowing analyzing millions of DNA fragments in parallel. One key application is the understanding of genetically heterogeneous and complex diseases where 50–100 different genes may converge to control the same pathways. These disorders cannot be studied using traditional approaches, based on gene-by-gene Sanger sequencing. We have set up an NGS protocol based on a specific selection of DNA regions belonging to about 900 genes of the autophagy-lysosomal (ALP) pathway. We here specify all the technical steps and challenges of our protocol, named LysoPlex. This is based on the Haloplex technology and together with high-coverage sequencing empowers a high and uniform coverage of ALP genes. LysoPlex outplays other NGS applications in sensitivity and specificity, providing an accurate picture of all variations in ALP genes.

Key words Next-generation sequencing, Sequence variant detection, Bioinformatic analysis, Lysosomal storage disorders, Neurodegenerative diseases, Targeted sequencing

1 Introduction

The physiological importance of autophagy-lysosomal (ALP) pathways is underlined by their involvement in human diseases. Lysosomal Storage Diseases (LSDs) [1, 2] are a group of approximately 50 rare, recessively inherited disorders, with an overall incidence of 1 in 5000, caused, in most cases, by mutations of genes encoding proteins that localize either to the lysosomal lumen or membrane. They are characterized by the progressive accumulation of un-degraded material in the lysosomes of most cells and tissues. Historically, LSDs have been classified on the basis of the type of material that accumulates in the lysosomes, such as mucopolysaccharides, sphingolipids, glycoproteins, glycogen, and lipofuscins. However, a precise molecular diagnosis is very

Karin Öllinger and Hanna Appelqvist (eds.), *Lysosomes: Methods and Protocols*, Methods in Molecular Biology, vol. 1594, DOI 10.1007/978-1-4939-6934-0_15,

important as new therapies are becoming available for several types of LSDs [3]. Unfortunately, the high extent of genetic heterogeneity renders molecular diagnosis a challenging effort for this group of diseases. This is also true for neuronal ceroid lipofuscinosis (NCL), a genetically heterogeneous subtype of LSDs with a devastating phenotype that can be caused by mutations of at least 13 different genes [4]. Mutations in the ALP pathway have also been linked to the pathogenesis of common neurodegenerative disorders, such as Parkinson and Alzheimer diseases [5–7]. It is very likely that mutations in the ALP pathway will be identified in an increasing number of diseases.

In the last 10 years, Next-Generation Sequencing (NGS) technologies have demonstrated a very powerful ability to identify DNA variants at unprecedented processivity. NGS enables the genetic diagnosis of highly heterogeneous diseases (50–100 different genes or large genes) that are far beyond the possibilities of single gene testing. In addition, Whole Exome Sequencing (WES) proved to be a cost-effective research tool to map new disease genes. However, accumulating data indicate that standard WES (at a mean coverage of 50–100×) may not be reliable in a diagnostic context when it is crucial to gain both high sensitivity and specificity. This WES is undermined by a fragmentary coverage that, for a significant fraction of the exome, may be lower than 20×. This is the minimal bioinformatics threshold to detect heterozygous variants and requires a labor-intensive gap-filling/validation steps based on PCR reactions and Sanger sequencing. For practical reasons, these manual diagnostic steps are usually restricted to a few genes, missing the remaining variability.

In this chapter, we describe a robust and cost-effective NGS platform, named LysoPlex, to identify the sequence variations in the ALP pathway: our protocol permits both high efficient capture of the coding sequences from about 900 genes. A specific procedure of capture based on the Haloplex technology together with high-coverage sequencing empowers a high and uniform coverage, improving both sensitivity and specificity.

2 Materials

2.1 Library Preparation

1. Target DNA in Tris EDTA 1× (*see* **Note 1**).
2. LysoPlex, HaloPlex Target Enrichment System Kit (Agilent Technologies).
3. Herculase II Fusion Enzyme with dNTPs (Agilent Technologies).
4. Nuclease-free water, not diethylpyrocarbonate (DEPC)-treated.
5. Agencourt AMPure XP Kit (Beckman Coulter Genomics).
6. 10 M NaOH.

7. 2 M acetic acid.
8. 10 mM Tris–HCl, pH 8.0.
9. 100% ethanol.
10. 96-well plate magnetic separator (Agencourt SPRIPlate Super Magnet Plate or equivalent).
11. 1.5 mL tube-compatible magnetic separator (DynaMag-2 magnet or equivalent).

2.2 Library Validation

1. Quant-iT dsDNA BR Assay Kit, for the Qubit fluorometer (Life Technologies).
2. High Sensitivity DNA Kit (Agilent).
3. Qubit assay tubes (Life Technologies).

2.3 Library Sequencing

1. cBot Barcoded Strip Tubes (8 wells) (Illumina).
2. 0.2 mL 8-tube strip.
3. 1.0 N NaOH.
4. HT1 (Hybridization Buffer; Illumina).
5. Tris–HCl 10 mM, pH 8.5 with 0.1% Tween 20.
6. [Optional] Illumina PhiX Control (Illumina).

2.4 Equipment

1. Qubit 2.0 Fluorometer (Life Technologies).
2. 2100 Bioanalyzer Laptop Bundle (Agilent).
3. 2100 Bioanalyzer Electrophoresis Set (Agilent).
4. cBot (Illumina).
5. cBot manifold (Illumina).
6. Illumina HiSeq 1000.
7. SBS reagent kits.
8. Thermal cycler with 96-well blocks.

3 Methods

3.1 Selection of LysoPlex Genes

LysoPlex includes:

1. Disease genes related to the different lysosomal storage disorders (LSDs) and involved in neurodegenerative disorders;
2. Genes encoding proteins with either established or predicted lysosomal localization;
3. A group of genes playing a role in the autophagic or endocytic pathways.

In particular, 98 genes already described for lysosomal localization, 9 genes co-localizing with the lysosomal marker LAMP2, the four

MIT family members, 10 genes involved in hereditary forms of Parkinson's disease, 4 involved in hereditary forms of Alzheimer disease, and all the known genes involved in the different LSDs selected by using the Online Mendelian Inheritance in Man (OMIM) were included. Moreover, for lysosomal genes, the gene list was selected using the "lysosome" term in AmiGO, Netview and UniProt databases and comparing the results with each other: genes predicted to be lysosomal by at least two tools were selected, for a total of 83 genes. The selection of the genes playing a role in the autophagic pathway was made to study genes already described in literature and using bioinformatics tools (GO hierarchy and Netview): in total, 106 genes were selected. For the genes involved in endocytic pathway and trafficking, both literature analysis and the Molecular Signatures Database (MSigDB) were used.

As a result, the LysoPlex gene list includes 891 genes that span 2054 megabases (Mb) of human DNA coding sequence: for the list of genes please visit the following URL: www.tandfonline.com/doi/suppl/10.1080/15548627.2015.1043077.

3.2 LysoPlex Design

The enrichment probes were designed using the Haloplex system (Agilent Technologies), based on the restriction enzyme digestion of genomic DNA and hybridization capture with specific oligonucleotide probes.

Using the SureDesign algorithm (https://earray.chem.agilent.com/suredesign), it is possible to create the LysoPlex design for HaloPlex target enrichment system. The algorithm selects the probe sequences based on the target genomic regions provided by the user. Create and finalize the Haloplex design as instructed in the SureDesign user guide.

3.3 Sample Preparation

1. Determine the concentration of your gDNA samples (*see* **Note 2**).
2. In 0.2-mL PCR tubes, dilute each gDNA sample in 45 μL nuclease-free water for a final concentration of 5 ng/μL (225 ng in total).
3. In a separate 0.2-mL PCR tube, dispense 45 μL of the Enrichment Control DNA (ECD; supplied in Agilent HaloPlex Target Enrichment System Kit).

3.4 Digestion of Genomic DNA

Each DNA sample is digested in eight different reactions, each containing two restriction enzymes. The restriction enzymes are provided in two 8-well strip tubes (marked by red and green color).

1. Combine RE Buffer and BSA Solution (supplied by Agilent HaloPlex kit) as indicated in Table 1 in a 1.5-mL tube.
2. Mix by vortexing briefly.
3. Dispense 4 μL for 1 reaction (56 μL for 12 reactions) of the RE Buffer/BSA mixture to each well of an 8-well strip tube.

Table 1
RE buffer/BSA mix

Reagent	Volume for 1 reaction	Volume for 12 reactions (included excess)
RE buffer	34 μL	476 μL
BSA solution	0.85 μL	11.9 μL
Total volume	34.85 μL	487.9 μL

Table 2
Digestion master mix

Reagent	Volume for 1 reaction (μL)	Volume for 12 reactions (included excess) (μL)
RE enzymes from green enzyme strip	0.5	7
RE enzymes from red enzyme strip	0.5	7

4. Using a multichannel pipette, add the appropriate volume of each enzyme from the Green Enzyme Strip and from the Red Enzyme Strip (Table 2). It is important to use the restriction enzyme tube strip in the proper orientation: the green or red color marker aligned with tube A, to corresponding tubes A to H of the 8-well strip previously prepared.
5. Mix by vortexing, spin briefly and keep on ice until it is used.
6. Align the strip along the vertical side of a 96-well PCR plate.
7. Using a multichannel pipette, carefully distribute 5 μL of each RE master mix row-wise into each well of the plate. Check the equal volumes in pipette tips before dispensing.
8. Each well of the plate now contains 5 μL of the same restriction enzyme combination.
9. Align the DNA samples (11 g DNA samples and the ECD sample), prepared in Subheading 3.1, along the horizontal side of the same 96-well PCR plate.
10. Using a multichannel pipette, carefully distribute 5 μL of each DNA sample column-wise into each well of the plate. Check the equal volumes in pipette tips before dispensing (*see* **Note 3**).
11. Seal the plate with an adhesive seal.
12. Vortex and spin briefly the plate.

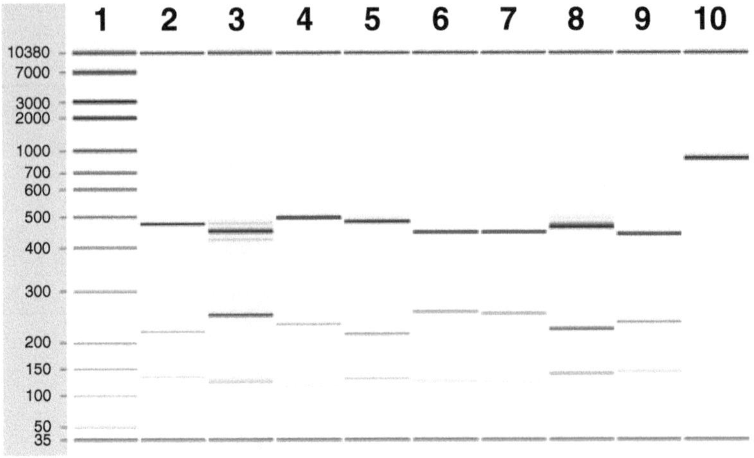

Fig. 1 Validation of restriction digestion by 2100 Bioanalyzer. *Lane 1*: 50-bp DNA ladder, *Lanes 2–9*: ECD digestion reactions A–H, *Lane 10*: Undigested Enrichment Control DNA

13. Each well of the plate now contains complete 10-μL restriction digestion reactions.
14. Place the plate in a thermal cycler with heated lid, and run the following program: 37 °C for 30 min and 8 °C hold.

3.5 Validation of the Restriction Digestion Reaction

To validate the digestion reaction, analyze the Enrichment Control DNA (ECD) reactions by electrophoresis. The ECD sample contains genomic DNA mixed with a PCR product that contains restriction sites for all the enzymes used in the digestion protocol.

1. Keep the plate on ice during validation.
2. Transfer 4 μL of each ECD digestion reaction from the wells of the plate to fresh 0.2-mL PCR tubes.
3. Incubate at 80 °C for 5 min to inactivate the restriction enzymes.
4. Prepare an undigested DNA gel control by combining 0.5 μL of the ECD stock solution and 3.5 μL of nuclease-free water.
5. Prepare the chip, samples, and ladder as instructed in the reagent kit guide.
6. Load the eight ECD digest samples and the undigested ECD sample.
7. Place the chip into the 2100 Bioanalyzer instrument and start the run.
8. Each of the eight digested ECD samples should have a smear of gDNA restriction fragments between 100 and 2500 bp (Fig. 1).
9. At this point, samples may be stored at −20 °C for long-term storage.

Table 3
Hybridization master mix

Reagent	Volume for 1 reaction (μL)	Volume for 12 reactions (included excess) (μL)
Hybridization solution	50	650
HaloPlex probe	20	260
Total volume	70	910

3.6 Hybridization of the Digested DNA to LysoPlex Probe

LysoPlex probes hybridize selectively to fragments originating from target regions containing genes of lysosomal-autophagic-endocytic pathway and to direct circularization of the DNA fragments. During this step, Illumina sequencing motifs and index sequences are incorporated into the targeted fragments.

1. Prepare the Hybridization Master Mix (Table 3).
2. Mix well by vortexing and spin the tube briefly.
3. Distribute 70 μL of the Hybridization Master Mix in 12 fresh 0.2-mL tubes.
4. Add 10 μL of the previously decided Index to each tube containing Hybridization Master Mix (*see* **Note 4**).
5. Using a multichannel pipette, transfer digested samples from the plate directly into the hybridization reaction tubes. Transfer each of the eight digestion reactions of one DNA sample into a tube (*see* **Note 5**).
6. After pooling, each reaction contains: 70 μL Hybridization Master Mix +10 μL Index + approximately 80 μL digested DNA sample (*see* **Note 6**).
7. Vortex and spin briefly.
8. Place the samples in a thermal cycler with heated lid, run the following program: 95 ° C for 10 min; 54 °C for at least 16 h (*see* **Note 7**).

3.7 Capture of the Target DNA

The circularized target DNA-probe hybrids contain biotin and, in this step, they are captured on streptavidin beads.

1. Resuspend the provided HaloPlex Magnetic Beads by vortexing.
2. Prepare 40 μL of HaloPlex Magnetic Beads per hybridization sample (520 μL for 12 reactions) in a 1.5-mL tube.
3. Put the tube into a compatible magnetic rack until the solution has cleared.
4. Remove and discard the supernatant using a pipette.

5. Add an equivalent volume of Capture Solution (40 μL per hybridization sample; 520 μL for 12 reactions) to the beads.
6. Resuspend by pipetting up and down.
7. Open the thermal cycler in which the hybridization reactions are going and immediately, add 40 μL of the prepared bead suspension to each 160-μL hybridization reaction and mix the solution by pipetting up and down (*see* **Note 8**).
8. Incubate at room temperature for 25 min.
9. Spin briefly and transfer the tubes to the Agencourt SPRIPlate Super Magnet magnetic plate.
10. As soon as the solution becomes clear, remove and discard the supernatant.
11. Remove the tubes from the magnetic plate and add 100 μL of Wash Solution to each tube.
12. Resuspend by pipetting up and down.
13. Incubate in a thermal cycler at 46 °C for 20 min, with heated lid (*see* **Note 9**).
14. Spin briefly and transfer the tubes to the magnetic plate.
15. Just the solution becomes clear, remove and discard the supernatant. Remove any residual liquid.

3.8 Ligation of the Circularized Fragments

DNA ligase closes nicks in the circularized target DNA.

1. Prepare the Ligation Master Mix (Table 4).
2. Mix the components by vortexing and spin briefly.
3. Add 50 μL of the Ligation Master Mix to the beads in each tube.
4. Resuspend by pipetting up and down.
5. Incubate in a thermal cycler at 55 °C for 20 min, with heated lid (*see* **Note 10**). During the incubation, prepare:
   (a) 2 M acetic acid, 0.5 μL per sample plus excess, for use in the PCR master mix preparation (*see* **Note 11**);

Table 4
Ligation master mix

Reagent	Volume for 1 reaction (μL)	Volume for 12 reactions (included excess) (μL)
Ligation solution	47.5	617.5
DNA ligase	2.5	32.5
Total volume	50	650

Table 5
PCR master mix

Reagent	Volume for 1 reaction (μL)	Volume for 12 reactions (included excess) (μL)
Nuclease-free water	16.1	209.3
5× Herculase II reaction buffer	10	130
dNTPs (25 mM for each dNTP)	0.4	5.2
Primer 1 (25 μM)	1	13
Primer 2 (25 μM)	1	13
2 M Acetic acid	0.5	6.5
Herculase II fusion DNA polymerase	1	13
Total volume	30	390

(b) 50 mM NaOH, 25 μL per sample plus excess, for use in the DNA elution step (*see* **Note 12**);

(c) PCR master mix.

3.9 Preparation of PCR Master Mix.

1. Prepare the PCR Master Mix as indicated in Table 5 (*see* **Note 13**).
2. Mix the master mix components by vortexing.
3. Distribute 30-μL of PCR master mix to fresh 0.2-mL reaction tubes.
4. Store the tubes on ice until they are used.

3.10 Elution with NaOH

1. When the ligation time is complete, briefly spin and transfer the tubes to the magnetic plate.
2. Just the solution becomes clear, remove and discard the supernatant.
3. Remove the tubes from the magnetic plate and add 100 μL of the SSC buffer (supplied by the Agilent HaloPlex kit) to each tube.
4. Resuspend by pipetting up and down.
5. Briefly spin and then return the tubes to the magnetic plate.
6. Just the solution becomes clear, remove and discard the supernatant. Remove any residual liquid.
7. Add 25 μL of 50 mM NaOH, freshly prepared, to each tube.
8. Resuspend by pipetting up and down.
9. Incubate at room temperature for 2 min.
10. Briefly spin and then return the tubes to the magnetic plate.

3.11 PCR Amplification

1. Transfer 20 μL of cleared supernatant from each tube on the magnetic plate to a PCR Master Mix tube.
2. Mix by vortexing and spin briefly.
3. Place the tubes in a thermal cycler and, with heated lid, run the following program: 98 °C for 2 min; followed by 16 cycles of 98 °C for 30 s, 60 °C for 1 min, 72 °C for 1 min and 30 s; with an extension step of 72 °C for 10 min.
4. At this point, PCR products may be stored at −20 °C for up to 72 h or at 8 °C overnight.

3.12 Purification of Amplified Libraries

1. Leave the AMPure XP beads at room temperature for at least 30 min.
2. Prepare 400 μL of 70% ethanol per sample, plus excess.
3. Transfer 40 μL of each PCR sample to a fresh 0.2-mL tube.
4. Resuspend the AMPure XP beads by vortexing, until the solution becomes homogeneous.
5. Prepare a master mix by combining 40 μL of nuclease-free water and 100 μL of the AMPure XP, for each sample as indicated in Table 6 (*see* **Note 14**).
6. Mix well, until the suspension appears homogeneous.
7. Add 140 μL of the master mix to each 40-μL amplified library sample.
8. Vortex thoroughly.
9. Incubate for 15 min at room temperature with continuous shaking.
10. Spin briefly and place the tubes in the magnetic plate.
11. Just the solution becomes clear, remove and discard the supernatant.
12. Keep the tubes in the magnetic plate and add 200 μL of 70% ethanol into the tubes (*see* **Note 15**).
13. Wait 30 s, then remove and discard the ethanol.
14. Keep the tubes in the magnetic plate and add 200 μL of 70% ethanol into the tubes.

Table 6
Post-amplification purification master mix

Reagent	Volume for 1 reaction (μL)	Volume for 12 reactions (included excess) (μL)
AMPure XP beads	100	1200
Nuclease-free water	40	480
Total volume	140	1680

15. Wait 30 s, then remove and discard the ethanol. Remove any residual ethanol.
16. Keeping the tubes in the magnetic plate, air-dry the tubes at room temperature until the ethanol completely evaporates.
17. Remove tubes from the magnetic plate.
18. Add 40 μL of 10 mM Tris-acetate (pH 8.0) to each sample.
19. Mix by pipetting up and down.
20. Incubate at room temperature for 5 min.
21. Put the tubes in the magnetic plate.
22. Just the solution becomes clear, remove the supernatant (approximately 40 μL) to a fresh tube.
23. Samples may be stored at −20 °C for long-term storage (up to 1 year) (*see* **Note 16**).

3.13 Validation and Quantification of LysoPlex Libraries

Each amplicon in the library contains one target insert surrounded by sequence motifs for sequencing using the Illumina platform. Use a Bioanalyzer High Sensitivity DNA Assay kit and the 2100 Bioanalyzer with 2100 Expert Software (See the reagent kit guide) (*see* **Note 17**) (Fig. 2).

To accurately quantify the LysoPlex libraries, use a fluorometry-based DNA quantitation method, such as Qubit fluorometry (*see* **Note 18**).

3.14 Sequencing and Analysis of LysoPlex Libraries

Once prepared the LysoPlex libraries, you need to proceed with the sequencing. It is clear that it is possible to perform the sequencing with any type of Next-Generation Sequencer, but in this chapter, we refer to Illumina HiSeq1000 system (Illumina Inc., San Diego, CA, USA). We recommend to set 32 libraries/lane to obtain a high-throughput sequencing.

Illumina sequencing platforms combine innovative engineering with widely used reversible terminator-based sequencing by synthesis (SBS) technology. To perform sequencing with these sequencers, specific adapters, containing Index 1 (i7) and Index 2

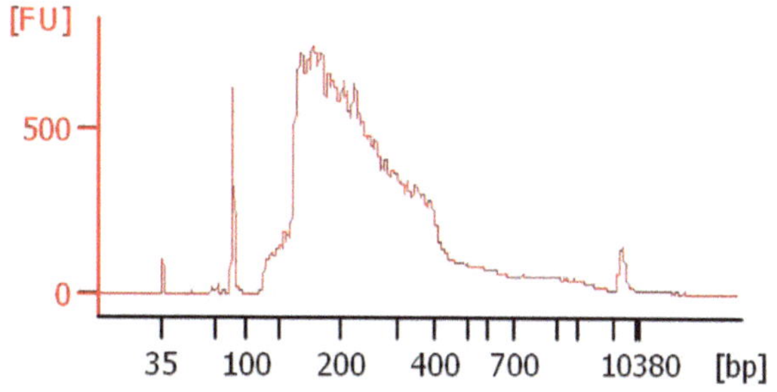

Fig. 2 Validation of LysoPlex libraries by 2100 Bioanalyzer. Using Bioanalyzer High Sensitivity DNA Assay kit, it is possible to verify the final result of each library

(i5), are added to libraries for bridge amplification. In LysoPlex libraries, the indexes 1 and 2 are added during the hybridization to LysoPlex probe. Through these adapters, denatured single-strand DNA molecules are attached to a solid surface, named flow cell and the molecules are amplified into clusters by solid-state bridge amplification. Each cluster is composed of about 1000 clonal copies of the molecule. For HiSeq1000 system, this step is performed by cBot, a revolutionary automated system that creates clonal clusters from single molecule DNA templates. After clusters generation, the sequencing by synthesis step is performed by HiSeq 1000: SBS technology uses four fluorescently labeled nucleotides placed on the flow cell and the fluorescent signal resulting from the incorporation of the nucleotides is detected during the imaging. In particular, the nucleotides are reversibly blocked: in this way, during each cycle, the primer is extended by a single base. After the imaging, the fluorophore is cleaved to allow incorporation of the next base.

The LysoPlex libraries are prepared before sequencing: in fact, it is necessary to obtain a pool containing all the LysoPlex libraries. Be sure to have a final Library pool concentration of 2 nM and that each library is represented in equimolar manner.

After that, it is necessary to denature and dilute the final pool for clustering on the cBot and subsequent sequencing on the HiSeq, obtaining a final DNA concentration of 12 pM for cluster generation on the cBot (*see* **Note 19**). See the reagent kit guide for detailed step. This final concentration is critical to obtain a good cluster density.

The generated sequences are analyzed using the appropriate tools available to the scientific community. Paired sequencing reads are aligned to the hg19 reference genome using BWA and sorted with SAMtools and Picard. The GATK tool is adapted with specific parameters to the Haloplex-generated sequences and it allows us to perform local realignment around single nucleotide variants (SNV) and small indel calling. Every produced calling is annotated using ANNOVAR: in this way, you obtain the relative position in genes using the RefSeq, amino acid change, presence in dbSNP and ClinVar database, frequency in NHLBI Exome Variant Server and prediction scores of damaging on protein activity.

With 32 libraries/lane, on average, on target reads represent more than 92% of total reads and the coverage depth is 40× for 95% and 100× for 80% of the target (*see* **Note 20**).

In conclusion, we obtain an excel file for each sample. The type of analysis you have to perform depends on the pattern of inheritance of the disease of your patients.

For the autosomal recessive diseases:

- select all genetic variants with a frequency in general population < 0.01 (1%);
- select all genetic variants in already known disease genes;

- focus your attention on homozygous variants, in case of evidence of consanguinity;
- focus your attention on heterozygous variants in the same gene, for compound heterozygosity;
- to simplify the analysis, concentrate on loss-of-function mutations (i.e., nonsense variants, indels) and already described as causative mutations;
- if relatives were sequenced, compare the detected variants, taking into account of the state of illness;
- if no causative variants in already known disease genes are found, deepen the analysis on the other genes in the LysoPlex and compare the observed phenotype to highlight candidate genes.

For the autosomal dominant diseases:

- select all genetic variants with a frequency in general population <0.01 (1%). In this case, to be causative, it is important for a variation not to be present in the general population;
- select all genetic variants in already known disease genes;
- focus your attention on heterozygous variant;
- to simplify the analysis, concentrate on already described as causative mutations;
- if relatives were sequenced, compare the detected variants, taking into account of the state of illness and focus on eventually de novo mutation;
- if no causative variants in already known disease genes are found, deepen the analysis on the other genes in the LysoPlex and compare the observed phenotype to highlight candidate genes.

4 Notes

1. High-quality DNA samples are required for the success of this protocol. Contaminants such as protein, salts, and RNA may interfere with accurate quantification and inhibit the digestion with restriction enzyme. Before you begin, verify that the OD 260/280 ratio is higher than 1.8 and the OD 260/230 ratio is higher than 2. Verify the concentration and the size distribution of each genomic DNA gel electrophoresis: the presence of smears indicates sample degradation.
2. To accurately quantify the gDNA samples, use a fluorometry-based DNA quantitation method, such as Qubit fluorometry. In this protocol, 200 ng genomic DNA is split among eight different digestions. Using less than 225 ng DNA can result in low yield and can potentiate rare allele dropouts.

3. Change tips for each different restriction reaction. It is important not to pool the restriction digests for a single DNA sample during the digestion time: if DNA samples are pooled before enzyme inactivation, restriction enzymes will catalyze inappropriate cleavage events. Restriction enzymes are inactivated during the hybridization.
4. Be sure to add only one specific Index to each hybridization tube, using different indexes for each sample. It is important to record the identity of the Index added to each tube.
5. When you add each digestion reaction to the hybridization solution, mix by pipetting to be sure that the enzymes are completely inactivated.
6. During the digestion, it is possible that a partial evaporation of samples occurs and for this reason, each restriction digest could be less than 10 μL. This will not modify hybridization performance.
7. During the hybridization program, do not include a low-temperature hold step in the thermal cycler program. It is important that the samples remain at 54 °C until they are processed.
8. When you add beads to the hybridization reactions, be sure that the bead preparation is homogenous without aggregated bead mass at the bottom of the tube. To avoid aggregation, pipette up and down the preparation.
9. During the wash of the bead-bound samples at 46 °C, do not include a low-temperature hold step in the thermal cycler program.
10. During the ligation step at 55 °C, it is possible to include a low-temperature hold step in the thermal cycler program.
11. Prepare 2 M acetic acid from high-quality glacial acetic acid.
12. Prepare the 50 mM NaOH solution fresh from a 10 M NaOH stock solution.
13. During the preparation of PCR Master Mix, it is critical to include 2 M acetic acid to be sure of the neutralization of the NaOH used for elution.
14. It is important to use the correct ratio between AMPure XP bead-H2O-sample to be sure optimal purification results.
15. Use fresh 70% ethanol for optimal results.
16. Avoid exposing the stored LysoPlex libraries to multiple freeze–thaw cycles.
17. Each enriched target DNA includes amplicons in the 175–625 bp size range.
18. During the pool of the LysoPlex libraries, it is important that all the libraries have the same concentration, obtaining a final

pool concentrated 2 nM. This step is critical because a sample more represented can cause an imbalance of the generated read. To accurately quantify the libraries, use a fluorometry-based DNA quantitation method, such as Qubit fluorometry.

19. Accurate quantification of final pool is a critical step to obtain an optimum cluster density, which means to achieve quality data.
20. Using 32 libraries/lane, the obtained coverage depth is significantly and this setting leads to a high specificity in variant calls.

References

1. Ballabio A, Gieselmann V (2009) Lysosomal disorders: from storage to cellular damage. Biochim Biophys Acta 1793(4):684–696. doi:10.1016/j.bbamcr.2008.12.001
2. Platt FM, Boland B, van der Spoel AC (2012) The cell biology of disease: lysosomal storage disorders: the cellular impact of lysosomal dysfunction. J Cell Biol 199(5):723–734. doi:10.1083/jcb.201208152
3. Parenti G, Andria G, Ballabio A (2015) Lysosomal storage diseases: from pathophysiology to therapy. Annu Rev Med 66:471–486. doi:10.1146/annurev-med-122313-085916
4. Schulz A, Kohlschutter A, Mink J, Simonati A, Williams R (2013) NCL diseases - clinical perspectives. Biochim Biophys Acta 1832(11):1801–1806. doi:10.1016/j.bbadis.2013.04.008
5. Aharon-Peretz J, Rosenbaum H, Gershoni-Baruch R (2004) Mutations in the glucocerebrosidase gene and Parkinson's disease in Ashkenazi Jews. N Engl J Med 351(19):1972–1977. doi:10.1056/NEJMoa033277
6. Sidransky E, Nalls MA, Aasly JO, Aharon-Peretz J, Annesi G, Barbosa ER, Bar-Shira A, Berg D, Bras J, Brice A, Chen CM, Clark LN, Condroyer C, De Marco EV, Durr A, Eblan MJ, Fahn S, Farrer MJ, Fung HC, Gan-Or Z, Gasser T, Gershoni-Baruch R, Giladi N, Griffith A, Gurevich T, Januario C, Kropp P, Lang AE, Lee-Chen GJ, Lesage S, Marder K, Mata IF, Mirelman A, Mitsui J, Mizuta I, Nicoletti G, Oliveira C, Ottman R, Orr-Urtreger A, Pereira LV, Quattrone A, Rogaeva E, Rolfs A, Rosenbaum H, Rozenberg R, Samii A, Samaddar T, Schulte C, Sharma M, Singleton A, Spitz M, Tan EK, Tayebi N, Toda T, Troiano AR, Tsuji S, Wittstock M, Wolfsberg TG, Wu YR, Zabetian CP, Zhao Y, Ziegler SG (2009) Multicenter analysis of glucocerebrosidase mutations in Parkinson's disease. N Engl J Med 361(17):1651–1661. doi:10.1056/NEJMoa0901281
7. Valente EM, Abou-Sleiman PM, Caputo V, Muqit MM, Harvey K, Gispert S, Ali Z, Del Turco D, Bentivoglio AR, Healy DG, Albanese A, Nussbaum R, Gonzalez-Maldonado R, Deller T, Salvi S, Cortelli P, Gilks WP, Latchman DS, Harvey RJ, Dallapiccola B, Auburger G, Wood NW (2004) Hereditary early-onset Parkinson's disease caused by mutations in PINK1. Science 304(5674):1158–1160. doi:10.1126/science.1096284

Chapter 16

Gelatin Zymography Using Leupeptin for the Detection of Various Cathepsin L Forms

Yoko Hashimoto

Abstract

Zymography is a highly sensitive method to assess the activities as well as molecular weights of enzymes in crude biological fluids and tissue extracts. Cathepsin L is a lysosomal cysteine proteinase that is optimally active at slightly acidic pH and is highly unstable in alkaline solutions such as electrode buffer (pH 8.3). Large amounts of cathepsin L are secreted by various cancer cells, where it promotes invasion and metastasis. Leupeptin is a tight-binding inhibitor of cysteine proteinases, and its complex with cathepsin L is stable in alkaline solutions. Moreover, leupeptin can be easily removed from the complex because it is a reversibly binding inhibitor. In addition, leupeptin is too small to influence the electrode migration distance of the complex with cathepsin L on a sodium dodecyl sulfate-polyacrylamide gel. Here, a novel gelatin zymography technique that employs leupeptin to detect pro-, intermediate, and mature cathepsin L forms on the basis of their gelatinolytic activities is described. Further, the differences in the glycosylation, phosphorylation, and processing statuses of lysosomal and secreted cathepsin L forms isolated from cultured HT 1080 cells are demonstrated using this method.

Key words Cathepsin L, Glycosylation, Leupeptin, Lysosome, Phosphorylation, Processing, Zymogram, Zymography

1 Introduction

Zymography is a kind of sodium dodecyl sulfate-polyacrylamide gel electrophoresis (SDS-PAGE) employed to detect possible hydrolytic activities using a substrate-embedded polyacrylamide gel. Zymography is extremely sensitive because the enzyme is in contact with the substrate. Enzymes can refold after the SDS is removed from the gel by washing with 2.5% Triton™ X-100. Moreover, proenzymes, which usually do not show enzymatic activities, show activities in this assay because the SDS changes their conformation. Even crude samples, which contain multiple proteinases and enzyme inhibitors, can be analyzed because they

Karin Öllinger and Hanna Appelqvist (eds.), *Lysosomes: Methods and Protocols*, Methods in Molecular Biology, vol. 1594, DOI 10.1007/978-1-4939-6934-0_16, © Springer Science+Business Media LLC 2017

are separated by SDS-PAGE. The principle of zymography is as follows: enzymes separated by SDS-PAGE hydrolyze a substrate (e.g., gelatin) at the migrated position during incubation at 37 °C for 20 h, resulting in the appearance of negatively stained bands upon staining of the gel. The activity is approximately proportional to the size of the unstained bands. However, this technique does not allow accurate quantification.

Cathepsin L is a lysosomal acid cysteine proteinase that is expressed at high levels in most cells and is presumed to play a major role in the degradation and turnover of both intracellular and extracellular proteins [1]. Cathepsin L exhibits a strong ability to degrade collagen, elastin, laminin, and other components of the basement membrane at acidic pH [2]. A large amount of cathepsin L is secreted from cancer cells, where it plays a role in invasion and metastasis [2–4]. Cathepsin L is synthesized as a preproform [5], and is subsequently processed into a 41-kDa proform in the Golgi apparatus, after which it is processed into a 28.5-kDa mature protein via intermediate forms [6]. Cathepsin L is the most unstable lysosomal proteinase at neutral and alkaline pH [7]. Leupeptin is a tight-binding, reversible competitive inhibitor of cysteine proteinases. The aldehyde group of leupeptin forms a thioester bond with the active-site -SH group of cathepsin L [8]. This stable three-dimensional structure is maintained during SDS-PAGE, thereby preventing the irreversible denaturation of cathepsin L in the alkaline electrode buffer [1, 6] (*see* Fig. 1). In addition, because leupeptin binds to the enzymes reversibly, it can be removed easily by washing the gel after electrophoresis. Moreover, leupeptin does not influence the migration distance of the complex as compared to cathepsin L alone, because of its very small size (formula weight < 500 Da; *see* Fig. 1).

Here, a novel leupeptin-based gelatin zymography method using polyacrylamide gels (11.5–12.0%) containing gelatin (0.8 mg/mL) for detecting various forms of cathepsin L is described. Fig. 1 shows representative zymograms of human liver cathepsin L (Fig. 1A) and intracellular and secreted cathepsin L (Fig. 1B) isolated from cultured HT 1080 cells with and without leupeptin treatment. This method enables researchers to examine (i) the glycosylation and phosphorylation statuses, and (ii) the activation and processing of cathepsin L. Cathepsin L has the highest endopeptidase activity among the lysosomal cysteine proteinases. Even when high levels of cathepsin B and S activities coexist with a low level of cathepsin L activity, the detected gelatinolytic activity in the zymogram can be attributed to cathepsin L [9]. However, to reduce contamination by other acid cysteine proteinases, it is best to use the lowest possible sample volume; the use of leupeptin allows minimizing the sample loading volume.

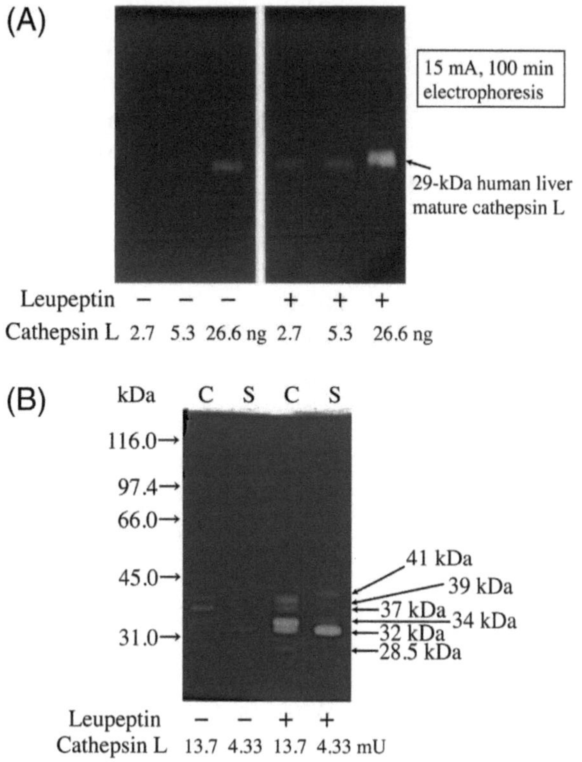

Fig. 1 Representative zymograms of (A) 29-kDa human liver mature cathepsin L (2.7–26.6 ng), and (B) intracellular (C, 13.7 mU) and secreted (S, 4.33 mU) cathepsin L from HT 1080 cells with (+) or without (−) leupeptin treatment. (**A**) Complexation with leupeptin did not influence the electrode migration distance of cathepsin L because of the very small size of leupeptin (formula weight < 500 Da). A higher sample quantity (several-folds) is required to detect enzyme activity without leupeptin treatment than with leupeptin treatment. (**B**) Six gelatinolytic active bands, including the 41-kDa proform and the 28.5-kDa mature form, were detected by this method with leupeptin treatment. All gelatinolytic active bands (with optimum pH 4.8) require thiol reagents to show the enzyme activities of the cathepsin L-derived enzymes [1] (*see* **Note 1**). (**B**) is reprinted from [6]

2 Materials

Prepare all solutions using deionized and distilled water and analytical grade reagents. Store all reagents at room temperature unless stated otherwise.

2.1 SDS-PA and Zymogram Gels

1. Acrylamide solution: 30% acrylamide, 0.8% bisacrylamide. Dissolve 30.0 g of acrylamide monomer and 0.8 g of *N,N′*-methylene-bisacrylamide (BIS) in 100 mL of water. Store at 4 °C in a dark-colored bottle (*see* **Note 2**).
2. Resolving gel buffer: 1.5 M Tris(hydroxymethyl)aminomethane (Tris)–HCl, pH 8.8. Dissolve 18.2 g of Tris base (FW

121.14) in water and adjust the pH with 6 N HCl. Bring the volume up to 100 mL with water.

3. Stacking gel buffer: 0.5 M Tris–HCl, pH 6.8. Dissolve 6.1 g of Tris base in water and adjust the pH with 6 N HCl. Bring the volume up to 100 mL with water.
4. 40% Glycerol (v/v): Dilute 40 mL of glycerol to 100 mL with water.
5. 5% SDS (w/v): Dissolve 5 g of SDS in 100 mL of water (*see* **Note 3**).
6. 10% Ammonium persulfate (APS; w/v): Dissolve 100 mg of APS in 1 mL of water (this stock can be kept for up to 1 week at 4 °C; *see* **Note 4**).
7. *N,N,N′,N′*-tetramethylethylenediamine (TEMED): 25 mL. Store at 4 °C.
8. Freshly prepared substrate (i.e., 8 mg/mL gelatin): Add 16 mg of gelatin to 2 mL of water. Heat to dissolve.
9. Water-saturated *n*-butanol (WSB): Add 50 mL each of *n*-butanol and water in a bottle and shake vigorously. Allow the mixture to stand until two layers separate. Use the top phase (WSB) only.

2.2 Zymogram Developing

1. SDS-PAGE running buffer: 25 mM Tris, 192 mM glycine buffer (pH 8.3), 0.1% SDS. Dissolve 3 g of Tris base, 14.4 g of glycine (FW 75.07), and 1 g of SDS in water and bring the volume up to 1 L. The pH need not be adjusted (*see* **Note 5**).
2. Washing and refolding solution: 2.5% Triton™ X-100 (v/v). Dilute 25 mL of Triton™ X-100 with water and bring the volume up to 1 L.
3. Zymogram incubation buffer: 0.1 M Sodium acetate (NaAc, pH 4.8), 1 mM ethylenediaminetetraacetic acid (EDTA), 20 mM cysteine-base (L-cysteine hydrochloride monohydrate). Dissolve 1.36 g of NaAc·3H_2O (FW 136.08), 37.2 mg of EDTA·2Na (FW 372.24), and 351.3 mg of cysteine-base (FW 175.64) in water, mix, adjust the pH with acetic acid, and then bring the volume up to 100 mL. It is best to prepare this fresh each time (*see* **Note 6**).
4. Fixing solution: 50% methanol, 10% acetic acid. Dilute 250 mL of methanol and 50 mL of acetic acid in water and bring the volume up to 500 mL.
5. Staining solution: 0.2% Coomassie brilliant blue R-250 (CBB), 50% methanol, 10% acetic acid. Dissolve 0.5 g of CBB in 250 mL of fixing solution. Filter through Whatman™ filter paper No. 1 to remove any insoluble CBB particles.
6. Destaining solution: 20% methanol, 10% acetic acid. Dilute 200 mL of methanol and 100 mL of acetic acid in water and bring the volume up to 1 L.

7. Drying solution: 5% glycerol (v/v). Dilute 5 mL of glycerol with water and bring the volume up to 100 mL.

2.3 Sample Preparation

1. Leupeptin stock solution (5 mM): Dissolve 21.3 mg of leupeptin (FW 426.55) in 10 mL of water. Store at −20 °C.
2. Sample dilution buffer: 0.1 M Tris–HCl (pH 6.8–7.2), 0.5 mM leupeptin. Mix 4 mL of 0.5 M Tris–HCl (pH 6.8–7.2, same as stacking gel buffer) and 2 mL of 5 mM leupeptin, and bring the volume up to 20 mL with water. Store at −20 °C.
3. SDS-PAGE sample buffer: 62.5 mM Tris–HCl (pH 6.8), 2% SDS, 25% glycerol, 0.02% bromophenol blue (BPB). Dissolve 12.5 mL of stacking gel buffer, 2 g of SDS, 25 mL of glycerol, 0.02 g of BPB in water, and bring the volume up to 100 mL.

2.4 Examination of the Glycosylation Status

1. Endoglycosidase H (endo-β-*N*-acetylglucosaminidase H; End H) from *Streptomyces griseus*: End H (0.1 U) is supplied as a lyophilized powder along with 10 mM Tris–HCl buffer, pH 7.2.
2. End H solution: Dissolve 0.1 U of End H in 100 μL of 0.1% bovine serum albumin (BSA).
3. End H incubation buffer (5×): 0.5 M NaAc (pH 5.0). Dissolve 1.36 g of NaAc·$3H_2O$ in water, adjust the pH with acetic acid, and bring the volume up to 20 mL.
4. Leupeptin stock solution (5 mM): Dissolve 21.3 mg of leupeptin (FW 426.55) in 10 mL of water. Store at −20 °C.

2.5 Examination of the Phosphorylation Status

1. Alkaline phosphatase from *Escherichia coli* (ALP): ALP (50 U) is supplied as a suspension in 110 μL of 2.5 M ammonium sulfate (ALP solution). Store at 4 °C.
2. ALP incubation buffer (10×): 50 mM Tris–HCl buffer (pH 9.3 at 25 °C), 1 mM $MgCl_2$, 0.1 mM $ZnCl_2$, 1 mM spermidine. Dissolve 121 mg of Tris base, 4.07 mg of $MgCl_2 \cdot 6H_2O$ (FW 203.3), 0.27 mg of $ZnCl_2$ (FW 136.32), and 2.91 mg of spermidine (FW 145.25) in water, adjust the pH with 1 N HCl, and bring the volume up to 20 mL.
3. Leupeptin stock solution (5 mM): Dissolve 21.3 mg of leupeptin (FW 426.55) in 10 mL of water. Store at −20 °C.

2.6 Examination of the Activation and Processing of Cathepsin L Forms

1. Activation and processing incubation buffer: 0.1 M NaAc (pH 4.8), 0.1 mM EDTA. Dissolve 680 mg of NaAc ·$3H_2O$ and 1.86 mg of EDTA·2Na in water, adjust the pH with acetic acid, and then bring the volume up to 50 mL. Store at 4 °C.
2. Dithiothreitol solution (DTT; 10 mM): Dissolve 1.54 mg of DTT (FW 154.2) in 1 mL of water. It is best to prepare this fresh each time.
3. Leupeptin stock solution (10 mM): Dissolve 21.3 mg of leupeptin in 5 mL of water. Store at −20 °C.

3 Methods

Carry out all procedures at room temperature unless specified otherwise.

3.1 Gelatin Zymography

1. Seal the mini slab gel plates (8.5 cm [width] × 8.0 cm [height] × 0.1 cm [depth]; *see* Fig. 2) at the bottom and both sides with a tube (1.0-mm diameter), mark the position for pouring the separating gel solution (approximately 58 mm from the bottom), and assemble the plates in the casting stand.
2. Prepare the separating gel (11.8%): Add 2.5 mL of resolving gel buffer, 3.84 mL of acrylamide solution, 1.0 mL of 8 mg/mL gelatin, 2.38 mL of water, 0.2 mL of 5% SDS, and 83.5 μL of 10% APS into a 25-mL conical flask. Mix well, add 10 μL of TEMED and mix again, and immediately pour the solution into the mini-slab gel plates until the mark. Allow space for the stacking gel and gently overlay with WSB to exclude oxygen to ensure good polymerization (*see* **Note 7**).
3. Pour off the WSB and wipe the area between the glass plates with Whatman™ filter paper after the separating gel has polymerized (0.5–1 h).
4. Prepare the stacking gel (3%): Add 2.0 mL of stacking gel buffer, 0.8 mL of acrylamide solution, 1.0 mL of 40% glycerol, 3.95 mL of water, 160 μL of 5% SDS, and 80 μL of 10% APS into a 25-mL conical flask. Mix well, add 10 μL of TEMED and mix again, and immediately pour the solution into the stacking gel space. Immediately insert a 12-well comb without introducing air bubbles (*see* Fig. 2).

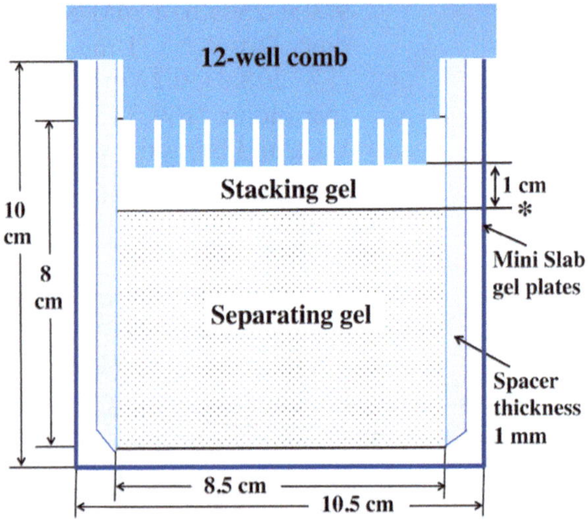

Fig. 2 Schematic representation of the mini-slab gel plates

5. After the stacking gel has solidified, remove the comb, arrange the plates in the electrode apparatus, and pour the SDS-PAGE running buffer.
6. SDS-PAGE sample preparation: Dilute the sample solutions with sample dilution buffer to prepare appropriate enzyme concentrations (*see* **Note 8**). Then, add the SDS-PAGE sample buffer without reducing reagents and mix well.
7. Apply aliquots of the sample solution (preferably <10 μL; without heat treatment) to the stacking gel wells.
8. Perform SDS-PAGE at a constant current of 25 mA for 80–90 min (unless stated otherwise) in SDS-PAGE running buffer at room temperature; maintain the temperature of the electrode buffer at approximately 2 °C by mixing (by hand) with a stick-type cold insulator during electrophoresis.
9. Following electrophoresis, open the gel plates with the aid of a spatula. The gel remains stuck to one of the glass plates. Remove the stacking gel by attaching and tearing off the Kimwipe paper. Transfer the separating gel into 25 mL (or a volume in which the gel is completely immersed) of 2.5% Triton™ X-100 washing and refolding solution in a plastic container (10 cm × 8 cm × 2.5 cm) with a cover with shaking at room temperature to remove leupeptin and SDS from the gel, and to allow the enzymes to refold (15 min × 4 times; *see* **Note 9**).
10. Change the solution to zymogram incubation buffer, wash the gel with shaking (×2 times), and incubate it in fresh zymogram incubation buffer at 37 °C for 20 h (*see* **Note 10**).
11. Change the buffer to fixing solution, wash the gel once with shaking, and incubate it in fresh fixing solution for 1 h at room temperature with shaking.
12. Change the solution to staining solution, wash the gel once with shaking, and incubate it in fresh staining solution for 1 h at room temperature with shaking.
13. Change the solution to destaining solution, wash the gel once with shaking, and destain the gel with fresh destaining solution with shaking until negatively stained bands appear; change the solution (check the gel every 10 min). Add Kimwipes to the destaining solution to adsorb the released dye.
14. After taking a photograph, equilibrate the gel in 5% glycerol, dry, and save it (*see* **Note 11**).

3.2 Comparison of Lysosomal and Secreted Cathepsin L Forms

1. Preparation of lysosomal and intracellular cathepsin L [1, 6, 10, 11]: Collect phosphate-buffered-saline-washed cultured HT 1080 cells by centrifugation at 700 × *g* for 10 min. For lysosomal cathepsin L, homogenize the cells in 0.25 M sucrose,

and remove the mitochondrial fraction by centrifugation at 3300 × *g* for 10 min at 4 °C. Precipitate the lysosomes by centrifugation at 25,000 × *g* for 10 min at 4 °C. Dissolve the lysosomes in 20 mM Tris–HCl buffer (pH 6.8; *see* **Note 12**), and remove the insoluble fraction by centrifugation (8000 × *g* for 10 min at 4 °C). The upper fraction contains lysosomal cathepsin L. For intracellular cathepsin L, homogenize the cells in 20 mM Tris–HCl buffer (pH 6.8), and remove the insoluble fraction by centrifugation (8000 × *g* for 10 min at 4 °C). This fraction corresponds to intracellular cathepsin L.

2. Preparation of a secreted cathepsin L [6]: Separate the culture medium from the HT 1080 cells by centrifugation (8000 × *g* for 10 min at 4 °C). This fraction contains secreted cathepsin L.
3. Perform gelatin zymography for each sample (*see* Fig. 3).

3.3 Examination of the Glycosylation Status of Cathepsin L

1. End H incubation mixture: Mix aliquots of the samples (intracellular, lysosomal, or secreted cathepsin L; *see* **Note 13**) with or without 10 μL of End H solution (10 mU), 6 μL of 5× End H incubation buffer (pH 5.0), and 1 μL of 5 mM leupeptin, and make up to 30 μL with water.
2. Incubate at 37 °C for 30 min.
3. Perform gelatin zymography for each sample (*see* Fig. 4).

3.4 Examination of the Phosphorylation Status of Cathepsin L

1. ALP incubation mixture: Mix aliquots of the samples (intracellular, lysosomal, or secreted cathepsin L; *see* **Note 13**) with or without 2.5 μL of ALP solution (1.14 U), 3 μL of 10× ALP incubation buffer, 1 μL of 5 mM leupeptin, and make up to 30 μL with water.
2. Incubate at 37 °C for 30 min.
3. Perform gelatin zymography for each sample (*see* Fig. 5).

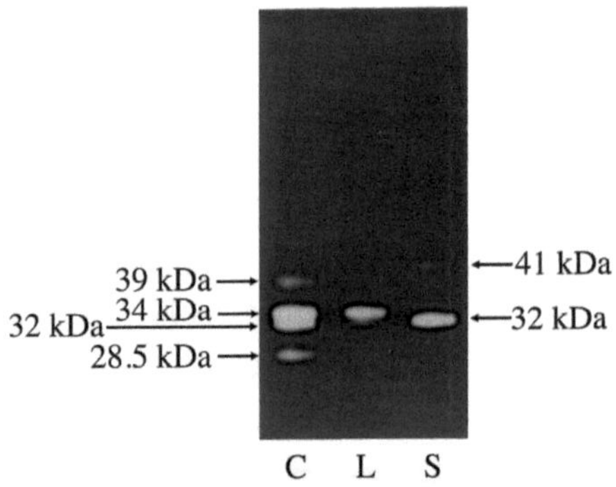

Fig. 3 Comparison of intracellular (C), lysosomal (L), and secreted (S) cathepsin L from HT 1080 cells. The molecular masses of lysosomal and secreted cathepsin L were 34 kDa and 32 kDa, respectively. The intracellular fraction contained both forms. Reprinted with modification from [6]

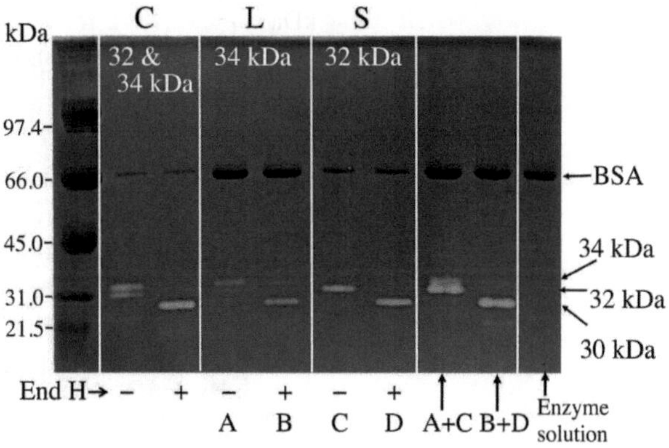

Fig. 4 Examination of the glycosylation status of the samples (32- and 34-kDa cathepsin L) by treatment with endoglycosidase H (End H). Both forms are End H-sensitive as they both shifted to 30 kDa after End H treatment (C, intracellular; L, lysosomal; and S, secreted cathepsin L). This suggests that both forms harbor *N*-linked oligosaccharides. Both core proteins are approximately 30 kDa and show gelatinolytic activity. The following molecular markers were used: 97.4 kDa, phosphorylase b; 66.0 kDa, BSA; 45.0 kDa, ovalbumin; 31.0 kDa, carbonic anhydrase; and 21.5 kDa, soybean trypsin inhibitor. Reprinted with modification from [6]

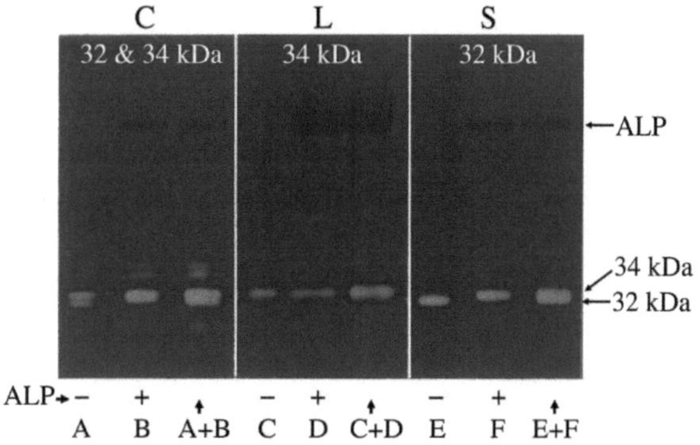

Fig. 5 Examination of the phosphorylation status of samples (32- and 34-kDa cathepsin L) by treatment with alkaline phosphatase (ALP). The 32-kDa form shifted to the 34-kDa position, while the position of the 34-kDa form did not change after ALP treatment, suggesting that 32-kDa cathepsin L is phosphorylated and 34-kDa cathepsin L is not (C, intracellular; L, lysosomal; and S, secreted cathepsin L). This discrepancy may be explained as follows: the 32-kDa form might migrate to the anode faster than the non-phosphorylated 34-kDa form because of the retention of the negative charge by the phosphate moiety. Reprinted with modification from [6]

3.5 Examination of the Activation and Processing of Cathepsin L

1. Activation and processing incubation mixture: Add the sample (6 μL of lysosomal or secreted cathepsin L) and 2 μL of 10 mM DTT to 20 μL of activation and processing incubation buffer, and incubate for 0, 2, and 4 h at 37 °C. Add 2 μL of 10 mM leupeptin after incubation.

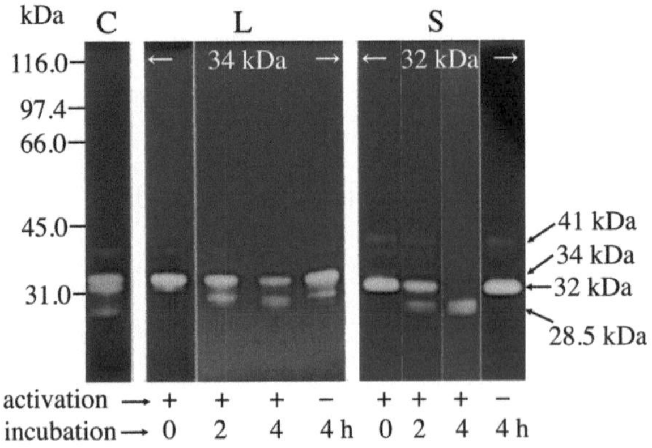

Fig. 6 Examination of the activation and processing of the samples (32- and 34-kDa single-chain cathepsin L). The 34-kDa cathepsin L is the lysosomal proteinase (L) and the 32-kDa cathepsin L is the secreted one (S). The intracellular cathepsin L includes both 32- and 34-kDa cathepsin L (C). Both the 32- and 34-kDa single-chain cathepsin L forms were processed further to a 28.5-kDa double-chain mature cathepsin L upon incubation at pH 4.8, suggesting that both are intermediate forms. An acidic pH of 4.8 is sufficient to initiate the auto-activation of single-chain cathepsin L intermediates into the mature double-chain form; leupeptin completely (32-kDa form) or partially (34-kDa form) inhibits this processing. Reprinted with modification from [6]

2. Non-activation control mixture with leupeptin and without DTT: Add the sample (6 μL of lysosomal or secreted cathepsin L), 2 μL of 10 mM leupeptin, and 2 μL of water to 20 μL of activation and processing incubation buffer, and incubate for 4 h at 37 °C.
3. Perform gelatin zymography for each sample (*see* Fig. 6).

4 Notes

1. All gelatinolytic activity bands in Fig. 1B were inhibited by synthetic inhibitors of cathepsin L (CLICKs-148 and -181), but not by inhibitors of cathepsin B, S, or K (CA-074, CLICK-60, or CLICK-164, respectively) [1]. In addition, the presence of the 32- and 41-kDa forms in the concentrated medium was confirmed by immunoblot analysis using antibodies against human amino-terminal and carboxyl-terminal cathepsin L [1]. Immunoblot analysis requires a larger amount of sample than synthetic inhibition analysis.
2. Wear a mask and gloves when weighing acrylamide. Unpolymerized acrylamide is a neurotoxin and care should be exercised to avoid skin contact.

3. Wear a mask when weighing SDS. Be careful not to breathe it. SDS is a protein denaturant and a fine powder that is easily dispersed through the air.
4. Keep the original powdery reagent under dry conditions with silica gel in a desiccator or bottle.
5. It is better to make a 10× native stock buffer (0.25 M Tris, 1.92 M glycine, pH 8.3) and 10% SDS separately. Dissolve 30.3 g of Tris base and 144 g of glycine in water, and bring the volume up to 1 L. Dissolve 10 g of SDS in 100 mL of water. For a working solution, dilute 100 mL of 10× native stock buffer with 890 mL of water, and add 10 mL of 10% SDS. Care should be taken when adding the SDS solution, since it forms bubbles.
6. Stock solution without reducing reagents (5×; 0.5 M NaAc, pH 4.8, 5 mM EDTA) is convenient. Dissolve 13.6 g of NaAc·3H_2O and 372 mg of EDTA·2Na in water, adjust the pH with acetic acid, and bring the volume up to 200 mL. Store at 4 °C. Cysteine-base reduces the pH of the solution. Therefore, when adding 20 mM cysteine-base to 5× stock solution, the pH of the solution should be adjusted with 1 N NaOH every time. Prepare fresh. Instead of cysteine-base, 5 mM DTT can be used; it does not influence the pH of the solution, but it is more expensive.
7. Oxygen inhibits polymerization of the gel. Therefore, cover the upper surface of the gel with WSB to prevent contact with air. Wait for 0.5–1 h for gel polymerization. When the gel has solidified, the boundary between the gel and WSB will become visible.
8. In this step, leupeptin interacts with cathepsin L to form a complex, which protects cathepsin L from irreversible denaturation in alkaline solutions during SDS-PAGE.
9. Leupeptin and SDS are removed from the gel, and the enzymes are allowed to refold in this step.
10. This step can be performed without shaking with equally good results.
11. In gels containing gelatin, crazing occurs easily, which can be prevented by adding 5% glycerol.
12. Buffer without leupeptin is used to prepare lysosomal and secreted cathepsin L because these fractions are used to examine the activation and processing of the cathepsin forms. When performing zymography, leupeptin should be added to the sample.
13. The sample loading volume depends on the enzyme activity (use preferably <10 μL).

References

1. Hashimoto Y, Kondo C, Kojima T, Nagata H, Moriyama A, Hayakawa T, Katunuma N (2006) Significance of 32-kDa cathepsin L secreted from cancer cells. Cancer Biother Radiopharm 21(3):217–224. doi:10.1089/cbr.2006.21.217
2. Chauhan SS, Goldstein LJ, Gottesman MM (1991) Expression of cathepsin L in human tumors. Cancer Res 51(5):1478–1481
3. Sudhan DR, Siemann DW (2013) Cathepsin L inhibition by the small molecule KGP94 suppresses tumor microenvironment enhanced metastasis associated cell functions of prostate and breast cancer cells. Clin Exp Metastasis 30(7):891–902. doi:10.1007/s10585-013-9590-9
4. Sudhan DR, Siemann DW (2015) Cathepsin L targeting in cancer treatment. Pharmacol Ther 155:105–116. doi:10.1016/j.pharmthera.2015.08.007
5. Ishidoh K, Imajoh S, Emori Y, Ohno S, Kawasaki H, Minami Y, Kominami E, Katunuma N, Suzuki K (1987) Molecular cloning and sequencing of cDNA for rat cathepsin H. Homology in pro-peptide regions of cysteine proteinases. FEBS Lett 226(1):33–37
6. Hashimoto Y, Kondo C, Katunuma N (2015) An active 32-kDa cathepsin L is secreted directly from HT 1080 fibrosarcoma cells and not via lysosomal exocytosis. PloS ONE 10(12):e0145067. doi:10.1371/journal.pone.0145067
7. Turk B, Dolenc I, Turk V, Bieth JG (1993) Kinetics of the pH-induced inactivation of human cathepsin L. Biochemistry 32(1):375–380
8. Umezawa H, Aoyagi T, Morishima H, Kunimoto S, Matsuzaki M, Hamada M, Takeuchi T (1970) Chymostatin, a new chymotrypsin inhibitor produced by actinomycetes. J Antibiotics 23:425–427
9. Hashimoto Y, Kakegawa H, Narita Y, Hachiya Y, Hayakawa T, Kos J, Turk V, Katunuma N (2001) Significance of cathepsin B accumulation in synovial fluid of rheumatoid arthritis. Biochem Biophys Res Commun 283 (2):334–339. doi:10.1006/bbrc.2001.4787
10. De Duve C, Pressman BC, Gianetto R, Wattiaux R, Appelmans F (1955) Tissue fractionation studies. 6. Intracellular distribution patterns of enzymes in rat-liver tissue. Biochem J 60(4):604–617
11. Novikoff AB, Beaufay H, De Duvo C (1956) Electron microscopy of lysosome-rich fractions from rat liver. J Biophys Biochem Cytol 2(4 Suppl):179–184

Chapter 17

Methods for Determination of α-Glycosidase, β-Glycosidase, and α-Galactosidase Activities in Dried Blood Spot Samples

Eser Yıldırım Sozmen and Ebru Demirel Sezer

Abstract

The lysosomal storage diseases (LDSs) are a heterogeneous group of inherited genetic disorders caused by defects of lysosomal proteins. The accumulation of undigested substrates from different catabolic pathways leads to cellular dysfunction. LSDs generally presents during early childhood and have a devastating impact on the families and on public health. Over the years, approaches for treatment of some LSDs have been developed with different strategies. Increasing availability of treatments of these diseases has accelerated the development of new methods and techniques for rapid diagnosis in patients with clinical indication.

The use of dried blood spot (DBS) test has been proposed as a first tier test to identify patients with Gaucher, Pompe, and Fabry diseases. DBS usage is advantageous for the purpose of screening as it is non-invasive, sensitive, has low-cost and fast turnaround time compared to measurements in leucocyte and/or fibroblast culture. This chapter focuses on the activity measurement of three lysosomal enzymes (α-glucosidase, β-glucosidase, and α galactosidase) in DBS samples by using fluorescent substrates and by the LC-MS/MS (liquid chromatography-mass spectrometry) method. All steps of the methods, from preparation of the solutions to calculation of the enzyme activity, will be explained in detail.

Key words α-Glycosidase, β-Glycosidase, α-Galactosidase, Dry blood spot analysis, Lysosomal storage disorders

1 Introduction

Lysosomal storage diseases comprise of more than 50 diseases, often related to a deficiency of a specific lysosomal protein, which results in clinical aspects due to accumulation of substrates in different tissues. Recent screening studies in newborns show that the prevalence of these diseases is higher than what was published in 1990s [1]. There is a risk of late/misdiagnosis due to the highly variable symptoms and characteristics of the diseases and physicians who are unfamiliar with these relatively rare diseases.

Pompe disease is an autosomal recessive hereditary disorder, which is characterized by deficiency of the lysosomal enzyme acid

Karin Öllinger and Hanna Appelqvist (eds.), *Lysosomes: Methods and Protocols*, Methods in Molecular Biology, vol. 1594, DOI 10.1007/978-1-4939-6934-0_17, © Springer Science+Business Media LLC 2017

α-glucosidase (GAA) and accumulation of glycogen in the lysosomes. The incidence has been reported to be between 1:4500 and 1:33,000 in live births screening studies within different populations [2, 3]. Fabry disease is an X-linked disorder, which is characterized by deficiency of the lysosomal enzyme α-galactosidase A and accumulation of its substrates (especially globotriaosylceramide and globotriaosylsphingosine (lysoGb3)) in tissues. The screening studies report the incidence of Fabry disease between 1:1200 and 1:13,000 of male live births [2, 3]. Gaucher disease is another autosomal recessive disorder with an incidence between 1:13,000 and 1:100,000 in newborn screenings [2, 3]. It is characterized by the deficiency of acid β-glucosidase (glucocerebrosidase) enzyme activity in lysosomes and the accumulation of cerebrosides in tissues.

In recent years enzyme replacement therapy has been developed for the above-mentioned diseases, which is why early diagnosis of patients is of critical importance to prevent fatal complications. Therefore, newborn screening programs are being developed over the last 10 years. The most advantageous method, to collect blood samples from a huge population for the screening, is the dried blood spot (DBS) test on filter paper. Since DBS is a non-invasive, sensitive and easy transportable method with acceptable enzyme stability compared to measurements in leucocyte and/or fibroblast culture [4–6], it is recommended as the first screening method, even though the false positive rate is higher than in other kinds of samples. It has been shown that the DBS samples can be stored at room temperature for about 1 month and at -20 °C for more than a year [7].

The most common methods are based on the measurement of enzyme activity using 4-methylumbelliferyl bound fluorescence substrates. These methods are reliable and valid for DBS samples, leucocytes as well as fibroblast culture.

The main factor that causes false-positive result in DBS samples is interference of hemoglobin with the fluorescent signals [8–10]. To overcome hemoglobin interference, other authors have proposed pre-analytical techniques based on the precipitation of hemoglobin with either 4% TCA or barium hydroxide/zinc sulfate before analysis [9, 10]. However, it was shown that sample blank correction provided accurate data in a simple but time-consuming way for the measurement of all three enzymatic activities explained in this chapter [11].

The 4-methyl umbelliferyl-α-D-glucoside synthetic substrate is used to determine α-glucosidase enzyme activity to diagnose Pompe disease. In order to prevent false-negative results due to interference of maltase-glucoamylase (MGA, E.C. 3.2.1.20), which is found in neutrophils, the specific MGA inhibitors acarbose and maltose are used [8, 12]. Acarbose selectively inhibits MGA activity even at low concentrations and is suitable both in fluorometric and tandem mass spectrometric detection methods.

Recently, the LC-MS/MS technique is recommended due to the accuracy of the analysis, low recall rate as well as the advantages of multiplex measurement (five or six enzymes in one sample). There are different MS/MS and LC-MS/MS methods currently used for lysosomal enzyme activity detections. Some of them do not involve UPLC/HPLC column separation but instead acquire offline liquid–liquid extractions and solid phase extraction (SPE) steps prior to MS/MS analysis. Such preparation steps are time-consuming and use of SPE plates for solid phase extraction is costly. The LC-MS/MS method explained in this chapter uses a UPLC column chromatographic step prior to the MS/MS analysis and consequently no offline extractions are required, saving time and money [13, 14].

2 Materials (See Notes 1–3)

2.1 Fluorometric Analysis Methods

2.1.1 Materials for α-Glucosidase Measurement

1. Buffer solution: 0.15 M Citric acid/phosphate buffer, pH 4.0, Citric acid mono hydrate ($C_6H_8O_7 \cdot H_2O$) 31.5 g/1000 ml distilled water, sodium phosphate dibasic anhydrous (Na_2HPO_4) 21.3 g/1000 ml distilled water. Mix about 620 ml of citric acid with 380 ml of sodium phosphate in an Erlenmeyer flask and adjust pH value to 4.0. This buffer might be stored at -20 °C up to 1 year.
2. Acarbose buffer: 6 μM Acarbose. Add 3.87 mg into a volumetric flask and solve in 1 l of citric acid/phosphate buffer, pH 4.0.
3. Substrate buffer: 2 mM 4-Methyl umbelliferyl (4-MU)-α-D-glucoside. Dissolve 0.68 mg in 1 ml of citric acid/phosphate buffer in a glass tube (*see* **Note 4**).
4. Stop buffer: 0.1 M Ethylenediamine. Dilute 660 μl in 100 ml of distilled water (*see* **Note 5**).

2.1.2 Materials for β-Glucosidase Measurement

1. Buffer solution: 0.4 M Citric acid/phosphate buffer, pH 5.2, Citric acid mono hydrate ($C_6H_8O_7 \cdot H_2O$) 8.41 g/100 ml distilled water, sodium phosphate dibasic anhydrous (Na_2HPO_4) 11.36 g/200 ml distilled water. Mix 48 ml citric acid monohydrate with 52 ml sodium phosphate in an Erlenmeyer flask and adjust pH value to 5.2.
2. Inhibitor buffer: 15 mM Sodium-taurodeoxycholate. Add 78.25 mg Na-taurodeoxycholate into a volumetric flask and solve in 10 ml of distilled water.
3. Substrate buffer: 20 mM 4-MU-β-D-glucopyranoside. Dissolve 6.8 mg in 1 ml of distilled water in a glass tube (*see* **Note 4**).
4. Stop buffer: 0.1 M Ethylenediamine. Dilute 660 μl of Ethylenediamine in 100 ml of distilled water (*see* **Note 5**).

2.1.3 Materials for α-Galactosidase Measurement

1. Buffer solution: 0.15 M Citric acid/phosphate buffer, pH 4.4, Citric acid mono hydrate ($C_6H_8O_7$)- 3.15 g/100 ml distilled water, Sodium phosphate dibasic anhydrous (Na_2HPO_4) 2.13 g/100 ml distilled water. Mix 56 ml of citric acid monohydrate with 44 ml of sodium phosphate in an Erlenmeyer flask and adjust pH value to 4.0.
2. NAG: 0.25 M N-acetyl-D-galactosamine (NAG). Add 55.3 mg into a glass tube and solve in 1 ml of distilled water.
3. Substrate buffer: 8 mM 4-MU-α-D-galactoside. Dissolve 2.706 mg in 1 ml of citric acid/phosphate buffer (0.15 M, pH 4.4) in a glass tube (*see* **Note 4**).
4. Stop buffer: 0.1 M Ethylenediamine. Dilute 660 μl in 100 ml of distilled water (*see* **Note 5**).

2.1.4 Materials for Calibration Curve

1. Stop buffer: 0.1 M Ethylenediamine. Dilute 660 μl in 100 ml of distilled water (*see* **Note 5**).
2. Standard stock solution: 25 mM 4-MU. Solve 11 mg in 2.5 ml of DMSO. Store at -20 °C for up to 1 year, protected from light and moisture.
3. Standard solutions: 25 μM standard solution. Add 10 μl of stock standard (25 mM) to 10 ml of distilled water.

2.2 LC-MS/MS Method [13, 14] (see Notes 1, 3, 4, and 6)

1. Triplex buffer: Sodium phosphate and citrate buffer, pH 4.4, 0.1 M Trisodium citrate dehydrate, 2.9 g/100 ml water (*see* **Note 6**), 0.2 M sodium phosphate monobasic anhydrous (NaH_2PO_4), 2.4 g/100 ml water. Mix about 90 ml of each solution in an Erlenmeyer flask and adjust pH value to 4.4.
2. Acarbose: 0.8 mM Acarbose. Add 5.16 mg into a volumetric flask and solve in 10 ml of water.
3. Sodium taurocholate solution: 0.18 M Na-taurodeoxycholate hydrate. Add 9.6 g into a volumetric flask and solve in 100 ml of water.
4. Triplex solution (*see* **Note** 7):

Completely solve the content of a β-glycosylcerebrosidase (ABG) vial including substrate and internal standard in 2 ml of methanol.

Completely solve the content of lysosomal α-glycosidase (GAA) vial including substrate and internal standard in 2 ml of methanol.

Mix GAA and ABG and evaporate to remove the methanol.

Solve the α-galactosidase (GLA) substrate and the internal standard in 1.8 ml of sodium taurocholate solution.

Add the GLA solution into the vial of GAA + ABG.

Add 0.3 ml of acarbose buffer and 15.9 ml of triplex buffer to the GAA + ABG + GLA mixture to prepare the triplex solution.

Vortex and heat the reagent solution to solve it completely.

5. Mobile phase: 80% acetonitrile, 20% water, 0.2% formic acid.
6. UHPLC column: HPLC analytical column equipped with a UPLC BEH C18 Uniguard pre-column(2.1 mm × 50 mm, 1.7 μm).
7. LC-MS/MS analysis is carried out on a triple quadrupole mass spectrometry with a flow-through-needle sample manager.

3 Methods

3.1 Fluorometric Measurement Methods

3.1.1 α-Glucosidase Measurement [5, 11]

1. Place punches of DBS (3.2 mm diameter) (*see* **Note 8**) (for each sample and blank) in four wells of a 96-well microplate, these are referred to as sample, sample blank, acarbose, and acarbose blank.
2. Add 50 μl of Buffer solution to the sample and sample blank wells while adding 50 μl of acarbose buffer to the acarbose and acarbose blank wells.
3. Add 50 μl of Substrate buffer to sample and acarbose wells.
4. Incubate at 37 °C for 2 h (*see* **Note 9**).
5. Add 200 μl of Stop buffer to all the wells.
6. Add 50 μl of Substrate Buffer (*see* **Note 10**) to sample blank and acarbose blank wells.
7. Measure the fluorescence of the wells in a microplate fluorometer (excitation 365 nm and emission 450 nm) (*see* **Note 11**).

3.1.2 β-Glucosidase Measurement [5, 11]

1. Place punches of DBS (3.2 mm diameter) (*see* **Note 8**) in two wells of a 96-well microplate, these are referred to as sample and sample blank.
2. Add 40 μl of Inhibition buffer to sample and sample blank wells.
3. Add 30 μl of Buffer solution to sample and sample blank wells.
4. Add 50 μl of Substrate buffer only to sample well.
5. Incubate at 37 °C for 20 h (*see* **Note 9**).
6. Add 200 μl of Stop buffer to sample and sample blank wells.
7. Add 50 μl of Substrate buffer (*see* **Note 10**) only to sample blank well.
8. Perform the readings (excitation 365 nm and emission 450 nm) on a microplate fluorometer (*see* **Note 11**).

3.1.3 α-Galactosidase Measurement [4, 6, 11]

1. Place punches of DBS (3.2 mm diameter) (*see* **Note 8**) in two wells of a 96-well microplate, these are referred to as sample and sample blank.

2. Add 20 μl of NAG to Sample and Sample Blank wells.
3. Add 30 μl of Buffer solution to Sample and Sample Blank wells.
4. Add 50 μl of Substrate buffer only to Sample wells.
5. Incubate at 37 °C for 20 h (*see* **Note 9**).
6. Add 200 μl of Stop buffer to Sample and Sample Blank wells.
7. Add 50 μl of Substrate buffer (*see* **Note 10**) only to Sample Blank well.
8. Perform the readings (excitation 365 nm and emission 450 nm) on a microplate fluorometer (*see* **Note 11**).

3.1.4 Calculation of Enzyme Activities

1. Prepare at least triplicate for each standard concentration.
2. Add 100 μl of distilled water into all wells.
3. Add 100 μl of standard (25 μM) in first well by pipetting up and down to get a final concentration of 12.5 μM.
4. Dilute the standards serially by transferring 100 μl from the first well to the consecutive well by pipetting up and down. After 6 transfer steps, discard 100 μl solution from the 7th well (Table 1).

   The last well (8th) should contain only water as a blank.
5. Add 200 μl of Stop buffer to all wells.
6. Perform the readings (excitation 365 nm and emission 450 nm) on a microplate fluorometer (*see* **Note 9**).
7. Plot the fluorescence readings (after subtraction of blank) against the corresponding concentrations and calculate the linear regression and first-degree polynomial equation

Table 1
Dilution of standards for the standard curve

	Water	Standard	Concentration after dilution (nmol/100 μl)	Stop buffer	Final concentration (nmol/well)
A	100 μl	100 μl of 25 μM	1.25	200 μl	0.417
B	100 μl	100 μl of 12.5 μM	0.625	200 μl	0.208
C	100 μl	100 μl of 6.25 μM	0.3125	200 μl	0.104
D	100 μl	100 μl of 3.125 μM	0.156	200 μl	0.052
E	100 μl	100 μl of 1.56 μM	0.078	200 μl	0.026
F	100 μl	100 μl of 0.78 μM	0.039	200 μl	0.013
G	100 μl	100 μl of 0.39 μM[a]	0.0195	200 μl	0.0065
H	100 μl	–	0	200 μl	0

[a]After mixing standard with water, 100 μl of solution will be discarded

($y = ax + b$) using a statistical package program (i.e., MS Excel). The r^2 value for curve fit should be greater than 0.98.

8. Read the enzyme activity using the calibration curve or the polynomial equation. This calculation gives the enzyme activity (nmol/well or nmol/punch) during the total incubation time.
9. Divide the enzyme activity by the incubation time, to calculate the enzyme activity as nmol/punch/hour.

3.2 3.2. LC-MS/MS Method [13, 14] (see Notes 1, 3, 4, and 6)

3.2.1 Sample Preparation Procedures

1. Place punches of DBS (3.2 mm diameter) from each sample (containing about 3.3–3.4 μl of blood if obtained from new born or about 2.8–3.0 μl if from adults) into the wells of a 96-well microplate (*see* **Note 3**).
2. Perform the blank reactions in duplicate by using 3.2 mm diameter disks from blank filter paper.
3. Add 20 μl of triplex solution (containing ABG, GAA, and GLA) into the wells.
4. Cover the plate with a lid and aluminum foil.
5. Incubate for 20 h on a rocking table at 250 rpm set at 37 °C.
6. After incubation quench reaction mixtures in the plates with 60 μl of methanol containing 0.1% formic acid.
7. Mix each well by aspirating and dispensing with a pipette 3 times.
8. Transfer the content to 2 ml-eppendorf tubes.
9. Centrifuge the mixtures at 750 × *g* for 3 min to remove any suspended paper particles.
10. Transfer 60 μl of the supernatant into wells of the loading plate for MSMS.

3.2.2 Analysis Procedure

1. Separate the analytes on a UPLC BEH C_{18} column (2.1 mm × 50 mm, 1.7 μm).
2. Set the column temperature to 30 °C while the auto-sampler temperature is maintained at 10 °C.
3. Inject 2 μl while operating with a variable composition of mobile phase A (methanol) and mobile phase B (0.05% formic acid in 98:2% water/methanol).
4. Elute at 0.5 ml/min using the following gradient:

0% A (0–1.0 min)
0% A → 100% A (1.0–2.0 min)
100% A → 100% A (2.0–4.0 min)
100% A → 0% A (4.0–4.5 min)

MS/MS conditions are summarized in Tables 2 and 3 and the graphical is presented in Fig. 1.

Table 2
The MS/MS conditions for each enzyme

Analyte	MRM transition	Cone voltage (V)	Collision energy (eV)	Retention time (min)
β-Glycosylcerebrosidase (ABG)	482.3>264.2	18	20	3.12
ABG-Internal Standard (IS)	510.3>264.2	18	20	3.39
α-Galactosidase (GLA)	484.3>384.2	20	20	2.33
GLA-Internal Standard (IS)	489.3>389.2	20	20	2.31
α-Glycosidase (GAA)	498.3>398.2	18	20	2.35
GAA-Internal Standard (IS)	503.3>403.3	18	20	2.35

Table 3
Ion mode and ESI condition for enzyme activity measurement

Ion mode	ESI (+)
Capillary Voltage(V)	3
Cone (V)	18 (20 GLA)
Extractor(V)	3
RF Lens (V)	2.5
Source temperature (°C)	150
Desolvation temprature (°C)	150
Cone Gas flow (L/h)	50
Desolvation Gas flow (L/h)	600
LM 1 Resolution	15
HM 1 Resolution	15
Ion Energy 1	0.5
MS MS mode Entrance	30
Collision	20
MS MS mode Exit	30
LM 2 Resolution	15
HM 2 Resolution	15
Ion Energy 2	0.5
Collision gas	Argon

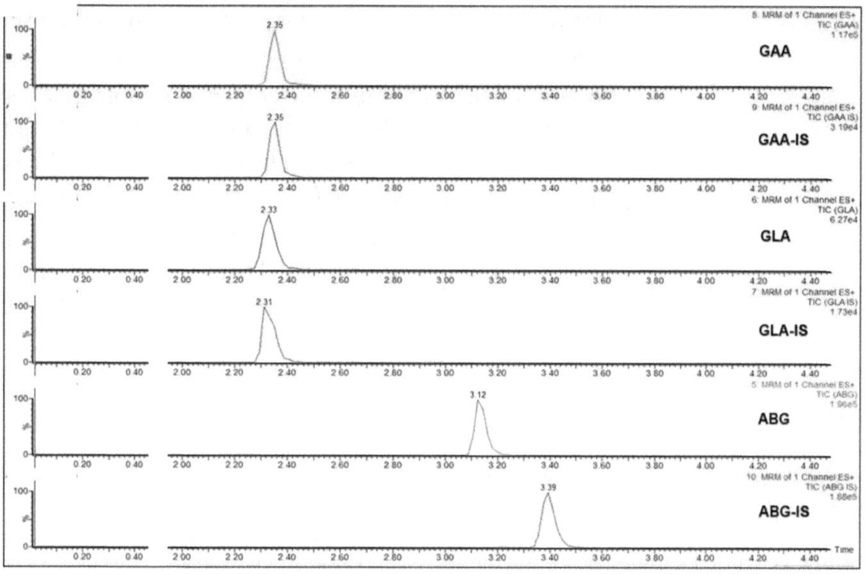

Fig. 1 Total chromatogram profiles of internal standard (IS) and patient samples with triple Quadrupole MRM (multiple reaction monitoring)-MS showing the peaks for β-glycosylcerebrosidase (ABG), ABG-IS, α-galactosidase (GLA), GLA-IS, α-glycosidase (GAA), GAA-IS

4 Notes

1. All solutions, especially substrates, should be prepared in glass tubes/Erlenmeyer flasks. Other materials (i.e., polyethylene, polyurethane) might affect the fluorescence of solutions, especially after the ultra-sonication process.
2. Use analytical grade chemicals and prepare the solutions in ultrapure distilled water.
3. Use powder-free gloves in order to prevent interference with the powder.
4. The substrate should be heated to +50 °C and sonicated in an ultrasonic bath until completely dissolved. Substrate solutions should be prepared once and aliquoted into small volumes (approximately 1 ml) in eppendorf tubes and stored at -20 °C until used. It is best to use all solution freshly prepared. Once thawed, the substrate should not be refrozen.
5. The stop solution can be kept at +4 °C in a dark bottle up to 6 months.
6. Use LC-MS/MS grade water.
7. The substrates and internal standards are collected from Center for Disease Control and Prevention.
8. If not analyzed immediately, the DBS samples should be kept in the freezer (-20 or -80 °C) and kept dry. If a manual puncher is used, the puncher should be cleaned by clean paper at least three times between each patient sampling.

9. During the incubation period, the wells should be sealed with a plate sealer in order to prevent evaporation.
10. Because of hemoglobin interference, a sample blank should be run with each sample and the substrate should be added into sample blank in the last step. The substrate should be kept in the same conditions as sample (i.e., incubation at 37 °C for 2–20 h).
11. Any type of microplate, white or black plate may be used since it does not affect the results. However, the calibration curve should be prepared in accordance with the same type of plate as is used to calculate the enzyme activity.

References

1. Pastores GM (2010) Lysosomal storage disorders, principles and practice. World Scientific Publishing Co, Singapore
2. Matern D, Gavrilov D, Oglesbee D, Raymond K, Rinaldo P, Tortorelli S (2015) Newborn screening for lysosomal storage disorders. Semin Perinatol 39:206–216
3. Kingma SDK, Bodamer OA, Wijburg FA (2015) Epidemiology and diagnosis of lysosomal storage disorders; challenges of screening. Best Pract Res Clin Endocrinol Metab 29:145–157
4. Lukacs Z, Hartung R, Beck M, Keil A, Mengel E (2007) Direct comparison of enzyme measurements from dried blood and leukocytes from male and female Fabry disease patients. J Inherit Metab Dis 30:614
5. Lukacs Z, Cobos PN, Mengel E, Hartung R, Beck M, Deschauer M, Keil A, Santer R (2010) Diagnostic efficacy of the fluorometric determination of enzyme activity for Pompe disease from dried blood specimens compared with lymphocytes—possibility for newborn screening. J Inherit Metab Dis 33:43–50
6. Chamoles NA, Blanco M, Gaggioli D (2001) Fabry disease: enzymatic diagnosis in dried blood spots on filter paper. Clin Chim Acta 308:195–196
7. de Jesus VR, Zhang XK, Keutzer C, Bodamer OA, Mühl A, Orsini JJ, Caggana M, Vogt FV, Hannon WH (2009) Development and evaluation of quality control dried blood spot materials in newborn screening for lysosomal storage disorders. Clin Chem 55(1):158–164
8. Reusera AJ, Verheijena FW, Balib D, van Diggelena OP, Germain DP, Hwud WL, Lukacs Z, Mühlf A, Olivovag P, Piraudh M, Wuytsi B, Zhang K, Keutzerg J (2011) The use of dried blood spot samples in the diagnosis of lysosomal storage disorders — Current status and perspectives. Mol Genet Metab 104:144–148
9. Shigeto S, Katafuchi T, Okada Y, Nakamura K, Endo F, Okuyama T, Takeuchi H, Kroos MA, Verheijen FW, Reusera AJ (2011) Improved assay for differential diagnosis between Pompe disease and acid α-glucosidase pseudodeficiency on dried blood spots. Mol Genet Metab 103:12–17
10. Oemardien LF, Boer AM, Ruijter GJ, van der Ploeg AT, de Klerk JB, Reuser AJ, Verheijen FW (2011) Hemoglobin precipitation greatly improves 4-methylumbelliferonebased diagnostic assays for lysosomal storage diseases in dried blood spots. Mol Genet Metab 102:44–48
11. Sözmen EY (2012) Letter to Editor. Sample blank subtraction outreachs Hemoglobin interferences in flurorometric methods for DBS. Mol Genet Metab 105(3):530
12. Zhang H, Kallwass H, Young SP, Carr C, Dai J, Kishnani PS, Millington DS, Keutzer J, Chen YT, Bali D (2006) Comparison of maltose and acarbose as inhibitors of maltase-glucoamylase activity in assaying acid alpha-glucosidase activity in dried blood spots for the diagnosis of infantile Pompe disease. Genet Med 8(5):302–306
13. Orsini JJ, Martin MM, Showers AL, Bodamer OA, Zhang XK, Gelb MH, Caggana M (2012) Lysosomal storage disorder 4 + 1 multiplex assay for newborn screening using tandem mass spectrometry: Application to a small-scale population study for five lysosomal storage disorders. Clin Chim Acta 413:1270–1273
14. La Marca G, Casetta B, Malvagia S, Guerrini R, Zammarchi E (2009) New strategy for the screening of lysosomal storage disorders: the use of the online trapping-and-cleanup liquid chromatography/mass spectrometry. Anal Chem 81:6113–6121

Chapter 18

Prenatal Diagnosis of Lysosomal Storage Disorders Using Chorionic Villi

Jyotsna Verma, Sunita Bijarnia-Mahay, and Ishwar C. Verma

Abstract

Prenatal enzymatic diagnosis for an array of lysosomal storage disorders (LSDs) can be performed accurately, provided that a confirmed diagnosis by biochemical/molecular study in the index case is available and a strict defined protocol, specific to each individual disorder is followed. The present chapter describes the protocols for reliable and accurate prenatal enzymatic diagnoses by fluorometric and spectrophotometric methods of lysosomal storage disorders: Gaucher, Fabry, Pompe, Niemann Pick A/B, Tay Sach, Sandhoff, GM1, Mucoplysaccharidoses, Wolman, Krabbe, Metachromatic leukodystrophy, and Batten diseases using uncultured chorionic villi samples. The biological reference intervals for enzyme levels in normal and affected fetuses are given for interpretation of prenatal results. It is imperative to establish normal reference interval in each laboratory to take into account the local environment, technical variations, and different ethnicities. Besides, enzyme activity in the fetus should be represented as percentage of the mean activity of enzyme of normal fetuses. The pitfalls and challenges in prenatal diagnosis as well as technical problems in performing enzyme assays are also discussed to help the reader in standardization and performing the assays for correct diagnosis.

Key words Prenatal diagnosis, Lysosomal storage disorders, Enzymatic diagnosis, Chorionic villi, Reference intervals

1 Introduction

Lysosomal storage disorders (LSDs) remain a significant cause of morbidity in the world, especially in developing countries, as the cost of treatment is either prohibitive for most patients or many of these disorders have currently no treatment. This renders genetic counseling and prenatal diagnosis (PND) immensely significant. Majority of LSDs have autosomal recessive inheritance with a 25% chance of penetrance except for Fabry and Hunter (MPS II) diseases, which are X-linked, and therefore have a 50% chance of penetrance in male offspring [1–3]. An essential prerequisite for PND is a confirmed diagnosis in the index patient, either by enzyme assays or molecular studies. The latter provide a specific diagnosis, however the identification of mutations in the proband is not

Karin Öllinger and Hanna Appelqvist (eds.), *Lysosomes: Methods and Protocols*, Methods in Molecular Biology, vol. 1594,
DOI 10.1007/978-1-4939-6934-0_18,

always possible because of the cost and lack of expertise. At times, the observed variant is difficult to classify as pathogenic or non-pathogenic. As a consequence, enzyme assays remain the most preferable approach for prenatal diagnosis. The inter- and intra-assay variations remain less than the threshold value (coefficient of variation: 15%). These assays have high sensitivity (97.7%) and specificity (99.5%) as reported by us recently [1]. As cultured chorionic villi and amniotic fluid samples take long time before providing a diagnosis, such tests are important only for confirmation of initial results. Overgrowth of maternal cells in comparison to fetal cells may also give incorrect results. Therefore, uncultured chorionic villi (CV) are the preferred samples for PND [4–6]. However, in cases where a high residual activity of the enzyme is detected (enzyme value is lower than the normal range but higher than the affected range) and mutation studies are not possible, the final confirmation is sometimes achieved after culturing chorionic villi or amniocytes.

1.1 Quality Management

Compliance with the internal quality control criteria and participation in external quality control schemes have now become essential to improve the diagnostic proficiency of the laboratory. It is necessary to establish the limit of detection and testing range for each enzyme in order to validate the assay. To minimize error, coefficient of variation of the analyses should be monitored regularly. European research network for evaluation and improvement of screening, diagnosis, and treatment of inherited disorders of metabolism (ERNDIM) in UK operates a quality assurance program for enzyme assays for the diagnosis of inborn errors of metabolism that provides scope to improve the testing quality [7]. Biological reference interval should be set up by analyzing one's own population data to interpret the results of these critical disorders.

1.2 Clinical Challenges

Since a definitive diagnosis is an essential prerequisite for any prenatal testing, the clinician is faced with the responsibility of providing an accurate diagnosis leading to correct enzyme assay followed by prenatal testing. Practically, this is not always possible. The most commonly encountered reason is the non-availability of the proband (i.e., death of proband prior to testing). Less often, there is a clinical overlap or atypical presentation of symptoms that leads to the delay in diagnosis. The presence of pseudo-deficiency also poses a problem, which can lead to a false diagnosis. This is resolved only after genetic testing and simultaneous parental testing. The correlation of the enzymatic results with clinical symptoms sometimes poses a challenge, especially in rare conditions where they are presented with atypical symptoms. For this reason, mutation analysis is recommended for the absolute confirmation of the diagnosis.

1.3 Basic Principle of the Assay of Enzyme

In acidic condition, hydrolysis of chromogenic/fluorogenic synthetic substrate by the enzyme results in the release of colored/fluorescent products. Enzyme activity can be determined in nmol/h/mg of protein by measuring the absorbance of the chromogenic or the fluorescence of the fluorogenic products, after stopping the reaction with alkaline buffer.

Though various advanced technologies have emerged for the diagnosis of lysosomal storage disorders, PND using uncultured CV is possible by uncomplicated bench-based spectrophotometric/fluorometric techniques [1–6, 8]. This chapter presents the enzymatic protocols for the prenatal diagnoses of LSDs using uncultured CV samples. For other details, readers can also refer to a recent article by the authors [1]. Table 1 lists the various lysosomal storage disorders, gene inheritance, the specific enzyme, accumulated material, clinical features in brief and provides biological reference intervals in normal and affected fetuses.

2 Materials

2.1 Samples

Collect CV sample at gestational week 11–13 from pregnant woman with confirmed diagnosis in the index case either by enzyme assay or mutation studies (*see* **Note 1**).

2.2 Standards

4-Methylumbelliferone (4-MU): Prepare a 50 μM stock solution of 4-MU in distilled water. The stock solution can be stored at −80 °C for 2 years.

p-Nitrocatechol (4-NC): Prepare stock solution of 4-NC in distilled water (0.2 mM) and store at −80 °C for 2 years.

p-Nitrophenol (4-NP): Prepare a 10 μM stock solution of 4-NP in distilled water and store at −80 °C for 2 years.

7-amino-4-methyl coumarin: Prepare a 59.4 μM stock solution of 5.2 mg 7-amino-4-methyl coumarin by solving in 1 mL DMSO and then make up the volume to 500 mL with 0.1 M phosphate buffer, pH 6.5.

2.3 Protein Estimation

1. Reagent a: prepare 2% sodium carbonate in 0.1 M sodium hydroxide.
2. Reagent b: prepare 1% copper sulfate in distilled water.
3. Reagent c: prepare 2% sodium potassium tartrate in distilled water.
4. Reagent d (copper tartrate reagent): Mix 100 μL of reagent b with 100 μL of reagent c, make up to 10 mL with reagent a.
5. Reagent e (Folin Ciocalteau reagent): Dilute 2:3 with distilled water.
6. Standard BSA (bovine serum albumin): 0.5 mg/mL in distilled water (*see* **Note 2**).

Table 1
Lysosomal disorders, inheritance, enzyme deficiency, accumulated material in the cells/organs, clinical features, and biologicalreference intervals for affected and normal fetuses

Disorders	Gene inherit-ance[a]	Enzyme deficiency (*see* Note 29)	Accumulated material	Clinical features in brief	Biological reference interval in uncultured chorionic villi samples[b] (*see* Notes 26–28)	
					Unaffected fetus	Affected fetus
Gaucher disease	GBA, *Autosomal recessive*	β-Glucocerebrosidase or β-glucosidase	Accumulation of Glucocerebrosidase in the cells of monocytes-macrophage system	Splenomegaly, anemia, or thrombocytopenia, chronic fatigue, hepatomegaly, bone pain, or pathologic fractures. Neuronopathic form is also known	33.0–294.0 (*n* = 30), Mean: 70.8	0.062–10.0 (*n* = 10), Mean: 7.5
Niemann Pick A/B disease	SMPD1, *Autosomal recessive*	Sphingomyelinase	Accumulation of Sphingomyelin (ceramide phosphoryl choline) in systemic organs (liver, spleen) and brain	Swelling of the abdomen due to hepatoslenomegaly. This usually occurs around 3–6 months of age. Swollen lymph nodes and macular cherry-red spot. Neuronopathic form is also known	70.0–120.0 (*n* = 19), Mean: 83.7	0.062–50.0 (*n* = 5), Mean: 32.1
GM1 gangliosidose/ MPS IVb/Morquio IVb disease	GLB1, *Autosomal recessive*	β-Galactosidase	Accumulation of sphingolipid GM1 ganglioside, glycol-protein derived oligosaccharides and keratan sulfate in cells of the central and peripheral nervous system	Coarse facial *features*, hepatosplenomegaly, dysostosis multiplex, macular cherry-red spots, and developmental delay/ arrest (followed by progressive neurologic deterioration)	200.0–2457.0 (*n* = 70), Mean: 693.5	0.062–100.0 (*n* = 17), Mean: 79.5

Tay Sach (GM2 gangliosidosis) disease	HEX A, *Autosomal recessive*	β-Hexosaminidase A	GM2 ganglioside accumulates abnormally in the nerve cells of the brain	Deafness, progressive blindness, macular cherry-red spots, decreased muscle strength, increased startle response, paralysis or loss of muscle function, seizure, spasticity, delayed mental and social development	200.0–1578.0 (*n* = 38), Mean: 689.0	0.062–150.0 (*n* = 8), Mean: 79.5
Sandhoff-Jatzkewitz disease	HEX B, *Autosomal recessive*	Total Hexosaminidase	GM2 ganglioside, meganeurites, asialo-GM2 accumulate in brain. Globoside and oligo-saccharides accumulate in liver and other non-neurogenic tissues	Seizures, vision and hearing loss, intellectual disability, and paralysis. Some have organomegaly or bone abnormalities. In late onset form, muscle weakness, loss of muscle coordination (ataxia), speech problems, and mental illness can be seen	2700.0–14,900.0 (*n* = 10), Mean: 11700.0	0.062–200.0 (*n* = 4), Mean: 112.6
Metachromatic leukodystrophy (MLD) disease	ARSA, *Autosomal recessive*	Aryl sulfatase A	Sulfatides and other sulfated glycolipids build up in many tissues of the body	Progressive motor and cognitive decline, difficulty walking, behavior problems, decreased mental function, difficulty eating or feeding, and incontinence can be seen	18.5–85.6 (*n* = 30), Mean: 32.5	0.062–6.0 (*n* = 5), Mean: 4.8 (*see***Note 29**)

(continued)

Table 1 (continued)

Disorders	Gene inherit-ance[a]	Enzyme deficiency (*see* Note 29)	Accumulated material	Clinical features in brief	Biological reference interval in uncultured chorionic villi samples[b] (*see* Notes 26–28)	
					Unaffected fetus	Affected fetus
Krabbe disease	GALC, *Autosomal recessive*	β-Galactocerebrosidase or galactosylceramidase	Accumulation of galactosylceramide in the patho-gnomonic "globoid cells" in the white matter of brain. Psychosine in oligo-dendrocytes and Schwann cells in the brain	Developmental regression, irritability in infancy, loss of head control, frequent vomiting, seizures, progressive loss of hearing and sight, rigid, constricted muscles, stiff, fixed posture, progressive loss of ability to swallow and breathe	18.0–49.0 ($n = 15$), Mean: 23.6	0.062–14.0 ($n = 5$), Mean: 10.6
Pompe (Glycogen storage disease type II or GSD type II) disease	GAA, *Autosomal recessive*	α-Glucosidase	Accumulation of glycogen in lysosome	Infantile form: Hypertrophic cardiomyopathy, cardiorespiratory failure and generalized marked muscle weakness (floppy infant). Late onset form: Skeletal and respiratory muscle disease	132.0–525.0 ($n = 24$), Mean: 186.4	0.062–50.0 ($n = 5$), Mean: 31.7

Fabry disease	GLA, X-linked	α-Galactosidase A	Accumulation of Gb3 in various tissues including vascular endothelial, perithelial and smooth muscle cells	Acroparesthesias, angiokeratomas, renal failure, hypohidrosis, corneal opacity, problems with the gastrointestinal system, tinnitus and hearing loss, hypertrophic cardiomyopathy, stroke	25–70 (n = 5), Mean: 42.5	NA
NCL1 (also known as Neuronal ceroid lipo-fuscinoses type 1 or Infantile NCL or Santavyori-Haltia disease or infantile Batten disease or CLN1)	PPT1, *Autosomal recessive*	Palmitoyl protein thioesterase (PPT)	Accumulation of autofluorescent ceroid lipopigments in lysosomes	Neuronal degeneration, visual impairment, progressive myoclonic epilepsy, decline in motor skills resulting in premature death	110.0–547.0 (n = 9), Mean: 322.9	0.062–2.5 (n = 3), Mean: 1.26
NCL2 (late infantile neuronal ceroid lipofuscinoses disease or LINCL or Jansky-Bielschowsky disease)	TPP 1 *Autosomal recessive*	Tripeptidyl peptidase 1 (TPP1)	Accumulation of subunit c of mitochondrial ATP synthase (SC-ATP) in neural tissues	The typical early signs are loss of muscle coordination (ataxia) and seizures, along with progressive mental deterioration	40.0–250.0 (n = 15), Mean: 107.7	0.062–10.0 (n = 7), Mean: 5.9
Mucopolysaccharidosis (MPS)	*Gene is given below for specific disease (a–g)*	*Specific enzyme is given below with the type of disease (a–g)*	*The accumulation of undegraded glycosoamino-glycans (GAGs) in the cells as given below*	*Each type of MPS differs clinically. Most of the patients generally experience a period of normal development followed by a decline in physical and/or mental function*	–	–

(continued)

Table 1
(continued)

Disorders	Gene inherit-ance[a]	Enzyme deficiency (*see* Note 29)	Accumulated material	Clinical features in brief	Biological reference interval in uncultured chorionic villi samples[b] (*see* Notes 26–28)	
					Unaffected fetus	Affected fetus
(a) Hurler (MPS I) disease (*Clinically classified into Hurler (MPS IH), Hurler-Scheie (MPS IHS) and non-neurological Scheie (MPS IS)*)	IDUA, *Autosomal recessive*	α-Iduronidase	Dermatan sulfate and chondroitin sulfate	Coarse facies, short stature, corneal clouding, joint stiffening, umbilical hernia, dysostosis multiplex, hepato-splenomegaly, severe airway obstruction, and little to no intellectual dysfunction	55.0–227.0 (*n* = 10), Mean: 184.8	0.062–25.0 (*n* = 6), Mean: 17.6
(b) Hunter (MPS II) disease	IDS, X-linked	Iduronate 2 sulfatase	Dermatan sulfate and heparan sulfate	Similar to MPS 1, except for corneal clouding	30.0–293.0 (*n* = 15), Mean: 139.5	0.062–8.0 (*n* = 8), Mean: 6.3
(c) Sanfilippo IIIa (MPS IIIa) disease	SGSH, *Autosomal recessive*	Heparan sulfamidase	Heparan sulfate	Severe neurological symptoms such as progressive dementia, aggressive behavior, hyperactivity, seizures, deafness, loss of vision, and an inability to sleep for more than a few hours at a time	23.0–120.0 (*n* = 5), Mean: 32.3	0.062–5.0 (*n* = 4), Mean: 3.8
(d) Sanfilippo IIIb (MPS IIIb) disease	NAGLU, *Autosomal recessive*	α-Hexosaminidase	Heparan sulfate	Similar to MPS IIIa	54.0–217.0 (*n* = 3), Mean: 424.8	0.062–15.0 (*n* = 2), Mean: 9.4

(e) Morquio Iva (MPS IVa) disease	GALNS, *Autosomal recessive*	Galactose 6-sulfatase	Keratan sulfate	Extreme, progressive skeletal changes leading to neurological complications include spinal nerve and nerve root compression, conductive and/or neurosensitive loss of hearing and clouded corneas	95.0–350.0 (n = 7), Mean: 124.1	0.062–12.0 (n = 4), Mean: 6.1
(f) Maroteaux Lamy (MPS VI) disease	ARSB, *Autosomal recessive*	Aryl sulfatase B	Dermatan sulfate, chondroitin sulfate	Clouded corneas, deafness, thickening of the dura, and pain caused by compressed or traumatized nerves and nerve roots. No intellectual disability	54.0–452.0 (n = 100), Mean: 238.6	0.062–25.0 (n = 5), Mean: 15.7
(g) Sly syndrome (MPS VII) disease	GUSB, *Autosomal recessive*	β-Glucuronidase	Dermatan sulfate, heparan sulfate, and chondroitin 6 sulfate	Intellectual impairment, communicating hydrocephalus, nerve entrapment, corneal clouding, and some loss of peripheral and night vision	85.0–399.0 (n = 15), Mean: 178.05	0.062–10.0 (n = 13), Mean: 5.9
Wolman disease	LIPA, *Autosomal recessive*	Lysosomal acid lipase/ esterase	Cholesterol esters abnormally accumulate in the tissues and organs of the body	Severe failure to thrive, diarrhea, abdominal distention, vomiting, and hepato-splenomegaly, death in infancy	75.0–250.0 (n = 5), Mean: 101.2	0.062–10.0 (n = 2)

[a]To have an ***autosomal recessive*** disorder, you inherit two mutated genes, one from each parent. These disorders are usually passed on by two carriers. Their health is rarely affected, but they have one mutated gene (*recessive* gene) and one normal gene (***dominant*** gene) for the condition

[b]Refer section 3.4 for the unit of enzyme activity.

2.4 Estimation of Lysosomal Enzyme Activities

2.4.1 β-Glucosidase for Gaucher Disease [9–11]

1. Substrate: 4-MU-β-D-glucopyranoside (5 mM) in substrate buffer (*see* **Note 3**).
2. Substrate buffer: 0.2 M sodium acetate, pH 5.5 containing 0.2% sodium tauroglycholate (*see* **Note 4**).
3. Stopping buffer: 0.1 M 2-amino-2-methyl-1-propanol (AMP) buffer, pH 10.3 (*see* **Note 5**).

2.4.2 Acid Sphingomyelinase (ASM) for Niemann Pick A/B [12]

1. Substrate: 6-hexadecanoylamino 4-MU-phosphorylcholine (0.66 mM) in substrate buffer. Heat for a while at 60 °C to dissolve (*see* **Notes 3** and **6**).
2. Substrate buffer: 0.1 M sodium acetate, pH 5.2 containing 0.2% sodium taurocholate and 0.02% sodium azide.
3. Stop buffer: 0.1 AMP buffer, pH 10.3 (*see* **Note 5**).

2.4.3 β-Galactosidase for GM1/MPS IVb/Morquio IVb Disease [13]

1. Substrate: 4-MU-β-D-galactopyranoside (1 mM) in substrate buffer (*see* **Note 3**).
2. Substrate buffer: 0.1 M citric acid/0.2 M di-sodium hydrogen phosphate buffer, pH 4.0 containing 0.1 M sodium chloride and 0.02% (w/v) sodium azide (*see* **Note 7**).
3. Stop buffer: 0.1 M AMP buffer, pH 10.3 (*see* **Note 5**).

2.4.4 β-Hexosaminidase A for Tay Sach/GM2 Gangliosidosis Disease [14]

1. Substrate: 4-MU-β-N-Ac-glucosaminide-6-sulfate (1 mM) in substrate buffer (*see* **Notes 3** and **8**).
2. Substrate buffer: 0.1 M citric acid/0.2 M di-sodium hydrogen phosphate buffer, pH 4.4 containing 0.02% (w/v) sodium azide.
3. Stop buffer: 0.1 M AMP buffer, pH 10.3 (*see* **Note 5**).

2.4.5 Total Hexosaminidase for Sandhoff-Jatzkewitz Disease [1]

1. Substrate: 4-MU-2-acetamido-2-deoxy-β-D-glucopyranoside (4.4 mM) in substrate buffer (*see* **Notes 3** and **8**).
2. Substrate buffer: 0.2 M citric acid/0.4 M di-sodium hydrogen phosphate buffer, pH 4.5 containing 0.02% (w/v) sodium azide.
3. Stop buffer: 0.1 M AMP buffer, pH 10.3 (*see* **Note 5**).

2.4.6 Aryl Sulfatase A for Metachromatic Leukodystrophy (MLD) Disease [15–17]

1. Substrate: p-nitocatechol sulfate di-potassium salt (40 mM) in distilled water (*see* **Notes 3** and **9**).
2. Buffer: 1 M sodium acetate, pH 5.0.
3. Stop solution: 1 M sodium hydroxide.

2.4.7 β-Galactocerebrosidase for Krabbe Disease [18]

1. Substrate: 6-hexadecanoylamino-4-MU-β-D-galactoside (0.9 mM) in chloroform:methanol (C/M), 2:1 (*see* **Note 3**).
2. Take 6.7 μL oleic acid (density: 0.895) and dilute to 1 mL with hexane (*see* **Note 10**).

3. Prepare sodium taurocholate (30 mg/mL in C/M) (*see* **Note 10**).
4. Make a substrate film: Mix 3 mL of oleic acid, 2 mL of taurocholic acid and 8.8 mL of 6-hexadecanoylamino 4-MU-β-D-galactoside in a glass bottle. Aliquot 175 μL of this mixture into 1.5-mL eppendorf tubes and incubate the tubes at 37 °C to evaporate all the solvents. Close the lids and seal the tubes (*see* **Notes 3** and **11**).
5. Substrate buffer: 0.2 M citric acid/0.4 M di-sodium hydrogen phosphate buffer, pH 5.2 containing 0.02% sodium azide.
6. Reconstitute the frozen substrate (**step 4**) in 250 μL substrate buffer (**step 5**), sonicate for 10 s at room temperature and centrifuge at 12,857 × *g* for 10 min. Collect the aqueous phase for use in the reaction.
7. Stop buffer: 0.1 M AMP buffer, pH 10.3 (*see* **Note 5**).

2.4.8 α-Glucosidase for Pompe/GSD Type II Disease [19]

1. Substrate: 4-MU-α-D-glucopyranoside (4 mM), dissolve 0.135 g substrate in 50–100 μl of di-methyl sulfoxide (DMSO) and make the volume to 100 mL with substrate buffer. To dissolve the substrate completely, keep in boiling water for a few minutes (*see* **Note 3**).
2. Substrate buffer: 0.1 M citric acid/0.2 M di-sodium hydrogen phosphate buffer, pH 4.0 containing 0.02% (w/v) sodium azide.
3. Stop buffer: 0.1 AMP buffer, pH 10.3 (*see* **Note 5**).

2.4.9 α-Galactosidase for Fabry Disease [20]

1. Substrate: 4-MU-α-D-galactopyranoside (5 mM) in substrate buffer containing N-acetyl-D-galactosamine (125 mM) (*see* **Notes 3** and **12**).
2. Substrate buffer: 0.1 M citric acid/0.2 M di-sodium hydrogen phosphate buffer, pH 4.5 containing 0.02% (w/v) sodium azide.
3. Saline: 0.9% sodium chloride.
4. Stop buffer: 0.1 M AMP buffer, pH 10.3 (*see* **Note 5**).

2.4.10 Palmitoyl Protein Thioesterase (PPT) for NCL 1/Santavyori-Haltia/ Infantile NCL Disease [21]

1. Chloroform:Methanol (C/M), prepare fresh 2:1 (v/v).
2. Triton X-100: Prepare 5% fresh in distilled water.
3. Buffer: 0.2 M citric acid/0.4 M di-sodium hydrogen phosphate buffer, pH 4.0 containing 0.02% (w/v) sodium azide.
4. β-Glu-A-stock: Prepare 100 U/mL β-glucosidase from almonds fresh in distilled water (*see* **Note 13**).
5. Dithiothreitol (33 mM): Prepare fresh in distilled water.
6. Substrate stock solution: Dissolve 2 mg 4-MU-6-thiopalmitoyl-β-D-glucoside (4-MU-6S-Palm-βGlc) in 1 mL C/M (3.4 mM).

7. Preparation of substrate working solution (0.64 mM):
   (a) Mix 150 μL of substrate stock with 60 μL Triton (5%) in 1.5-mL eppendorf tubes.
   (b) Evaporate the solvent in a dry block at 25–30 °C. (Prepare 5–6 such tubes of working substrate each time).
   (c) When dry, add 400 μL of citrate/phosphate buffer, pH 4.0 to each tube and sonicate for 10 s.
   (d) Then add 360 μL of DTT solution and 40 μL β-Glu-A solution. Vortex and pool the content of all tubes. Finally mix the whole content and aliquot in 100 μL portions in small eppendorf tubes and store at −80 °C (*see* **Note 3**).
8. Stop buffer: 0.1 M AMP buffer, pH 10.3 (*see* **Note 5**).

2.4.11 Tripeptidyl Peptidase (TPP) for NCL2/Jansky-Bielschowsky/Late Infantile NCL Disease [1]

1. Substrate: Ala-Ala-Phe-7-amido-4 methyl coumarin (500 μM) in substrate buffer (*see* **Note 3**).
2. Substrate buffer: 0.1 M sodium acetate buffer, pH 4.0.
3. Stop buffer: 0.1 M AMP buffer, pH 10.3 (*see* **Note 5**).

2.4.12 α-Iduronidase for MPS I/Hurler Disease [22]

1. Substrate: 4-MU-α-L-iduronide (2 mM) in substrate buffer containing 0.9% sodium chloride (*see* **Note 3**).
2. Substrate buffer: 0.4 M formate buffer, pH 3.5. Take 1.85 mL formic acid in 80 mL distilled water and adjust pH to 3.5 with 1 M sodium hydroxide. Make up the volume to 100 mL and then add 0.9 g sodium chloride.
3. Inhibitor: 20 mM saccharo-1,4-Lactone in distilled water (*see* **Note 14**).
4. Stop buffer: 0.1 M AMP buffer, pH 10.3 (*see* **Note 5**).

2.4.13 α-Iduronate 2 Sulfatase for MPS II/Hunter Disease [23]

1. Substrate: 4-MU-α-L-Iduronide-2-sulfate sodium salt (1.25 mM). Add 3 mL of substrate buffer to a vial containing 5 mg of the substrate. Dissolve and transfer to a 10-mL tube. Rinse the vial with 3 mL substrate buffer and add to the tube, and add additional 2.33 mL substrate buffer to the 10-mL tube. Divide it in aliquots for storage (*see* **Note 3**).
2. Substrate buffer: 0.1 M sodium acetate buffer, pH 5.0 containing 10 mM lead acetate (*see* **Note 15**).
3. LEBT (lysosomal enzymes purified from bovine testis). Reconstitute one vial with 2.2 mL water and store at −80 °C. It can be thawed and flash frozen a couple of times.
4. Buffer: 0.1 M citric acid/0.2 M di-sodium hydrogen phosphate buffer, pH 4.5 containing 0.02% (w/v) sodium azide.
5. Stop buffer: 0.1 M AMP buffer, pH 10.3 (*see* **Note 5**).

2.4.14 Heparan Sulfamidase for MPS IIIa/ Sanfilippo IIIa Disease [24]

1. Substrate: dissolve 4-MU-α-D-N-sulfoglucosaminide sodium salt (5 mM) in substrate buffer (*see* **Note 3**).
2. Substrate buffer:
   - (a) Sodium barbital/sodium acetate buffer, pH 6.5 containing 0.7% sodium chloride and 0.02% (w/v) sodium azide.
   - (b) Barb/Ac stock A buffer (143 mM): Make 5-mL solution by mixing 130 mg sodium barbitone and 85 mg sodium acetate.
   - (c) Barb/Ac stock B buffer: Dissolve 85 mg sodium chloride in 1 mL of Barb/Ac stock A.
   - (d) Sodium azide 20% (w/v): 200 mg/mL.
   - (e) Prepare the substrate buffer by mixing 1000 μL Barb/Ac stock B with 2500 μL Barb/Ac stock A and 5000 μL distilled water. Adjust pH to 6.5 by adding approx. 3500 μL HCl (0.1 M) and make the volume to 12.5 mL with distilled water. Add 250 μL of sodium azide (20% stock).
3. α-Glucosidase (Sigma from Bacillus stearothermophilus): Prepare a 10 U/mL working solution of α-glucosidase of in distilled water (*see* **Note 16**).
4. Buffer: 0.2 M citric acid/0.4 M di-sodium hydrogen phosphate buffer, pH 6.7 containing 0.02% (w/v) sodium azide.
5. Stop buffer: 0.1 M AMP buffer, pH 10.3 (*see* **Note 5**).

2.4.15 α-N-Acetylglucosaminidase for MPS IIIb/ Sanfilippo IIIb Disease [25]

1. Substrate: 4-Mu-α-N-acetyl glucosaminide (0.25 mM) in substrate buffer (*see* **Note 3**).
2. Substrate buffer: 0.1 M citric acid/0.2 M di-sodium hydrogen phosphate buffer, pH 7 containing 0.02% (w/v) sodium azide.
3. Stop buffer: 0.1 M AMP buffer, pH 10.3 (*see* **Note 5**).

2.4.16 Galactose 6-Sulfatase for MPS IVa/ Morquio IVa Disease [26, 27]

1. Substrate: 4-MU-β-Gal-6-sulfate-triethyl ammonium (10 mM) in 1 mL substrate buffer.
2. Substrate buffer: 0.1 M sodium acetate buffer, pH 4.3 containing 0.1 M sodium chloride and 5 mM lead acetate (*see* **Note 15**).
3. Phosphate buffer: 0.9 M sodium hydrogen phosphate buffer, pH 4.3 containing 0.02% (w/v) sodium azide.
4. β-Galactosidase stock solution (β-Gal-Ao-stock): Add 10 mL distilled water directly to the batch of lyophilized 25,000 U β-D-galactoside galactohydrolase (from Aspergillus oryze) and dissolve by vortexing. From this stock, make a solution of 10 units/mL (4 μL stock/mL distilled water). Store at −80 °C for <2 years (*see* **Note 17**).
5. Stop buffer: 0.1 M AMP buffer, pH 10.3 (*see* **Note 5**).

2.4.17 Aryl Sulfatase B for MPS VI/Mourtaux Lamy Disease [28, 29]

1. Substrate: 4-MU-sulfate (10 mM) in substrate buffer containing 0.6% taurocholic acid (*see* **Note 3**). Dissolve by stirring at 37 °C for 15 min and then cool it.
2. Substrate buffer: 0.2 M sodium acetate buffer, pH 6.0 containing 5 mM barium acetate (*see* **Note 18**).
3. Stop buffer: 0.1 M AMP buffer, pH 10.3 (*see* **Note 2**).

2.4.18 β-Glucuronidase for MPS VII/Sly Disease [1, 28]

1. Substrate: 4-MU-β-D-glucuronidase (5 mM) in substrate buffer (*see* **Note 3**).
2. Substrate buffer: 0.1 M sodium acetate buffer, pH 4.0.
3. 5% trichloro-acetate (TCA) in distilled water.
4. Stop buffer: 0.1 M AMP buffer, pH 10.3 (*see* **Note 5**).

2.4.19 Esterase/Acid Lipase for Wolman Disease [30, 31]

1. Substrate: p-nitro-phenylpalmitate (0.6 μM) in 10 μM glycine–HCl buffer, pH 4.0 (*see* **Note 3**).
2. Substrate buffer: Prepare 1 mM glycine–HCl buffer, pH 4.0 (adjust pH with 1 M HCl). Dilute this buffer 100 times to bring the concentration to 10 μM for the preparation of substrate.
3. 20% perchloric acid in distilled water.
4. Stop buffer: 1 M Tris–HCl buffer, pH 9.0.

3 Methods

3.1 Preparation of Chorionic Villi

Homogenize the chorionic villi sample in distilled water using a sonicator with metal tip and dilute the cell lysate to the desired protein concentration in distilled water. Calculate the protein content in chorionic villi extracts (as given in Subheading 3.3) by the modified method of Lowry et al [32] for further use in the enzyme assays.

3.2 Construction of Standard Curves

3.2.1 4-Methylumbelliferone (4-MU)

1. Make standard solutions of concentrations between 0.062 and 1 nmol/mL in stop buffer (0.1 M AMP).
2. Analyze fluorescence at λ_{ex}365 nm and λ_{em}460 nm.
3. Construct a standard curve by extrapolating the fluorescence count versus concentration (*see* **Note 19**).

3.2.2 p-Nitrocatechol (4-NC)

1. Make standard solutions of concentrations between 0.62 and 10 nmol 4-NC/300 μL of 1 M sodium hydroxide.
2. Analyze the absorbance at λ 500 nm.
3. Construct a standard curve by extrapolating the absorbance versus concentration.

3.2.3 p-Nitrophenol

1. Make standard solutions of concentrations between 10 and 50 nmol/mL in distilled water.
2. Analyze the absorbance at λ 400 nm.
3. Construct a standard curve by extrapolating the absorbance versus concentration.

3.2.4 7-Amino-4-Methyl Coumarin

1. Take 2.5 mL from the stock solution and dilute 8 times to make the concentration 7.4 μM. Take 200 μL of this 7.4 μM standard and add 800 μL of 0.1 M sodium chloroacetate prepared in 30 mM sodium acetate buffer, pH 4.3.
2. Make standard solutions of concentrations between 0.1 and 1.5 nmol/mL in stop buffer (0.1 M AMP).
3. Analyze the fluorescence at λ_{ex}365 nm and λ_{em}440 nm.
4. Construct a standard curve by extrapolating the fluorescence versus concentration.

3.3 Protein Analysis

1. Take 10 μL of sample (cell lysate) and dilute it with distilled water to the final volume of 100 μL.
2. Add 0.5 mL of reagent d, mix and incubate for 10 min.
3. Add 0.05 mL reagent e, mix and incubate for 30 min.
4. Prepare a standard curve of protein every month by taking different concentrations of BSA (0–50 μg/100 μL) to check the quality of reagents. Run one standard (50 μg/100 μL BSA) with each assay to estimate the protein in the samples.
5. For blank, take 100 μL distilled water and treat like the sample.
6. Measure the absorbance at λ 500 nm after 30 min using water as a blank to set the spectrophotometer at zero.
7. Calculate the protein concentration according to the below given formula:

$$\frac{(\text{Sample}-\text{blank})\,\text{absorbance} \times 0.5 \times 10}{(\text{Standard}-\text{blank})\,\text{absorbance}} = (\text{mg protein}/\text{mL})$$

3.4 Estimation of Lysosomal Enzyme Activities

3.4.1 β-Glucosidase for Gaucher Disease [9–11]

1. Take 25 μL cell lysate (1 mg/mL protein) of patient/controls and incubate with 25 μL substrate for 30 min at 37 °C in shaking water bath.
2. For blank, incubate 25 μL of substrate in a separate tube and treat as the sample.
3. After 30 min of incubation stop the reaction by adding with 950 μL of AMP buffer.
4. After stopping the reaction, add sample in the blank tube.
5. Measure fluorescence at λ_{ex}355 nm and λ_{em}460nm by transferring 200 μL of the reaction mixture into a microplate (*see* **Notes 20–22**).

6. In the standard tube, take 20 μL of 4-MU of the concentration 50 μM and dilute it with stopping buffer to 1 mL (no incubation is required).
7. Calculate the activity in nmol/h/mg of protein using 4-MU (1 nmol/mL) as a standard using the following formula:

$$\frac{(\text{Sample}-\text{blank})\ \text{fluorescence} \times 40 \times 2}{(\text{Standard}-\text{blank})\ \text{fluorescence} \times \text{protein}\ (\text{mg/ml})} = \text{enzyme activity}\ (\text{nmol/h/mg})$$

3.4.2 Sphingomyelinase for Niemann Pick A/B [12]

1. Take 10 μL of cell lysate (10 μg protein) of patient and control samples.
2. Add 20 μL of substrate and incubate for 1 h at 37 °C. For blank, incubate substrate and samples in separate tubes.
3. Add 200 μL stop buffer in all tubes except blanks.
4. In blank tubes, add 200 μL stop buffer just before mixing incubated substrate with samples (**step 2**).
5. Analyze the fluorescence of MU at λ_{ex}355 nm and λ_{em}460 nm using a fluorometer (*see* **Notes 20–22**).
6. Calculate the activity of enzyme in nmol/17 h/mg by using 4-MU (1 nmol/mL) as a standard:

$$\frac{(\text{Sample}-\text{blank})\ \text{fluorescence} \times 23 \times 4.3}{(\text{Standard}-\text{blank})\ \text{fluorescence} \times \text{protein}\ (\text{mg/ml})} = \text{enzyme activity}\ (\text{nmol/17 h/mg})$$

3.4.3 β-Galactosidase for GM1/MPS IVb/Morquio IVb Disease [13]

1. Take 10 μL cell lysate (1 mg/mL protein) of patient and controls and incubate with 20 μL substrate for 5 min at 37 °C in a shaking water bath.
2. For blank, incubate substrate and samples in separate tubes.
3. After 5 min incubation, stop the reaction with 970 μL of AMP stop buffer. In blank tubes, add 970 μL stop buffer just before mixing the incubated substrate with samples (**step 2**).
4. Read the fluorescence at λ_{ex}355 nm and λ_{em}460 nm by transferring 200 μL of the reaction mixture into a microplate (*see* **Notes 20–22**).
5. Calculate the activity in nmol/h/mg of protein by using 4-MU (1 nmol/mL) as a standard:

$$\frac{(\text{Sample}-\text{blank})\ \text{fluoresence} \times 23 \times 4.3 \times 12}{(\text{Standard}-\text{blank})\ \text{fluoresence} \times \text{protein}\ (\text{mg/mL})} = \text{enzyme activity}\ (\text{nmol/h/mL})$$

3.4.4 β-Hexosaminidase A for Tay Sach/GM2 Gangliosidosis Disease [14]

1. Take 10 μL diluted cell lysate (3 μg protein) of patient and controls. Add 20 μL of substrate and incubate for 1 h at 37 °C.
2. For blank, incubate substrate and samples in separate tubes.

3. After incubation, add 200 μL stop buffer in all tubes except the blank. In blank tubes, add 200 μL stop buffer just before mixing incubated substrate with samples (**step 2**).
4. Read the fluorescence of MU at λ_{ex}355 nm and λ_{em}460 nm using a fluorometer (*see* **Notes 20–22**).
5. Calculate the activity of enzyme in nmol/h/mg of protein by using 4-MU (1 nmol/mL) as a standard:

$$\frac{(\text{Sample}-\text{blank})\,\text{fluorescence}\times 23}{(\text{Standard}-\text{blank})\,\text{fluorescence}\times\text{protein}\,(\text{mg}/\text{mL})}=\text{enzyme activity}\,(\text{nmol}/\text{h}/\text{mL})$$

3.4.5 Total Hexosaminidase for Sandhoff-Jatzkewitz Disease [1]

1. Take 5 μL (1 mg/mL) cell lysate of patient and control samples and dilute with 500 μL substrate buffer.
2. Take 100 μL of this diluted enzyme in two tubes (one for sample and other for blank) separately.
3. Add 100 μL of citrate-phosphate buffer to all tubes.
4. Add 100 μL of substrate to all tubes except blank. For blank, incubate the substrate separately.
5. Incubate all tubes at 37 °C for 15 min.
6. Stop the reaction by adding 800 μL stopping solution and in blank tubes, add the incubated substrate just before making the reading using a fluorometer at λ_{ex}355 nm and λ_{em} 460 nm.
7. Calculate the activity in nmol/h/mg by taking the fluorescence of the blank, samples and 4-MU (1 nmol/mL) standard (*see* **Notes 20–22**):

$$\frac{(\text{Sample}-\text{blank})\ \text{fluorescence}\times 4}{(\text{Standard}-\text{blank})\,\text{fluorescence}}=\text{enzyme activity}\,(\text{nmol}/\text{h}/\text{mg})$$

3.4.6 Aryl sulfatase A for Metachromatic Leukodystrophy (MLD) Disease [15–17]

1. Use two tubes for each assay, sample and sample blank.
2. Use two tubes for standards, standard and standard blank.
3. Add 100 μL of buffer in sample blank, sample and standard tubes.
4. Add 60 μL (approx. 20 μg protein) cell lysate in the tubes set for the patient/control samples.
5. Add 60 μL distilled water in standard blank and 50 μL standard with 10 μL distilled water in standard tube.
6. Incubate 60 μL cell lysate with 100 μL buffer in a separate tube (blank).
7. Add 40 μL of substrate in all tubes except the tube in **step 6**.
8. Incubate for 16–18 h at 0 °C (*see* **Notes 9** and **23**).
9. Add 400 μL stopping solution in all the tubes after incubation.
10. In blank tube (**step 6**), now add 40 μL of substrate and take the readings for all tubes at 500 nm (*see* **Notes 20–22**).

11. Calculate the activity in nmol/h/mg of protein using 4-NC (10 nmol/300 μL) as a standard:

$$\frac{(\text{Sample}-\text{blank})\,\text{absorbance} \times 1000}{(\text{Standard}-\text{blank})\,\text{absorbance} \times 6 \times \text{protein}\,(\text{mg/mL}) \times T\,(\text{h})} = \text{enzyme activity}\,(\text{nmol/h/mg})$$

3.4.7 β-Galactocerebrosidase for Krabbe Disease [18]

1. Take 10 μL cell lysate (20 μg protein) of patient and controls.
2. Add 20 μL of substrate and incubate for 17 h at 37 °C. For blank, incubate substrate and samples in separate tubes (*see* **Note 23**).
3. Add 200 μL stop buffer in all tubes except blanks. Add 200 μl stop buffer in incubated substrate and then add incubated samples (set for blanks) just before taking the readings.
4. Read the fluorescence of MU at λ_{ex}355 nm and λ_{em}460 nm using a fluorometer (*see* **Notes 20–22**).
5. Calculate the activity of enzyme in nmol/17 h/mg by taking 4-MU (1 nmol/mL) as a standard:

$$\frac{(\text{Sample}-\text{blank})\ \text{fluorescence} \times 23 \times 4.3}{(\text{Standard}-\text{blank})\ \text{fluorescence} \times \text{protein}\,(\text{mg/mL})} = \text{enzyme activity}\,(\text{nmol/17 h/mg})$$

3.4.8 α-Glucosidase for Pompe/GSD Type II Disease [19]

1. Take 50 μL of cell lysate (approx. 50 μg of protein) of patient and control samples and add 50 μL of substrate.
2. Incubate at 37 °C for 60 min.
3. For blank, incubate substrate and samples in separate tubes. Mix both tubes after adding stop solution in **step 4** just before making the readings.
4. Terminate the reaction by adding 900 μL of stopping buffer (*see* **Notes 20–22**).
5. Analyze the fluorescence at λ_{ex}355 nm and λ_{em}460 nm against the standard (1 nmol/mL) and calculate the enzyme activity in nmol/h/mg of protein by using 4-MU (1 nmol/mL) as a standard:

$$\frac{(\text{Sample}-\text{blank})\,\text{fluorescence} \times 1000 \times 1}{(\text{Standard}-\text{blank})\,\text{fluorescence} \times 50 \times \text{protein}\,(\text{mg/mL})} = \text{enzyme activity}\,(\text{nmol/h/mg})$$

3.4.9 α-Galactosidase for Fabry Disease [20]

1. Take 5 μL cell lysate (approx. 10 μg of protein) of patient and control samples.
2. Add 15 μL saline and 80 μL substrate.
3. Incubate for 1 h at 37 °C.
4. For blank, incubate substrate and samples in separate tubes. Mix both tubes after adding stop solution in **step 5** just before making readings.

5. After 1 h incubation, stop the reaction by adding 900 μL stop buffer and read the fluorescence of MU at λ_{ex}355 nm and λ_{em}460 nm using a fluorometer (*see* **Notes 20–22**).
6. Calculate the activity of enzyme in nmol/h/mg using 4-MU (1 nmol/mL) as a standard:

$$\frac{(\text{Sample}-\text{blank})\ \text{fluorescence} \times 200}{(\text{Standard}-\text{blank})\ \text{fluorescence} \times \text{protein}\ (\text{mg/mL})} = \text{enzyme activity}\ (\text{nmol/h/mg})$$

3.4.10 Palmitoyl Protein Thioesterase for NCL 1/ Santavyori-Haltia/Infantile NCL Disease [21]

1. Take 10 μL cell lysate (30 μg protein) of patient and control samples and add 20 μL of substrate.
2. Incubate for 1 h at 37 °C.
3. For blank, incubate substrate and samples in separate tubes. Mix both tubes after adding stop solution in **step 4** just before making readings.
4. Stop the reaction by adding 200 μL stop buffer and read the fluorescence of MU at λ_{ex}355 nm and λ_{em}460 nm using a fluorometer (*see* **Notes 20–22**).
5. Calculate the activity of enzyme in nmol/h/mg using 4-MU as a standard:

$$\frac{(\text{Sample}-\text{blank})\ \text{fluorescence} \times 23 \times 4.3}{(\text{Standard}-\text{blank})\ \text{fluorescence} \times \text{protein}\ (\text{mg/mL})} = \text{enzyme activity}\ (\text{nmol/h/mg})$$

3.4.11 Tripeptidyl Peptidase for NCL2/ Jansky-Bielschowsky/Late Infantile NCL Disease [1]

1. Take 10 μL cell lysate (30 μg protein) and add 90 μL of buffer.
2. Add 100 μL of substrate and incubate for 1 h. For blank, incubate substrate and samples in separate tubes. Mix both tubes after adding stop solution in **step 3** just before making readings.
3. Stop the reaction by adding 800 μL stop buffer and read the fluorescence of 7-amino-4-methylcoumarin at λ_{ex}365 nm and λ_{em}440 nm using a fluorometer (*see* **Notes 20–22**).
4. Calculate the activity of enzyme in nmol/h/mg using 7-amino-4-methyl coumarin (1.5 nmol/mL) as a standard:

$$\frac{(\text{Sample}-\text{blank})\ \text{fluorescence} \times 1.5 \times 100}{(\text{Standard}-\text{blank})\ \text{fluorescence} \times \text{protein}\ (\text{mg/mL})} = \text{enzyme activity}\ (\text{nmol/h/mg})$$

3.4.12 α-Iduronidase for MPS I/Hurler Disease [22]

1. Take 20 μL cell lysate (1 mg/mL protein) of patient and control samples and add 20 μL of saccharo lactone.
2. Add 20 μL of substrate and incubate for 30 min at 37 °C.
3. For blank, incubate substrate and samples in separate tubes. Mix both tubes after adding stop solution in **step 4** just before making readings.

4. Add 940 μL of stop buffer and read the fluorescence of MU at λ_{ex}355 nm and λ_{em}460 nm using a fluorometer (*see* **Notes 20–22**).
5. Calculate the enzyme activity in nmol/h/mg using 4-MU (1 nmol/mL) as a standard using the equation:

$$\frac{(\text{Sample}-\text{blank})\ \text{fluorescence}\times 100\times 2}{(\text{Standard}-\text{blank})\ \text{fluorescence}\times \text{protein}\ (\text{mg/mL})}=\text{enzyme activity}\ (\text{nmol/h/mg})$$

3.4.13 α-Iduronate-2-Sulfatase for MPS II/Hunter Disease [23]

1. Take 10 μL cell lysate (15 μg protein) of patient and control samples, add 20 μL of substrate and incubate for 4 h at 37 °C (*see* **Note 23**).
2. After 4 h, add 20 μL of citrate/phosphate buffer to all samples.
3. For blank, incubate substrate and samples in separate tubes.
4. Add 10 μL LEBT in all tubes except incubated samples (**step 3**) and mix (*see* **Note 24**).
5. Incubate for 24 h at 37 °C and then stop the reaction by adding 200 μL stop buffer. Mix both tubes set for the purpose of blank (described in **step 3**) after adding stop solution just before making readings.
6. Read the fluorescence of MU at λ_{ex}355 nm and λ_{em}460 nm using a fluorometer and calculate the enzyme activity in nmol/4 h/mg using 4-MU (1 nmol/mL) as a standard (*see* **Notes 20–22**) and the equation:

$$\frac{(\text{Sample}-\text{blank})\ \text{fluorescence}\times 26}{(\text{Standard}-\text{blank})\ \text{fluorescence}\times \text{protein}\ (\text{mg/mL})}=\text{enzyme activity}\ (\text{nmol/4 h/mg})$$

3.4.14 Heparan Sulfamidase for MPS IIIa/ Sanfilippo IIIa Disease [24]

1. Take 10 μL cell lysate (10 μg protein) of patient and control samples, add 20 μL of the substrate 4-MU-α-D-N--sulfoglucosaminide sodium salt (5 mM) and incubate for 17 h at 37 °C (*see* **Note 23**).
2. For blank, incubate substrate and samples in separate tubes and treat them as the samples.
3. Stop the incubation with 6 μL citrate/phosphate buffer and mix (*see* **Note 25**).
4. Add 10 μL α-glucosidase and mix.
5. Incubate for 24 h at 37 °C and then add 200 μL stop buffer in all tubes except the blank tubes in **step 2**.
6. Add 200 μL stop buffer and mix substrate and sample set for blank before making readings.
7. Read the fluorescence of MU at λ_{ex}355 nm and λ_{em}460 nm using a fluorometer and calculate the activity of enzyme in

nmol/17 h/mg by using 4-MU (1 nmol/mL) as a standard (*see* **Notes 20–22**) and the equation:

$$\frac{(\text{Sample}-\text{blank})\ \text{fluorescence} \times 23 \times 4.3}{(\text{Standard}-\text{blank})\ \text{fluorescence} \times \text{protein}\ (\text{mg/mL})} = \text{enzyme activity}\ (\text{nmol/17 h/mg})$$

3.4.15 α-N-Acetylglucosaminidase for MPS IIIb/Sanfilippo IIIb Disease [25]

1. Take 10 μL cell lysate (10 μg protein), add 20 μL of substrate and incubate 17 h at 37 °C (*see* **Note 23**).
2. For blank, incubate substrate and samples in separate tubes. Mix both tubes after adding stop solution in **step 3** just before making readings.
3. Add 200 μL stop buffer and read fluorescence of MU at λ_{ex}355 nm and λ_{em}460 nm using a fluorometer (*see* **Notes 20–22**).
4. Calculate the activity of enzyme in nmol/17 h/mg by using 4-MU (1 nmol/mL) as a standard and the equation:

$$\frac{(\text{Sample}-\text{blank})\ \text{fluorescence} \times 23 \times 4.3}{(\text{Standard}-\text{blank})\ \text{fluorescence} \times \text{protein}\ (\text{mg/mL})} = \text{enzyme activity}\ (\text{nmol/17 h/mg})$$

3.4.16 Galactose-6-Sulfatase for MPS IVa/Morquio IVa Disease [26, 27]

1. Take 10 μL cell lysate (5 μg protein) of patient and controls, add 20 μL of substrate and incubate for 17 h at 37 °C (*see* **Note 23**).
2. After 17 h, add 5 μL phosphate buffer to inhibit the sulfatase reaction (*see* **Note 25**).
3. Then add 10 μL of β-gal-Ao (10 U stock solution resulting in the final concentration 0.01 U) and incubate for 2 h at 37 °C.
4. Stop the reaction by adding 200 μL stop buffer and read the fluorescence of MU at λ_{ex}355 nm and λ_{em}460 nm using a fluorometer.
5. Calculate the activity of enzyme in nmol/17 h/mg using 4-MU (1 nmol/mL) as a standard and the following equation:

$$\frac{(\text{Sample}-\text{blank})\ \text{fluorescence} \times 22.5}{(\text{Standard}-\text{blank})\ \text{fluorescence} \times \text{protein}\ (\text{mg/mL})} = \text{enzyme activity}\ (\text{nmol/17 h/mg})$$

3.4.17 Aryl-Sulfatase B for MPS VI/Mourtaux Lamy Disease [28, 29]

1. Take 10 μL cell lysate (1 mg/mL of protein) of patient and controls, add 50 μL of substrate and incubate at 37 °C for 30 min in a shaking water bath.
2. Take half of the reaction mixture (30 μL) and stop the reaction with 200 μL of stop buffer. Read the fluorescence at λ_{ex}355 nm and λ_{em}460 nm (incubation time 30 min).
3. Continue the incubation of the remaining reaction mixture (30 μL) at 37 °C for additional 60 min (total incubation time for these samples is 90 min).

4. Stop the reaction with 200 μL of stop buffer for the 90-min samples.
5. For blank, incubate substrate and samples in separate tubes. Mix both tubes after adding stop solution just before making readings.
6. Similar to the samples, blank tubes will be read in two steps, at 30 min and then at 90 min.
7. Read the fluorescence of 4-MU using a fluorometer (*see* **Notes 20–22**).
8. Calculate the activity of enzyme in nmol/h/mg using 4-MU (1 nmol/mL) as a standard:

$$\frac{\left[(90\text{'Sample}-\text{blank})-(30\text{'Sample}-\text{blank})\right]\text{fluorescence}\times 100}{(\text{Standard}-\text{blank})\,\text{fluorescence}\times\text{protein}\,(\text{mg/mL})}=\text{enzyme activity}\,(\text{nmol/h/mg})$$

3.4.18 β-Glucuronidase for MPS VII/Sly Disease [1, 28]

1. Take 10 μL cell lysate of patient and control samples, add 20 μL of substrate and incubate at 37 °C for 15 min in a shaking water bath.
2. Terminate the reaction by adding 100 μL TCA (5%).
3. For blank, incubate substrate and samples in separate tubes. Mix both tubes after adding stop solution just before making readings.
4. Add 200 μL of stop buffer and read the fluorescence of MU at λ_{ex}355 nm and λ_{em}460 nm using a fluorometer (*see* **Notes 20–22**).
5. Calculate the activity of enzyme in nmol/h/mg using 4-MU (1 nmol/mL) as a standard using the equation:

$$\frac{(\text{Sample}-\text{blank})\,\text{fluorescence}\times 113\times 4}{(\text{Standard}-\text{blank})\,\text{fluorescence}\times\text{protein}\,(\text{mg/mL})}=\text{enzyme activity}\,(\text{nmol/h/mg})$$

3.4.19 Esterase/Acid Lipase for Wolman Disease [30, 31]

1. Take 10 μL cell lysate (10–50 μg protein) and add 150 μL of substrate.
2. Incubate for 30 min at 37 °C.
3. For blank, incubate substrate and samples in separate tubes. Mix both tubes after adding stop solution just before making readings.
4. Stop the reaction by adding 50 μL perchloric acid and after 5 min, centrifuge at 3214 × *g* for 10 min.
5. Take 50 μL supernatant and add 350 μL of 1 M Tris–HCl, pH 9.0.
6. Read the absorbance at 400 nm and calculate the activity of enzyme in nmol/h/mg using p-nitrophenol (10 nmol/mL) as a standard (*see* **Notes 20–22**) and the equation:

$$\frac{(\text{Sample}-\text{blank})\,\text{absorbance}\times 1000}{(\text{Standard}-\text{blank})\,\text{absorbance}\times\text{protein}\,(\text{mg/mL})}=\text{enzyme activity}\,(\text{nmol/h/mg})$$

4 Notes

1. Maternal tissue contamination can give false results. Therefore, the CV sample has to be selected carefully to isolate fetal cells under a dissecting microscope [33].
2. Store all the protein estimation reagents in refrigerator at 2–8 °C for up to 3 months.
3. After preparation, store all the enzyme substrates at −80 °C for 2 years or as per instructions for each substrate in Subheading 2. At the time of assay, keep the substrate on ice after thawing.
4. Cytosolic β-glucosidase and saposin C have been the main reasons for high residual enzyme activity in the diagnosis of Gaucher patients. Therefore, add taurocholate in the substrate buffer to inhibit the cytosolic enzyme and pro-saposin. Perform a genetic study to rule out the deficiency of saposin C.
5. Use 0.1 M 2-amino-2-methyl-1-propanol (AMP), pH 10.3 as a stopping solution for the reactions. Take 943 μL AMP and dilute to the final volume of 100 mL with distilled water. It is easy to prepare and is stable for at least 6 months at room temperature without the change in pH and microbial growth. Moreover, the fluorescence remains stable for up to 1 h at room temperature after stopping the reaction [34].
6. For the estimation of acid sphingomyelinase (ASM), 4-MU bound fluorogenic substrate is used since radio-labeled substrates are accompanied by health hazards. Performing the ASM assay with the artificial substrate 6-hexadecanoylamino-4-methylumbelliferyl-phosphorylcholine (HMU-PC) without addition of the lysosphingomyelin (lysoSM) inhibitor could give false-negative result in Niemann-Pick A/B patients with mutation Q292K [12].
7. β-Galactosidase has two isoenzymes (I and II), which differ in molecular size and pH activity. Chloride ions provide stability to the lysosomal β-galactosidase isoenzyme and therefore sodium chloride should be added in the buffer.
8. For the diagnosis of Tay Sach/Sandoff disease, hexosaminidase B and hexosaminidase S enzymes can interfere in the measurement. It is recommended to adhere to the protocol, as the heat and acid labile enzyme hexosaminidase A can hydrolyze both the substrates 4-MU β-D-N-acetyl glucosamine 6-sulfate and 4-MU-β-D-N-acetyl glucosamine while the heat and acid stable hexosaminidase B enzyme can hydrolyze 4-MU-β-D-N-acetyl glucosamine only.
9. Aryl sulfatase A (ASA) and aryl sulfatase B (ASB) are soluble enzymes while aryl sulfatase C (ASC) is membrane bound (a microsomal enzyme). Therefore, the activity of ASC remains

high in CV samples in comparison to other aryl sulfatases. Even though ASC has minimal activity at acidic pH, there might be high residual activity if the assay is performed at 37 °C. In addition, both the isoenzymes ASB and ASC can hydrolyze the substrate p-nitrocatecol sulfate. To get maximum activity of ASA and minimum activity of ASB and ASC, the enzyme assay for PND of metachromatic leukodystrophy should be done at 0 °C. Besides, ASB can be inhibited by adding sodium chloride and sodium pyrophosphate in the buffer. Assay for PND at 0 °C has been found more reliable than at 37 °C. At this temperature, ASA remains active while ASB and ASC are inactivated [17]. It is also essential to assay ASA in the parents before offering PND. DNA must be checked for the presence of pseudo-deficiency allele, if parents show low ASA enzyme activity. In addition, other sulfatase group enzymes should be checked in patient to rule out the possibility of multiple sulfatidosis [17, 29, 34].

10. Concentration of oleic acid, taurocholate as well as the ratio of detergents greatly influence the enzyme activity. It is recommended to perform the assay carefully and adhere to the protocol strictly.
11. Removal of all organic solvents by evaporation in the last step of the preparation of the substrate film should be done thoroughly, as traces of solvents in the substrate film can reduce the enzyme activity considerably.
12. N-acetyl galactosamine is an inhibitor of α-galactosidase B. It should be added in the assay to measure α-galactosidase A activity specifically for the diagnosis of Fabry disease.
13. Store the stock solution of β-glucosidase (100 U/mL) in aliquots at -80 °C until deterioration is noted. Enzyme activity should be validated quarterly by comparing the results with known previous samples. The enzyme activity varies from batch to batch due to the different specifications on the vial (units and concentration), therefore these should be checked before making stock solution.
14. To get maximum activity of α-iduronidase, add saccharolactone in the buffer to inhibit the activity of the β-glucuronidase enzyme [34].
15. Disposal of all waste should be in accordance with local regulations. Since lead is an environmental hazard, wear gloves, eye protection and laboratory coat while using the solution/buffer containing lead. As per waste management guidelines, discard these wastes (after diluting with water) in leak proof plastic container labeled with the words "hazardous waste". Tightly closed container should be kept according to local regulations and send to disposal plant for destruction.

16. Test each new batch of α-glucosidase for its efficacy with normal control and MPS IIIa patient samples.
17. Test each new batch of substrate for efficacy with both control and patient samples before using for further testing. Similarly, if additional enzyme (like β-galactosidase in Gal-6-sulfatase (MPS IVa) and α-glucosidase in heparin sulfamidase (MPS IIIa)) is used in the assay from outside, its performance should be checked with known samples.
18. Barium ions are a potent inhibitor of aryl sulfatase A. Substrate (4-MU sulfate) specificity for aryl sulfatase B was enhanced by including barium acetate in the buffer [29].
19. FluorAce™ ß-galactosidase reporter assay kit can be purchased from Bio-Rad Laboratories.
20. Quality control is essential. Enzyme deficiency is one of the characteristics for the diagnosis of the lysosomal storage disorder. It is absolutely necessary to validate the assay in one's own laboratory with the samples from the patients with proven enzyme deficiency. At least one negative control (CV sample with normal enzyme activity) and one positive control (CV sample with deficient enzyme activity) should be run with each assay [7].
21. The laboratory should participate in external quality assurance program to improve the quality of testing and provide reliable results [7].
22. If the value of the enzyme activity is in the deficient range, a second control enzyme should be analyzed to check the sample integrity [1]. Sometimes degradation of enzymes can occur due to the high temperature and humidity during transportation.
23. In every assay where the incubation time of ≥17 h is required, perform the assay in 200 μL PCR tubes which are well insulated during incubation to avoid evaporation.
24. Formation of fluorochromes in the assay of Hunter (MPS II) disease requires sequential hydrolysis of both iduronate-2-sulfatase (IDS) and α-iduronidase. Therefore, α-iduronidase is required exogenously to prevent underestimation of IDS. Enzyme activity should be checked for each batch of substrate to prevent underestimation of IDS activity [23].
25. Sometimes, reaction mixtures do not remain homogeneous during incubation (e.g., iduronate-2-sulfatase and galactose-6-sulfatase enzymes analyses) and a precipitate is formed. This has no adverse effect on activity and measurement.
26. Due to the complexity of CV sample, some of the enzymes have wide/narrow range. Therefore, reference values for affected and normal fetus should be established in one's own population and updated annually.

27. Higher enzyme activities than the normal values have no clinical significance as there could be a genetic variability among the individuals (except in Mucolipidosis II where significantly high plasma enzyme values are diagnostic).
28. Biochemical assays cannot reliably establish carrier status on the account of indistinguishable values of enzyme activity in carriers and affected individuals. Although with experience, these results can be interpreted but should be confirmed by mutation studies. Where mutation data are available for index case, isolate the DNA from CV samples by phenol-chloroform method [35] and perform mutation analysis by relevant techniques like, gap-PCR, ARMS-PCR and sequencing for the diagnoses of different LSDs.
29. In case of low enzyme activity, parents should be checked for pseudo-deficiency status with consent before attempting for PND [36].

Acknowledgement

We would like to thank our colleagues Ratna D. Puri for clinical suggestions, Renu Saxena and Sudha Kohli for their cooperation in molecular analysis, Divya C. Thomas for scientific discussion, Sandeepika, Geetu and Azad for technical assistance and the Chairman of the Hospital for his support in setting up the genetic facility.

References

1. Verma J, Thomas DC, Sharma S (2015) Inherited metabolic disorders: prenatal diagnosis of lysosomal storage disorders. Prenat Diagn 35:1–11
2. Sheth J, Sheth F (2013) Genetics in clinical practice—symptoms, diagnosis and therapy. Jaypee Brothers, Ahmedabad, Gujarat. ISBN: 978-93-5152-153-2
3. Fletcher JM (2006) Screening for lysosomal storage disorders: a clinical perspective. J Inherit Metab Dis 29(2–3):405–408
4. Beck M (2010) Therapy for lysosomal storage disorders. IUBMB Life 62:33–40
5. Kaminsky P, Lidove O (2014) Current therapeutic strategies in lysosomal disorders. Presse Med 43(11):1174–1184
6. Winchester B, Young E (1991) Prenatal diagnosis of enzyme defect—an update. Arch Dis Child 66:451–454
7. Verma J, Thomas DC, Sharma S et al (2015) Inherited metabolic disorders: quality management in laboratory diagnosis. Clin Chim Acta 447:1–7
8. Verma J, Thomas DC, Verma IC (2016) New generation techniques in rapid diagnosis of lysosomal storage disorders. Eur J Mol Biol Biochem 3(1):42–47
9. Peter SP, Lee RE, Glew RH (1975) A micro assay for Gaucher disease. Clin Chim Acta 60:391–396
10. Daniels LB, Glew RH (1982) Beta glucosidase assays in the diagnosis of Gaucher disease. Clin Chem 28:569–577
11. Arora D, Kochar SPS (2010) Prenatal diagnosis of Gaucher disease. Med J Armed Forces India 66:170–171
12. van Diggelen OP, Voznyi YV, Keulemans JLM et al (2005) A new fluorimetric enzyme

assay for the diagnosis of Niemann–Pick A/B, with specificity of natural sphingomyelinase substrate. J Inherit Metab Dis 28(5): 733–741
13. Suzuki K, Glew R, Peters SP (1977) Practical enzymology of the sphingolipidosis. AR Liss, New York, pp 101–136
14. Grabowski GA, Kruse JR, Goldberg JD et al (1984) First-trimester prenatal diagnosis of Tay-Sachs disease. Am J Hum Genet 36(6): 1369–1378
15. Percy AK, Brady RO (1968) Metachromatic leukodystrophy: diagnosis with samples of venous blood. Science 161(3841):594–595
16. Galjaard H, van Hoogstraten JJ, de Jong JE et al (1974) Methodology of the quantitative cytochemical analysis of single or small numbers of cultured cells. Histochem J 6(4): 409–429
17. Giles L, Coope A, Fowler B et al (1987) Aryl sulfatase isoenzymes of chorionic villi: implications for prenatal diagnosis. Prenat Diagn 7(4):245–252
18. Wiederschain G, Raghvan S, Kolodny E (1992) Characterization of 6-hexadecanoylamino-4-methylumbelliferyl-beta-D-galactopyranoside as fluorogenic substrate of galactocerebrosidase for the diagnosis of Krabbe disease. Clin Chim Acta 205(1–2):87–96
19. PrajnyaR RC, Phadke SR et al (2011) Prenatal diagnosis of Pompe disease—enzyme assay or molecular testing? Indian Pediatr 48(11): 901–902
20. Chamoles NA, Blanco M, Gaggioli D (2001) Fabry disease: enzymatic diagnosis in dried blood spots on filter paper, letter to the editor. Clin Chim Acta 308:195–196
21. Van Diggelen OP, Keulemans JLM, Winchester B et al (1999) A rapid fluorogenic palmitoyl protein thioesterase assay: pre and post natal diagnosis of INCL. Mol Genet Metab 66: 240–244
22. Hopwood JJ, Muller V, Smithson A et al (1979) A flourometric assay using 4 methylumbelliferyl alpha-l-iduronide for the estimation of alpha-l-iduronidase activity and the detection of Hurler and Schie syndromes. Clin Chem Acta 92(2):257–265
23. Voznyi YV, Keulemans JLM, van Digglen OP (2001) A flourogenic assay for the diagnosis of hunter disease (MPS II). J Inherit Metab Dis 24:675–680
24. Karpova EA, Voznyi YV, Keulemans JLM et al (1996) A fluorometric assay for the diagnosis of Sanfilippo disease type A (MPS IIIA). J Inherit Metab Dis 19:278–285
25. Marsh J, Fensom AH (1985) 4-Methylumbelliferyl-α-N-acetylglucosaminidase activity for diagnosis of Sanfilippo B disease. Clin Genet 27:258–262
26. Zao H, van Diggelen OP, Thoomes R et al (1990) Prenatal diagnosis of Morquio disease type A using a simple fluorometric enzyme assay. Prenat Diagn 10:85–91
27. van Diggelen OP, Zhao H, Kleijer WJ et al (1990) A fluorometric enzyme assay for the diagnosis of Morquio disease type A. Clin Chem Acta 187:131–140
28. Kally S (1977) Biochemical methods in medical genetics. American lecture series 1977; Publication no. 1008. Harvard, 18th edn. Springfield III, Thomas, Chicago
29. Sanguinetti N, Marsh J, Jackson M et al (1986) The aryl sulfatases of chorionic villi: potential problems in the first trimester diagnosis of Metachromatic leukodystrophy and Maroteaux-lamy disease. Clin Genet 30(4):302–308
30. Patrick D, Willcox P, Stephens R et al (1976) Prenatal diagnosis of Wolman's disease. J Med Genet 13:49–51. doi:10.1136/jmg.13.1.49
31. Huggins C, Lapides J (1947) Chromogenic substrates IV. Acyl esters of p-nitrophenol as substrates for the colorimetric determination of esterase. J Biol Chem 170:467–482
32. Lowry OH, Rosebrough NJ, Farr AL et al (1951) Determination of protein concentration. J Biol Chem 193:265–275
33. Fowler B, Giles L, Cooper A et al (1989) Chorionic villi sampling: diagnostic uses and limitations of enzyme assays. J Inherit Metab Dis 12 suppl(1):105–117
34. Verma J, Thomas DC, Kasper DC et al (2016) Inherited metabolic disorders: efficacy of enzyme assays on dried blood spots for the diagnosis of lysosomal storage disorders. JIMD Rep. doi:10.1007/8904_2016_548
35. Miller SA, Dykes DD, Polesky HF (1988) A simple salting out procedure for extracting DNA from human nucleated cells. Nucleic Acids Res 16(3):1215
36. Sheth J, Mistri M, Sheth F et al (2014) Prenatal diagnosis of lysosomal storage disorders by enzymes study using chorionic villus and amniotic fluid. J Fetal Med 1:17–24

Chapter 19

Lysosomal Biology in Cancer

Colin Fennelly and Ravi K. Amaravadi

Abstract

Cells depend on the lysosome for sequestration and degradation of macromolecules in order to maintain metabolic homeostasis. These membrane-enclosed organelles can receive intracellular and extracellular cargo through endocytosis, phagocytosis, and autophagy. Lysosomes establish acidic environments to activate enzymes that are able to break down biomolecules engulfed through these various pathways. Recent advances in methods to study the lysosome have allowed the discovery of extended roles for the lysosome in various diseases, including cancer, making it an attractive and targetable node for therapeutic intervention. This review focuses on key aspects of lysosomal biology in the context of cancer and how these properties can be exploited for the development of new therapeutic strategies. This will provide a contextual framework for how advances in methodology could be applied in future translational research.

Key words Lysosome, Cancer, Drug therapy, Autophagy

1 Introduction to the Lysosome

The lysosome is a membrane-enclosed organelle that functions as an essential part of the digestive system of the cell. Christian de Duve first discovered lysosomes in 1955 and the name was derived from the Greek term for digestive body. This organelle contains over 60 different types of hydrolases that can break down biological polymers such as proteins, carbohydrates, lipids, and nucleic acids [1]. These enzymes require an acidic pH for optimal function, which is achieved by using ATP hydrolysis to pump protons against their electrochemical gradient into the lysosome by the vacuolar H^+ ATPase (V-ATPase) [2, 3]. Due to their pH dependence, these enzymes are also called acid hydrolases. They are produced in the endoplasmic reticulum (ER), trafficked to the golgi apparatus and tagged with mannose-6-phosphate to be targeted to the lysosome [4]. Vesicular exchange between the trans-golgi network (TGN) and endosomes is involved in the transport of newly synthesized proteins from the TGN to the endolysosomal compartment, and the reverse, where some proteins are trafficked from the endolysosomal compartment back to the TGN. Trafficking

Karin Öllinger and Hanna Appelqvist (eds.), *Lysosomes: Methods and Protocols*, Methods in Molecular Biology, vol. 1594, DOI 10.1007/978-1-4939-6934-0_19, © Springer Science+Business Media LLC 2017

vesicles are categorized by their internal pH, where early endosomes are around 6–6.6, late endosomes are at 5 and lysosomes are the most acidic at pH 4.5. The mannose-6-phosphate receptor marker is used to identify endosomes and lysosomes. Intracellular cargo is delivered to the lysosomes through autophagy, whereas exogenous material can be engulfed from outside the cell and delivered via endocytosis or phagocytosis. Endocytosis occurs in either a clathrin-dependent or -independent manner. The lysosome serves as the end organelle for these degradative endocytic pathways that begin at the plasma membrane. Once lysosomal substrates are broken down, their components can be recycled and reused by the cell as building blocks for macromolecules through various carrier-mediated transport channels back into the cytosol [5, 6]. Other than the ubiquitin-proteosome system, autophagy is the main degradation pathway for intracellular proteins. Unlike proteasomal degradation, autophagy can accommodate organelles and cytoskeletal components in addition to proteins. During chaperone-mediated autophagy, cytosolic proteins that contain specific motifs are localized to the lysosome through the action of chaperones and the lysosomal receptor LAMP-2A [7–9]. Another form of autophagy is microautophagy, which involves the direct engulfment of cytoplasmic cargo at the lysosomal membrane [10]. The most widely studied form of autophagy is macroautophagy, and therefore, hereafter our discussion of autophagy will focus on this form. Autophagic degradation is accomplished through the sequestration of soluble cargo into a double membrane structure—referred to as a phagophore—to form an autophagosome that eventually fuses with the lysosome to complete the degradation process [11]. In addition to the main autophagy pathway, there are the more recently recognized organelle-specific autophagy processes of lysophagy [12, 13], mitophagy [14], ER-phagy [15], nucleophagy [16], and pexophagy [17]. This review will outline how lysosomal biogenesis is regulated, our current understanding of the many roles lysosomes play in cancer progression and cell death, examples of tool compounds that can be used to modulate lysosomal function (Fig. 1), and a brief overview of efforts to translate some of these findings into clinical trials.

2 Control of Lysosomal Biogenesis at the Transcriptional Level

Lysosome formation is typically thought of in terms of simply the vesicular trafficking of key lysosomal proteins from the ER, golgi, endosomes, and eventually into lysosomes. However, recent evidence suggests that lysosomal biogenesis is coordinated at the transcriptional level in a sophisticated manner, and can even play a critical role in cancer cell metabolism [18, 19]. Transcription factor

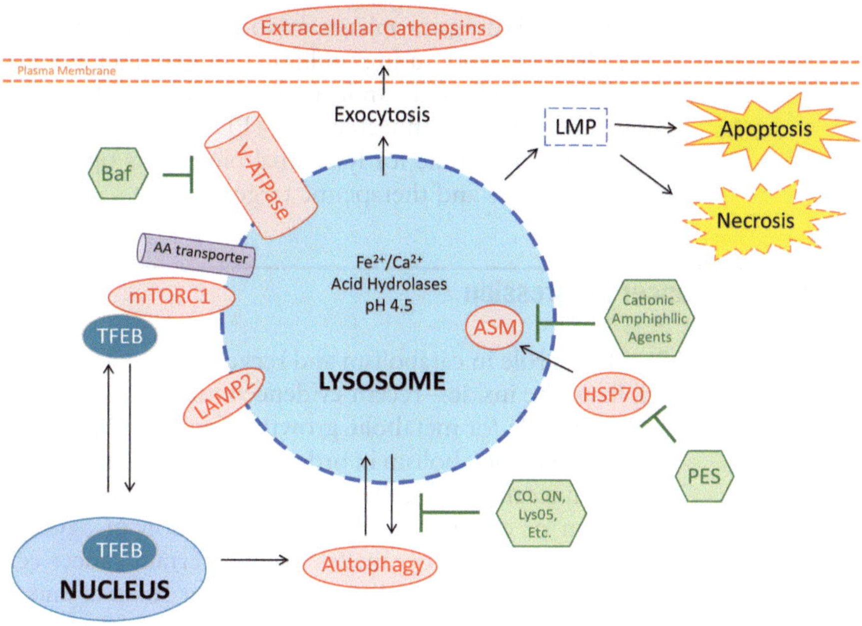

Fig. 1 The key elements involved in lysosomal biology in cancer. Factors that can play roles in cancer progression are in red. Vesicular trafficking funnels cellular contents into the lysosome from various pathways including endocytosis and autophagy. TFEB regulates expression of autophagy and lysosomal genes. Elevated autophagy has proven to aid in oncogenesis by providing macromolecules for sustained growth and eliminating damaged proteins, organelles, and hazardous waste including ammonia and ROS. Cellular contents can also be secreted to the extracellular space through exocytosis. Lysosomal cathepsins have been observed in the extracellular space where they are free to engage the ECM for remodeling to promote tumor invasion. Lysosomal membrane proteins such as the V-ATPase and LAMP2 have also been observed promoting metastasis by exerting an influence on the tumor microenvironment from the plasma membrane. MTORC1 is known to positively regulate growth and feed into other oncogenic signaling pathways. HSP70 supports lysosomal membrane integrity by activating ASM and can even be used as a biomarker for prognosis. Various drugs (*green*) that target the lysosome or cellular components involved with the lysosome can destabilize the organelle leading to LMP and activation of cell death pathways (*yellow*). *aa* amino acid, *ASM* acid sphingomyelinase, *Baf* Bafilomycin A 1, *CQ* chloroquine, *ECM* extracellular matrix, *HSP70* 70-kDa heat shock proteins, *LAMP* lysosome associated membrane glycoproteins, *LMP* lysosomal membrane permeabilization, *mTORC1* mammalian target of rapamycin complex 1/mechanistic target of rapamycin complex 1, *PES* phenylethynesulfonamide, *QN* quinacrine, *ROS* reactive oxygen species, *TFEB* transcription factor EB, *V-ATPase* Vacuolar-type H^+ATPase

EB (TFEB) is a transcription factor that acts as a master regulator for lysosomal biogenesis and drives the expression of over 500 genes related to autophagy and autophagosome-lysosome fusion [20]. Other family members of the TFE/MiTF family control this expression profile in different cellular contexts. Activation of this expression profile called the CLEAR (coordinated lysosomal expression and regulation) network occurs when TFEB translocates from the lysosomal membrane into the nucleus. This system controls the expression of lysosomal enzymes required for the breakdown of biomolecules and genes linked to the main trafficking pathways including autophagy, endo/exocytosis, and phagocytosis [21]. Recent work using unbiased global metabolite profiling revealed the MiT/TFE family critically supports the metabolism of pancreatic ductal adenocarcinoma (PDA) [18]. The discovery of this expression profile for lysosomal biogenesis opens the door to new biomarkers and therapeutic targets.

3 Lysosomes and Cancer Progression

Besides its role in catabolism and recycling—i.e. feeding the cancer cell from the inside—recent evidence indicates the lysosome is also a central node for metabolic growth signaling. Cancer cells deviate from normal metabolism in order to acquire their idiosyncratic feature of uncontrolled growth. This transformation results in rapid depletion of cellular nutrients, accumulation of aggregated proteins, and damaged organelles making certain cancer cells dependent on lysosomal recycling programs for survival and continued growth. Autophagic-lysosomal degradation of macromolecules and organelles serves as a coping mechanism for cancer cells to deal with these stresses while also providing a consistent supply of nutrients to promote further growth. Additionally, lysosomes are not just degradative vesicles, but signaling scaffolds for mTOR and AMPK signaling, as described later. They are arguably the main nutrient sensing organelle in the cell. Targeting lysosomes can have pleiotropic effects involving metabolism [22], reactive oxygen species (ROS) [23], DNA damage [24], cell death [25, 26], and protein secretion [27].

Cancer cells depend on lysosome function and demonstrate changes in lysosomal volume and subcellular localization during oncogenic transformation [28, 29]. Cathepsin proteases are lysosomal hydrolases that can play dual roles in promoting and suppressing tumor growth. They are observed as being upregulated and mislocalized in cancer [29, 30]. Intracellular cathepsins are able to activate the intrinsic apoptotic pathway, but in contrast, extracellular cathepsins promote tumor invasion through their ability to break down basement membranes and activate other oncogenic proteins. In addition, cathepsins B, E, and S have all been recognized as contributing to malignancy in different cancers [31–33]. Lysosomal membrane proteins like lysosome-associated membrane protein 1 (LAMP-1] have been observed on the cell surface of highly metastatic colon cells, indicating a role for these proteins in the extracellular matrix [34]. Other lysosomal membrane proteins such as the V-ATPase have been shown to exert an influence on the tumor microenvironment by pumping protons to the extracellular space [35]. The Na^+/H^+ exchanger has also been associated with extracellular acidification and cancer cell invasion [36]. Another intriguing aspect of lysosomes is their ability to secrete contents out of the cell by fusing with the plasma membrane [4, 37]. For example, cells can expel ATP to the extracellular space with this secretory pathway to mediate cellular signaling through ATP receptors [38, 39]. The observation that lysosomal exocytosis can play a role in cell signaling, proteolytic extracellular matrix (ECM) remodeling and tumor invasion suggests that targeting lysosomal exocytosis rather than individual cathepsins would be a more promising strategy [40].

The lysosome is an important signaling hub that responds to both external and internal stimuli to perceive the availability of nutrients, growth factor signals, and energy to maintain metabolic homeostasis. One of the main regulators of cell growth and proliferation is mammalian target of rapamycin complex 1 (mTORC1), which exerts its function directly from the lysosomal membrane surface. mTORC1 is a multicomponent protein kinase complex that includes mTOR, Regulatory Associated Protein of mTOR (RAPTOR), and mLST8/GβL [41]. mTORC1 and its regulatory complexes detailed below, together integrate various nutritional and environmental cues including the presence of amino acids, growth factors, glucose, hormones, and oxygen to drive anabolic processes such as protein, mRNA, and lipid biosynthesis [22, 42, 43]. Active mTORC1 also phosphorylates ULK1 and ATG13 to inhibit their activity and block autophagy [44]. Additionally, TFEB has been recognized as a target for mTORC1 suggesting this interaction directly influences expression of the CLEAR network genes [45]. This ability to control biosynthetic and catabolic states makes mTORC1 an important factor in metabolic signaling and mutations that lead to defective mTORC1 regulation are commonly observed in human cancers [41, 46]. Oncogenic transformation is significantly enabled by mutations that lead to inactivation of key tumor suppressor genes including phosphatase and tensin homolog (PTEN), tuberous sclerosis complex 1/2 (TSC1/TSC2), neurofibromin 1/2 (NF1/NF2), and liver kinase B1 (LKB1). In all of these cases the downstream consequence of this inactivation is promotion of mTORC1 signaling [47]. mTORC1 kinase activity is stimulated by direct interaction with the GTPase Ras homolog enriched in brain (RHEB) on the lysosome surface [48]. This interaction is negatively regulated by the heterodimer TSC1/TSC2 and promoted by amino acids that recruit mTORC1 to the lysosomal surface through Rag GTPases that are stabilized by the Ragulator complex [49]. Recent investigations have also revealed that the V-ATPase can mediate mTORC1 activation and autophagy [50]. This provides further evidence that mTORC1 localization to the lysosomal surface is essential for its activation. Molecular sensors for amino acids (Rags), growth factor inputs (Rheb), energy status (LKB1/AMPK1), and lysosomal health (V-ATPase) all have to be aligned for full activation of mTORC1.

Loss of the tumor suppressor PTEN has been shown to activate mTORC1 via protein kinase B (AKT), which inhibits TSC1/TSC2 through phosphorylation [41]. TSC1/TSC2 can also be suppressed when LKB1 is inactivated, preventing TSC1/TSC2 phosphorylation and activation [41]. LKB1 regulates AMPK directly, and recent evidence indicates that AMPK is also closely associated with the lysosomal surface. Another example of mTORC1-driven oncogenesis is activation of eukaryotic translation initiation factor 4B (eIF4E) through mTORC1-mediated inhibition of 4EBP1 [51]. This results in mRNA translation of cell cycle regulatory genes and pro-tumorigenic genes such as the anti-apoptotic protein Mcl-1 that can promote cancer cell survival in in vivo mouse models of lymphoma [52–54].

The lysosome's role in catabolic recycling and metabolic growth decisions suggests there may be therapeutic potential in targeting the lysosome. Great progress has been made in understanding the cell fates associated with lysosomal targeting, i.e. the role of the lysosome in eliciting cell death.

4 Lysosomes and Cell Death

Lysosomes can play a role in each of the three major types of cancer cell death that include apoptosis, autophagy, or necrosis [55]. A more recent form of cell death ferroptosis is also dependent on the lysosome [56]. For apoptosis there are intrinsic and extrinsic pathways that can be activated by different mechanisms. The intrinsic pathway involves mitochondrial outer membrane permeabilization (MOMP) and cytochrome c release into the cytoplasm, whereas the extrinsic is initiated by cell death receptors [57]. Both result in caspase signaling cascades that are governed by the Bcl-2 family of proteins that ultimately regulate MOMP. These proteins are classified in two categories: antiapoptotic (e.g. Bcl-2 and Bcl-xL) and proapoptotic (e.g. Bax and Bid) [58]. Damaged lysosomes allow proteolytic enzymes to be released into the cytosol and initiate apoptosis [59]. Cathepsins B and D are known to cleave Bid when ectopically in the cytosol, which results in MOMP followed by cytochrome c release [60].

Cancer metabolism can create harsh byproducts such as ammonia, ROS, and hypoxia [61–64]. To sustain oncogenic growth and cell survival autophagy can play a cytoprotective role that counteracts apoptosis by intercepting damaged mitochondria that could trigger apoptosis [65, 66]. Autophagy can have dual roles in the context of cancer and is also recognized as a cell death pathway [67]. High drug doses can initiate apoptosis-independent and autophagy-dependent cell death in vitro, although the relevance of autophagic cell death in vivo has been called into question [68].

Necroptosis is a form of programmed necrosis that serves as a backup role when apoptosis signaling is blocked by endogenous or exogenous factors including viruses or mutations. The receptor-interacting serine/threonine protein kinase 1 (RIPK1) and RIPK3 have been identified as regulators of apoptosis and necrosis. RIPK1/3 can induce necroptosis via sphingomyelinase-mediated lysosomal membrane permeabilization (LMP) [69]. Ferroptosis is an additional and unique form of cell death that is dependent on iron and ROS [56]. It is distinct in that it has its own morphological, genetic, and biochemical signatures. Misregulation of iron metabolism and lipid peroxidation has been implicated in various pathologies including cancer [70, 71].

One common theme that can impact multiple cell death pathways is LMP. This process has been the topic of intense study for decades, and steadily methods to measure LMP in cancer cells have become more reproducible and versatile. LMP permits the release of lysosomal hydrolases into the cytosol and can contribute to the forms of cell death discussed above [28, 60]. Depending on the degree of permeabilization, LMP can either induce lysosomal cell death through apoptosis or necrosis if the subsequent enzyme release is extensive enough [72, 73]. For instance, LMP can initiate caspase cascades via the intrinsic apoptosis pathway through cleavage of Bid and induction of Bax-mediated release of cytochrome c, but cathepsins are able to mediate cell death in a caspase-independent manner as well [74]. Identifying stimuli that can cause the release of these lysosomal enzymes such as cathepsins into the cytosolic lumen has potential applications for targeting the lysosome in cancer. LMP can be evaluated in cells by detecting functional enzyme activity of lysosomal hydrolases present in the cytosol or visually by either tracking lysosomes with fluorescent dextran or staining with antibody probes against galectins [75]. A small molecule screen done to identify compounds able to induce p53-independent cell death found that the ones that were effective worked through an LMP mechanism [76]. These findings and other studies have lead to the proposal that transformed cells are more sensitive to lysosomal cell death and further support the notion that targeting the lysosome can be an effective therapeutic strategy [77].

5 Drugs That Target the Lysosome

There are at least five categories of drugs that target the lysosome. These include lysosomal hydrolase inhibitors [78–82], heat shock protein 70 (HSP70) inhibitors, cationic amphiphilic compounds, V-ATPase inhibitors, and chloroquine derivatives that do not yet have a clear mechanism of action.

Acid sphingomyelinase (ASM) is located in the lysosome and breaks down sphingomyelin into ceramide, which is the substrate for the generation of other sphingolipids including sphingomyelin and sphingosine 1-phosphate (S1P) [83]. It has been observed that cancer cells display decreased levels of the proapoptotic lipid ceramide and increased levels of proliferation promoting lipid S1P [84]. Cancer cells also exhibit lower ASM activity leading to higher sphingomyelin levels. Blocking ASM activity has shown to further elevate sphingomyelin levels and interrupt the function of the lysosomal membrane [85]. HSP70 is expressed in many tumor types and can activate ASM, which is associated with increased lysosomal integrity. Targeting HSP70 thereby inactivating ASM with small molecule inhibitors like 2-phenylethynesulfonamide (PES) can increase LMP and cause cell death [86, 87]. Other drugs such as chloroquine (CQ), chlorpromazine, and amiodarone are cationic amphiphilic agents that displace ASM from vesicular membranes in the lysosome and result in lysosomal membrane permeabilization (LMP) and eventual tumor cell death [86].

HSP70 promotes tumor cell metastasis and survival by protecting lysosomal membrane integrity. It can serve as a biomarker for poor prognosis due to its higher expression in many cancers. The HSP70 modulator PES disrupts the protein interaction with p53 resulting in massive accumulation of autophagosomes loaded with undigested cargo and cellular apoptosis [88].

Bafilomycin A1 is the prototypical inhibitor of the V-ATPase and prevents lysosomal acidification and autophagic flux. It is similar to other compounds that are of microbial origin including archazolid and cleistanthin A. However, other mechanisms of action have been proposed for bafilomycin's effects on the lysosome and autophagy. Bafilomycin A1 has also been shown to prevent autophagosome formation by activation of mTOR signaling, suggesting that it may target both the early and late stages of autophagy [89]. This impairment is mediated by dissociation of the Beclin1-Vps34 complex and encourages Bcl-2 interaction to drive autophagy inhibition and apoptotic cell death [37]. Interestingly, Bafilomycin has also been shown to engage the mitochondria and induce translocation of apoptosis inducing factor to the nucleus and provoke caspase-independent apoptosis [90]. Archazolid is another V-ATPase inhibitor and myxobacterial agent that has shown the ability to reduce the activity of the protease cathepsin B both in vitro and in vivo [91]. A member of the manzamine alkaloids, manzamine A, was isolated from marine sponges of the genus *Haliclona*, and demonstrated to have inhibiting effects on autophagy and the V-ATPase in pancreatic cancer cells [92]. The diphyllin glycoside cleistanthin-A also has cytotoxic effects on various tumor cell lines and targets the V-ATPase [93].

Salinomycin is a monocarboxylic polyether antibiotic that was isolated from a Streptomyces albus strain and functions as an ionophore in the lysosome to facilitate the transport of cations across cellular membranes (including lysosomal) [94–96]. Salinomycin has been shown to impair autophagic flux in breast cancer cells [97] and even act in concert with Gefitinib to induce apoptosis in colorectal cancer cells [98]. For the latter, this process was dependent on ROS production and lead to loss of mitochondrial outer membrane potential and LMP. Other groups have also recognized oxidative stress as an important factor in salinomycin-induced cell growth inhibition in prostate cancer cells [99]. Co-treatment of salinomycin with doxorubicin or etoposide led to DNA damage and apoptosis in drug-resistant cancer cells. This was also associated with enhanced expression of p53 and H2AX as well as concurrent reduction in p21 [100]. In a different study, salinomycin suppressed elevated p21 resulting from radiation treatment and promoted activation of H2AX and p53 resulting in DNA damage and G2 arrest [101]. Salinomycin is suggested to have selective cytotoxic effects on cancer stem cells and also sensitize tumor cells to conventional chemotherapeutic drugs including methotrexate, adriamycin, and cisplatin in vitro and in vivo [102].

CQ accumulates in lysosomes and blocks autophagy by disrupting acidification and enzyme function [103]. However, a definitive mechanism for CQ in mammalian cells remains elusive. Other weak base compounds are known to also accumulate in lysosomes [104], but none are known to inhibit autophagy. Interestingly, other drugs have been observed to accumulate in lysosomes and it has been suggested that this mitigates the cytotoxic effect of these compounds and aids in drug resistance [105–108]. A series of novel monomeric CQ derivatives were tested in both lung and pancreatic cancer cells and proved to be eightfold more potent than CQ [109]. Other efforts have identified the antimalarial agent quinacrine (QN) as being much more effective at autophagy inhibition than CQ [110]. The synthesis of novel monomeric QN analogs led to the generation of improved lysomotropic agents that targeted the lysosome and elicited cell death in various cancer cell types in vitro [110]. Another lysosomal agent, lucanthone, has been reported to inhibit lysosomal function and induce apoptosis in a p53-independent manner [111]. Interestingly, these effects appear to be dependent on cathepsin D. However, this agent has been suggested to block topoisomerase II activity and inhibits AP endonuclease (APE1), an important enzyme in DNA base excision repair suggesting it may not be specific for the lysosome [112].

6 Clinical Trials and Future Outlook

Over 40 clinical trials using hydroxychloroquine (HCQ) are being conducted worldwide in humans and dogs [113]. Six phase I/II clinical trials have been performed in patients diagnosed with refractory myeloma, glioblastoma, melanoma, and other cancers [114–119]. These trials also include combination therapies that were designed from preclinical studies [120–124]. Clinical trials to date have demonstrated that autophagy inhibition could be achieved safely in patients. This was concluded from evidence of accumulated autophagic vesicles in peripheral blood mononuclear cells and tumor cells. Even though high doses were required to achieve this effect, treatment combinations were generally well tolerated and there were not any signs of liver damage, metabolic dysfunction, or neurological impairment [113]. However, there were some HCQ-cancer drug combinations that did result in dose-limiting toxicities. In phase II clinical trials, patients previously treated for metastatic pancreatic cancer were given HCQ alone and high doses were tolerated, but did not demonstrate high therapeutic efficacy [119]. This suggests more potent compounds are needed to generate the desired outcomes with the overall strategy of autophagy inhibition.

One of the limitations for clinical trials is biomarker availability for assessment of drug efficacy. In the case of the autophagy inhibitor HCQ, the current methods include EM visualization of autophagic vesicle accumulation in peripheral blood mononuclear cells and tumor cells along with LC3 western blotting and evaluation of total LC3 with immunohistochemistry. Studies have been done to characterize secreted factors of tumor cells exhibiting high autophagy and indicate that these could be potential candidates for biomarkers [27]. There is growing evidence to encourage the concept that more potent autophagy inhibitors could eventually be used synergistically with conventional chemotherapy or radiotherapy [125]. Recent work has led to the dimerization of CQ to generate a CQ derivative (Lys05), which has proven to be far more potent as a single agent in vivo and in combination with B-Raf proto-oncogene serine/threonine protein kinase (BRAF) inhibitors [126, 127]. Future studies will need to expand on the findings to date to further elucidate the role of lysosomal function in tumor biology. Fortunately, there are many opportunities to elicit an effect on lysosomal activity involving factors related to nutrient sensing, kinase signaling, death signaling, and cell trafficking. Coupling functional studies and molecular biology techniques will confirm the identification of new target candidates and potential biomarkers. Capitalizing on other approaches involving high-throughout readouts to analyze patient samples could also help detect what aspects of human cancer is prone to lysosomal inhibition and lead to new clinical therapies.

References

1. Schröder BA, Wrocklage C, Hasilik A, Saftig P (2010) The proteome of lysosomes. Proteomics 10(22):4053–4076
2. Mindell JA (2012) Lysosomal acidification mechanisms. Annu Rev Physiol 74:69–86
3. Ohkuma S, Moriyama Y, Takano T (1982) Identification and characterization of a proton pump on 'lysosomes by fluorescein isothiocyanate-dextran fluorescence. Proc Natl Acad Sci 79(9):2758–2762
4. Repnik U, Cesen MH, Turk B (2013) The endolysosomal system in cell death and survival. Cold Spring Harb Perspect Biol 5(1):a008755. doi:10.1101/cshperspect.a008755
5. Rong Y, McPhee CK, Deng S, Huang L, Chen L, Liu M, Tracy K, Baehrecke EH, Yu L, Lenardo MJ (2011) Spinster is required for autophagic lysosome reformation and mTOR reactivation following starvation. Proc Natl Acad Sci U S A 108(19):7826–7831
6. Lloyd JB (1996) Metabolite efflux and influx across the lysosome membrane. Subcell Biochem 27:361–386
7. Dice JF (1990) Peptide sequences that target cytosolic proteins for lysosomal proteolysis. Trends Biochem Sci 15(8):305–309
8. Cuervo AM, Dice JF (1996) A receptor for the selective uptake and degradation of proteins by lysosomes. Science 273(5274):501–503
9. Kaushik S, Cuervo AM (2012) Chaperone-mediated autophagy: a unique way to enter the lysosome world. Trends Cell Biol 8(22):407–417. doi:10.1016/j.tcb.2012.05.006
10. Li WW, Li J, Bao JK (2012) Microautophagy: lesser-known self-eating. Cell Mol Life Sci 69(7):1125–1136
11. Burman C, Ktistakis NT (2010) Autophagosome formation in mammalian cells. Semin Immunopathol 6(27):421–429
12. Maejima I, Takahashi A, Omori H, Kimura T, Takabatake Y, Saitoh T, Yamamoto A, Hamasaki M, Noda T, Isaka Y, Yoshimori T (2013) Autophagy sequesters damaged lysosomes to control lysosomal biogenesis and kidney injury. EMBO J 32(17):2336–2347. doi:10.1038/emboj.2013.171
13. Okamoto K (2014) Organellophagy: eliminating cellular building blocks via selective autophagy. J Cell Biol 205(4):435–445. doi:10.1083/jcb.201402054
14. Baumann K (2015) Mitophagy receptors unravelled. Nat Rev Mol Cell Biol 16(10):580
15. Khaminets A, Heinrich T, Mari M, Grumati P, Huebner AK, Akutsu M, Liebmann L, Stolz A, Nietzsche N, Koch S, Mauthe M, Katona I, Qualmann B, Weis J, Reggiori F, Kurth I, Hübner CA, Dikic I (2015) Regulation of endoplasmic reticulum turnover by selective autophagy. Nature 522(7556):354–358
16. Mijaljica D, Devenish RJ (2013) Nucleophagy at a glance. J Cell Sci 129(pt 19):4325–4330
17. Sakaia Y, Okua M, van der Klei IJ, Kiel JA (2006) Pexophagy: autophagic degradation of peroxisomes. Mol Cell Res 1763(12):1767–1775
18. Perera RM, Stoykova S, Nicolay BN, Ross KN, Fitamant J, Boukhali M, Lengrand J, Deshpande V, Selig MK, Ferrone CR, Settleman J, Stephanopoulosn G, Dyson NJ, Zoncu R, Ramaswamy S, Haas W, Bardeesy N (2015) Transcriptional control of autophagy–lysosome function drives pancreatic cancer metabolism. Nature 524(7565):361–365
19. Sardiello M, Palmieri M, di Ronza A, Medina DL, Valenza M, Gennarino VA, Di Malta C, Donaudy F, Embrione V, Polishchuk RS, Banfi S, Parenti G, Cattaneo E, Ballabio A (2009) A gene network regulating lysosomal biogenesis and function. Science 325(5939):473–477
20. Settembre C, Di Malta C, Polito VA, Garcia Arencibia M, Vetrini F, Erdin S, Erdin SU, Huynh T, Medina D, Colella P, Sardiello M, Rubinsztein DC, Ballabio A (2011) TFEB links autophagy to lysosomal biogenesis. Science 332(6036):1429–1433
21. Palmieri M, Impey S, Kang H, di Ronza A, Pelz C, Sardiello M, Ballabio A (2011) Characterization of the CLEAR network reveals an integrated control of cellular clearance pathways. Hum Mol Genet 19(20):3852–3866. doi:10.1093/hmg/ddr306
22. Settembre C, Fraldi A, Medina DL, Ballabio A (2013) Signals from the lysosome: a control centre for cellular clearance and energy metabolism. Nat Rev Mol Cell Biol 14(5):283–296. doi:10.1038/nrm3565
23. Lee J, Giordano S, Zhang J (2012) Autophagy, mitochondria and oxidative stress: cross-talk and redox signalling. Biochem J 441(2):523–540. doi:10.1042/BJ20111451
24. Kroemer G, Marino G, Levine B (2010) Autophagy and the integrated stress response. Mol Cell 40(2):280–293. doi:10.1016/j.molcel.2010.09.023
25. Amaravadi RK, Thompson CB (2007) The roles of therapy-induced autoph-

agy and necrosis in cancer treatment. Clin Cancer Res 13(24):7271–7279. doi:10.1158/1078-0432.CCR-07-1595
26. Yu F, Chen Z, Wang B, Jin Z, Hou Y, Ma S, Liu X (2016) The role of lysosome in cell death regulation. Tumour Biol 37(2):1427–1436
27. Kraya AA, Piao S, Xu X, Zhang G, Herlyn M, Gimotty P, Levine B, Amaravadi RK, Speicher DW (2015) Identification of secreted proteins that reflect autophagy dynamics within tumor cells. Autophagy 11(1):60–74. doi:10.4161/15548627.2014.984273
28. Kirkegaard T, Jäättelä M (2009) Lysosomal involvement in cell death and cancer. Biochem Biophys Acta 1793(4):746–754
29. Kallunki T, Olsen OD, Jäättelä M (2013) Cancer-associated lysosomal changes: friends or foes? Oncogene 32(16):1995–2004
30. Palermo C, Joyce JA (2008) Cysteine cathepsin proteases as pharmacological targets in cancer. Trends Pharmacol Sci 29(1):22–28
31. Withana NP, Blum G, Sameni M, Slaney C, Anbalagan A, Olive MB, Bidwell BN, Edgington L, Wang L, Moin K, Sloane BF, Anderson RL, Bogyo MS, Parker BS (2012) Cathepsin B inhibition limits bone metastasis in breast cancer. Cancer Res 72(5):1199–1209
32. Small DM, Burden RE, Jaworski J, Hegarty SM, Spence S, Burrows JF, McFarlane C, Kissenpfennig A, McCarthy HO, Johnston JA, Walker B, Scott CJ (2013) Cathepsin S from both tumor and tumor-associated cells promote cancer growth and neovascularization. Int J Cancer 133(9):2102–2112
33. Keliher EJ, Reiner T, Earley S, Klubnick J, Tassa C, Lee AJ, Ramaswamy S, Bardeesy N, Hanahan D, DePinho RA, Castro CM, Weissleder R (2013) Targeting cathepsin E in pancreatic cancer by a small molecule allows in vivo detection. Neoplasia 15(7):684–683. doi:10.1593/neo.13276
34. Furuta K, Ikeda M, Nakayama Y, Nakamura K, Tanaka M, Hamasaki N, Himeno M, Hamilton SR, August JT (2001) Expression of lysosome-associated membrane proteins in human colorectal neoplasms and inflammatory diseases. Am J Pathol 159(2):449–455
35. Dettmer J, Hong-Hermesdorf A, Stierhof YD, Schumacher K (2006) Vacuolar H+-ATPase activity is required for endocytic and secretory trafficking in Arabidopsis. Plant Cell 18(3):715–730. doi:10.1105/tpc.105.037978
36. Cardone RA, Casavola V, Reshkin SJ (2005) The role of disturbed pH dynamics and the Na+/H+ exchanger in metastasis. Nat Rev Cancer 5(10):786–795
37. Piao S, Amaravadi RK (2016) Targeting the lysosome in cancer. Ann N Y Acad Sci 1371(1):45–54. doi:10.1111/nyas.12953
38. Zhang Z, Chen G, Zhou W, Song A, Xu T, Luo Q, Wang W, Gu XS, Duan S (2007) Regulated ATP release from astrocytes through lysosome exocytosis. Nat Cell Biol 9(8):945–953
39. Ralevic V, Burnstock G (1998) Receptors for purines and pyrimidines. Pharmacol Rev 50(3):413–492
40. Hämälistö S, Jäättelä M (2016) Lysosomes in cancer-living on the edge (of the cell). Curr Opin Cell Biol 39:69–79
41. Sabatini DM (2006) mTOR and cancer: insights into a complex relationship. Nat Rev Cancer 6(9):729–734
42. Laplante M, Sabatini DM (2012) mTOR signaling in growth control and disease. Cell 149(2):274–293. doi:10.1016/j.cell.2012.03.017
43. Efeyan A, Zoncu R, Chang S, Gumper I, Snitkin H, Wolfson RL, Kirak O, Sabatini DD, Sabatini DM (2013) Regulation of mTORC1 by the Rag GTPases is necessary for neonatal autophagy and survival. Nature 493(7434):679–683. doi:10.1038/nature11745
44. Ganley IG, Lam du H, Wang J, Ding X, Chen S, Jiang X (2009) ULK-Atg13-FIP200 complexes mediate mTOR signaling to the autophagy machinery. J Biol Chem 20(7):1992–2003. doi:10.1091/mbc.E08
45. Roczniak-Ferguson A, Petit CS, Froehlich F, Qian S, Ky J, Angarola B, Walther TC, Ferguson SM (2012) The transcription factor TFEB links mTORC1 signaling to transcriptional control of lysosome homeostasis. Sci Signal 5(228):ra42. doi:10.1126/scisignal.2002790
46. Guertin DA, Sabatini DM (2007) Defining the role of mTOR in cancer. Cancer Cell 12(1):9–22. doi:10.1016/j.ccr.2007.05.008
47. Leone RD, Amaravadi RK (2013) Autophagy: a targetable linchpin of cancer cell metabolism. Trends Endocrinol Metab 24(4):209–217. doi:10.1016/j.tem.2013.01.008
48. Long X, Lin Y, Ortiz-Vega S, Yonezawa K, Avruch J (2005) Rheb binds and regulates the mTOR kinase. Curr Biol 15(8):702–713. doi:10.1016/j.cub.2005.02.053
49. Kim DH, Sarbassov DD, Ali SM, King JE, Latek RR, Erdjument-Bromage H, Tempst P, Sabatini DM (2002) mTOR interacts with raptor to form a nutrient-sensitive complex that signals to the cell growth machinery. Cell 110(2):163–175
50. Meo-Evoli N, Almacellas E, Massucci FA, Gentilella A, Ambrosio S, Kozma SC, Thomas G, Tauler A (2015) V-ATPase—a master

effector of E2F1-mediated lysosomal trafficking, mTORC1 activation and autophagy. Oncotarget 6(28):28057–28070
51. Zoncu R, Efeyan A, Sabatini DM (2011) mTOR: from growth signal integration to cancer, diabetes and ageing. Nat Rev Mol Cell Biol 12(1):21–35. doi:10.1038/nrm3025
52. Hsieh AC, Costa M, Zollo O, Davis C, Feldman ME, Testa JR, Meyuhas O, Shokat KM, Ruggero D (2010) Genetic dissection of the oncogenic mTOR pathway reveals druggable addiction to translational control via 4EBP-eIF4E. Cancer Cell 17(3):249–261. doi:10.1016/j.ccr.2010.01.021
53. Wendel HG, De Stanchina E, Fridman JS, Malina A, Ray S, Kogan S, Cordon-Cardo C, Pelletier J, Lowe SW (2004) Survival signalling by Akt and eIF4E in oncogenesis and cancer therapy. Nature 428(6980):332–337
54. Wendel HG, Silva RL, Malina A, Mills JR, Zhu H, Ueda T, Watanabe-Fukunaga R, Fukunaga R, Teruya-Feldstein J, Pelletier J, Lowe SW (2007) Dissecting eIF4E action in tumorigenesis. Genes Dev 21(24):3232–3237. doi:10.1101/gad.1604407
55. Duprez L, Wirawan E, Vanden Berghe T, Vandenabeele P (2009) Major cell death pathways at a glance. Microbes Infect 11(13):1050–1062
56. Dixon SJ, Lemberg KM, Lamprecht MR, Skouta R, Zaitsev EM, Gleason CE, Patel DN, Bauer AJ, Cantley AM, Yang WS, Morrison B 3rd, Stockwell BR (2012) Ferroptosis: an iron-dependent form of nonapoptotic cell death. Cell 149(5):1060–1072. doi:10.1016/j.cell.2012.03.042
57. Elmore S (2007) Apoptosis—a review of programmed cell death. Toxicol Pathol 35(4):495–516
58. Dupreza L, Wirawana E, Vanden Berghea T, Vandenabeele P (2009) Major cell death pathways at a glance. Microbes Infect 11(13):1050–1062
59. Guicciardi ME, Leist M, Gores GJ (2004) Lysosomes in cell death. Oncogene 23(16):2881–2890
60. Boya P, Kroemer G (2008) Lysosomal membrane permeabilization in cell death. Oncogene 27(50):6434–6451. doi:10.1038/onc.2008.310
61. Eng CH, Yu K, Lucas J, White E, Abraham RT (2010) Ammonia derived from glutaminolysis is a diffusible regulator of autophagy. Sci Signal 3(119):ra31
62. Zhang H, Bosch-Marce M, Shimoda LA, Tan YS, Baek JH, Wesley JB, Gonzalez FJ, Semenza GL (2008) Mitochondrial autophagy is an HIF-1-dependent adaptive metabolic response to hypoxia. J Biol Chem 283(16):10892–10903
63. Noman MZ, Janji B, Kaminska B, Van Moer K, Pierson S, Przanowski P, Buart S, Berchem G, Romero P, Mami-Chouaib F, Chouaib S (2011) Blocking hypoxia-induced autophagy in tumors restores cytotoxic T-cell activity and promotes regression. Cancer Res 18(71):5976–5986
64. Bellot G, Garcia-Medina R, Gounon P, Chiche J, Roux D, Pouyssegur J, Mazure NM (2009) Hypoxia-induced autophagy is mediated through hypoxia-inducible factor induction of BNIP3 and BNIP3L via their BH3 domains. Mol Cell Biol 29(10):2570–2581. doi:10.1128/MCB.00166-09
65. Green DR, Galluzzi L, Kroemer G (2011) Mitochondria and the autophagy-inflammation-cell death axis in organismal aging. Science 333(6046):1109–1112. doi:10.1126/science.1201940
66. Geisler S, Holmström KM, Skujat D, Fiesel FC, Rothfuss OC, Kahle PJ, Springer W (2010) PINK1/Parkin-mediated mitophagy is dependent on VDAC1 and p62/SQSTM1. Nat Cell Biol 12(2):119–131
67. White E, Mehnert JM, Chan CS (2015) Autophagy, metabolism, and cancer. Clin Cancer Res 21(22):5037–5046. doi:10.1158/1078-0432.CCR-15-0490
68. Amaravadi RK (2008) Autophagy-induced tumor dormancy in ovarian cancer. J Clin Invest 118(12):3837–3840. doi:10.1172/JCI37667
69. Vandenabeele P, Galluzzi L, Berghe TV, Kroemer G (2010) Molecular mechanisms of necroptosis: an ordered cellular explosion. Nat Rev Mol Cell Biol 11(10):700–714
70. Yang WS, Stockwell BR (2016) Ferroptosis: death by lipid peroxidation. Trends Cell Biol 26(3):165–176
71. Xie Y, Hou W, Song X, Yu Y, Huang J, Sun X, Kang R, Tang D (2016) Ferroptosis: process and function. Cell Death Differ 23(3):369–379. doi:10.1038/cdd.2015.158
72. Foghsgaard L, Wissing D, Mauch D, Lademann U, Bastholm L, Boes M, Elling F, Leist M, Jäättelää M (2001) Cathepsin B acts as a dominant execution protease in tumor cell apoptosis induced by tumor necrosis factor. J Cell Biol 5(153):999–1010
73. Vancompernolle K, Van Herreweghe F, Pynaert G, Van de Craen M, De Vos K, Totty N, Sterling A, Fiers W, Vandenabeele P, Grooten J (1998) Atractyloside-induced release of cathepsin B, a protease with caspase-processing activity. FEBS Lett 3(438):150–158
74. Imamura K, Ogura T, Kishimoto A, Kaminishi M, Esumi H (2001) Cell cycle regulation via p53 phosphorylation by a 5'-AMP activated protein kinase activator, 5-aminoimidazole-4-carboxamide-1-beta-D-ribofuranoside, in a

human hepatocellular carcinoma cell line. Biochem Biophys Res Commun 287(2):562–567
75. Aits S, Jäättelä M, Nylandsted J (2015) Methods for the quantification of lysosomal membrane permeabilization: a hallmark of lysosomal cell death. Methods Cell Biol 126:261–285
76. Erdal H, Berndtsson M, Castron J, Brunk U, Shoshan M, Linder S (2005) Induction of lysosomal membrane permeabilization by compounds that activate p53-independent apoptosis. Proc Natl Acad Sci U S A 102(1):192–197
77. Fehrenbacher N, Gyrd-Hansen M, Poulsen B, Felbor U, Kallunki T, Boes M, Weber E, Leist M, Jäättelä M (2004) Sensitization to the lysosomal cell death pathway upon immortalization and transformation. Cancer Res 64(15):5301–5310
78. Maynadier M, Vezenkov LL, Amblard M, Martin V, Gandreuil C, Vaillant O, Gary-Bobo M, Basile I, Hernandez JF, Garcia M, Martinez J (2013) Dipeptide mimic oligomer transporter mediates intracellular delivery of Cathepsin D inhibitors: a potential target for cancer therapy. J Control Release 171(2):251–257
79. Kos J, Mitrović A, Mirković B (2014) The current stage of cathepsin B inhibitors as potential anticancer agents. Future Med Chem 6(11):1355–1371
80. Duong LT, Wesolowski GA, Leung P, Oballa R, Pickarski M (2014) Efficacy of a cathepsin K inhibitor in a preclinical model for prevention and treatment of breast cancer bone metastasis. Mol Cancer Ther 13(12):2898–2909. doi:10.1158/1535-7163.MCT-14-0253
81. Tsai JY, Lee MJ, Chang MD, Wang HC, Lin CC, Huang H (2014) Effects of novel human cathepsin S inhibitors on cell migration in human cancer cells. J Enzyme Inhib Med Chem 29(4):538–546
82. Lankelma JM, Voorend DM, Barwari T, Koetsveld J, Van der Spek AH, De Porto AP, Van Rooijen G, Van Noorden CJ (2010) Cathepsin L, target in cancer treatment? Life Sci 86(7–8):225–233
83. Petersen NH, Olsen OD, Groth-Pedersen L, Ellegaard AM, Bilgin M, Redmer S, Ostenfeld MS, Ulanet D, Dovmark TH, Lonborg A, Vindelov SD, Hanahan D, Arenz C, Ejsing CS, Kirkegaard T, Rohde M, Nylandsted J, Jaattela M (2013) Transformation-associated changes in sphingolipid metabolism sensitize cells to lysosomal cell death induced by inhibitors of acid sphingomyelinase. Cancer Cell 24(3):379–393. doi:10.1016/j.ccr.2013.08.003
84. Savić R, Schuchman EH (2013) Use of acid sphingomyelinase for cancer therapy. Adv Cancer Res 117:91–115
85. Smith EL, Schuchman EH (2008) Acid sphingomyelinase overexpression enhances the antineoplastic effects of irradiation in vitro and in vivo. Mol Ther 16(9):1565–1571. doi:10.1038/mt.2008.145
86. Saftig P, Sandhoff K (2013) Cancer: killing from the inside. Nature 7471(502):312–313
87. Leu JI, Pimkina J, Pandey P, Murphy ME, George DL (2011) HSP70 inhibition by the small-molecule 2-phenylethynesulfonamide impairs protein clearance pathways in tumor cells. Mol Cancer Res 9(7):936–947. doi:10.1158/1541-7786.MCR-11-0019
88. Granato M, Lacconi V, Peddis M, Lotti LV, Di Renzo L, Gonnella R, Santarelli R, Trivedi P, Frati L, D'Orazi G, Faggioni A, Cirone M (2013) HSP70 inhibition by 2-phenylethynesulfonamide induces lysosomal cathepsin D release and immunogenic cell death in primary effusion lymphoma. Cell Death Dis 4:e730. doi:10.1038/cddis.2013.263
89. Yamamoto A, Tagawa Y, Yoshimori T, Moriyama Y, Masaki R, Tashiro Y (1998) Bafilomycin A1 prevents maturation of autophagic vacuoles by inhibiting fusion between autophagosomes and lysosomes in rat hepatoma cell line, H-4-II-E cells. Cell Struct Funct 1(23):33–42
90. Yuan N, Song L, Zhang S, Lin W, Cao Y, Xu F, Fang Y, Wang Z, Zhang H, Li X, Wang Z, Cai J, Wang J, Zhang Y, Mao X, Zhao W, Hu S, Chen S, Wang J (2015) Bafilomycin A1 targets both autophagy and apoptosis pathways in pediatric B-cell acute lymphoblastic leukemia. Haematologica 3(100):345–356. doi:10.3324/haematol.2014.113324
91. Kubisch R, Frohlich T, Arnold GJ, Schreiner L, von Schwarzenberg K, Roidl A, Vollmar AM, Wagner E (2014) V-ATPase inhibition by archazolid leads to lysosomal dysfunction resulting in impaired cathepsin B activation in vivo. Int J Cancer 134(10):2478–2488. doi:10.1002/ijc.28562
92. Kallifatidis G, Hoepfner D, Jaeg T, Guzman EA, Wright AE (2013) The marine natural product manzamine A targets vacuolar ATPases and inhibits autophagy in pancreatic cancer cells. Mar Drugs 11(9):3500–3516. doi:10.3390/md11093500
93. Zhao Y, Lu Y, Ma J, Zhu L (2015) Synthesis and evaluation of cleistanthin A derivatives as potent vacuolar H+-ATPase inhibitors. Chem Biol Drug Des 86(4):691–696
94. Zhou S, Wang F, Wong ET, Fonkem E, Hsieh T, Wu JM, Wu E (2014) Salinomycin—a

novel anti-cancer agent with known anticoccidial activities. Curr Med Chem 20(33):4095–4101
95. Fuchs D, Heinold A, Opelz G, Daniel V, Naujokat C (2009) Salinomycin induces apoptosis and overcomes apoptosis resistance in human cancer cells. Biochem Biophys Res Commun 390(3):743–749
96. Kim KY, Yu SN, Lee SY, Chun SS, Choi YL, Park YM, Song CS, Chatterjee B, Ahn SC (2011) Salinomycin-induced apoptosis of human prostate cancer cells due to accumulated reactive oxygen species and mitochondrial membrane depolarization. Biochem Biophys Res Commun 413(1):80–86
97. Yue W, Hamai A, Tonelli G, Bauvy C, Nicolas V, Tharinger H, Codogno P, Mehrpour M (2013) Inhibition of the autophagic flux by salinomycin in breast cancer stem-like/progenitor cells interferes with their maintenance. Autophagy 9(5):714–729. doi:10.4161/auto.23997
98. Zou ZZ, Nie PP, Li YW, Hou BX, Rui-Li, Shi XP, Ma ZK, Han BW, Luo XY (2015) Synergistic induction of apoptosis by salinomycin and gefitinib through lysosomal and mitochondrial dependent pathway overcomes gefitinib resistance in colorectal cancer. Oncotarget 1–19. doi: 10.18632/oncotarget.5628
99. Ketola K, Hilvo M, Hyotylainen T, Vuoristo A, Ruskeepaa AL, Oresic M, Kallioniemi O, Iljin K (2012) Salinomycin inhibits prostate cancer growth and migration via induction of oxidative stress. Br J Cancer 106(1):99–106. doi:10.1038/bjc.2011.530
100. Kim JH, Chae M, Kim WK, Kim YJ, Kang HS, Kim HS, Yoon S (2011) Salinomycin sensitizes cancer cells to the effects of doxorubicin and etoposide treatment by increasing DNA damage and reducing p21 protein. Br J Pharmacol 162(3):773–784. doi:10.1111/j.1476-5381.2010.01089.x
101. Kim WK, Kim JH, Yoon K, Kim S, Ro J, Kang HS, Yoon S (2012) Salinomycin, a p-glycoprotein inhibitor, sensitizes radiation-treated cancer cells by increasing DNA damage and inducing G2 arrest. Invest New Drugs 30(4):1311–1318
102. Tang QL, Zhao ZQ, Li JC, Liang Y, Yin JQ, Zou CY, Xie XB, Zeng YX, Shen JN, Kang T, Wang J (2011) Salinomycin inhibits osteosarcoma by targeting its tumor stem cells. Cancer Lett 311(1):113–121
103. Steinman RM, Mellman IS, Muller WA, Cohn ZA (1983) Endocytosis and the recycling of plasma membrane. J Cell Biol 96(1):1–27
104. Fu D, Zhou J, Zhu WS, Manley PW, Wang YK, Hood T, Wylie A, Xie XS (2014) Imaging the intracellular distribution of tyrosine kinase inhibitors in living cells with quantitative hyperspectral stimulated Raman scattering. Nat Chem 6(7):614–622
105. Zhitomirsky B, Assaraf YG (2015) Lysosomal sequestration of hydrophobic weak base chemotherapeutics triggers lysosomal biogenesis and lysosome-dependent cancer multidrug resistance. Oncotarget 6(2):1143–1156
106. Duvvuri M, Gong Y, Chatterji D, Krise JP (2004) Weak base permeability characteristics influence the intracellular sequestration site in the multidrug-resistant human leukemic cell line HL-60. J Biol Chem 279(31):32367–32372. doi:10.1074/jbc.M400735200
107. Gorden BH, Saha J, Khammanivong A, Schwartz GK, Dickerson EB (2014) Lysosomal drug sequestration as a mechanism of drug resistance in vascular sarcoma cells marked by high CSF-1R expression. Vasc Cell 6(20):1–14
108. Wang E, Lee MD, Dunn KW (2000) Lysosomal accumulation of drugs in drug-sensitive MES-SA but not multidrug-resistant MES-SA/Dx5 uterine sarcoma cells. J Cell Physiol 184(2):263–274
109. Nordstrom LU, Sironi J, Aranda E, Maisonet J, Perez-Soler R, Wu P, Schwartz EL (2015) Discovery of autophagy inhibitors with antiproliferative activity in lung and pancreatic cancer cells. ACS Med Chem Lett 6(2):134–139. doi:10.1021/ml500348p
110. Wang T, Goodall ML, Gonzales P, Sepulveda M, Martin KR, Gately S, MacKeigan JP (2015) Synthesis of improved lysomotropic autophagy inhibitors. J Med Chem 58(7):3025–3035
111. Carew JS, Espitia CM, Esquivel JA, Mahalingam D, Kelly KR, Reddy G, Giles FJ, Nawrocki ST (2011) Lucanthone is a novel inhibitor of autophagy that induces cathepsin D-mediated apoptosis. J Biol Chem 286(8):6602–6613
112. Luo M, Kelley MR (2004) Inhibition of the human apurinic/apyrimidinic endonuclease (Ape1) repair activity and sensitization of breast cancer cells to DNA alkylating agents with lucanthone. Anticancer Res 24(4):2127–2134
113. Rebecca VW, Amaravadi RK (2016) Emerging strategies to effectively target autophagy in cancer. Oncogene 35(1):1–11. doi:10.1038/onc.2015.99
114. Mahalingam D, Mita M, Sarantopoulos J, Wood L, Amaravadi RK, Davis LE, Mita AC, Curiel TJ, Espitia CM, Nawrocki ST, Giles FJ, Carew JS (2014) Combined autophagy and HDAC inhibition: a phase I safety, tolerability, pharmacokinetic, and pharmacodynamic

analysis of hydroxychloroquine in combination with the HDAC inhibitor vorinostat in patients with advanced solid tumors. Autophagy 10(8):1403–1414. doi:10.4161/auto.29231

115. Rangwala R, Chang YC, Hu J, Algazy KM, Evans TL, Fecher LA, Schuchter LM, Torigian DA, Panosian JT, Troxel AB, Tan KS, Heitjan DF, DeMichele AM, Vaughn DJ, Redlinger M, Alavi A, Kaiser J, Pontiggia L, Davis LE, O'Dwyer PJ, Amaravadi RK (2014) Combined MTOR and autophagy inhibition: phase I trial of hydroxychloroquine and temsirolimus in patients with advanced solid tumors and melanoma. Autophagy 10(8):1391–1402. doi:10.4161/auto.29119

116. Barnard RA, Wittenburg LA, Amaravadi RK, Gustafson DL, Thorburn A, Thamm DH (2014) Phase I clinical trial and pharmacodynamic evaluation of combination hydroxychloroquine and doxorubicin treatment in pet dogs treated for spontaneously occurring lymphoma. Autophagy 10(8):1415–1425. doi:10.4161/auto.29165

117. Rangwala R, Leone R, Chang YC, Fecher LA, Schuchter LM, Kramer A, Tan KS, Heitjan DF, Rodgers G, Gallagher M, Piao S, Troxel AB, Evans TL, DeMichele AM, Nathanson KL, O'Dwyer PJ, Kaiser J, Pontiggia L, Davis LE, Amaravadi RK (2014) Phase I trial of hydroxychloroquine with dose-intense temozolomide in patients with advanced solid tumors and melanoma. Autophagy 10(8):1369–1379. doi:10.4161/auto.29118

118. Rosenfeld MR, Ye X, Supko JG, Desideri S, Grossman SA, Brem S, Mikkelson T, Wang D, Chang YC, Hu J, McAfee Q, Fisher J, Troxel AB, Piao S, Heitjan DF, Tan KS, Pontiggia L, O'Dwyer PJ, Davis LE, Amaravadi RK (2014) A phase I/II trial of hydroxychloroquine in conjunction with radiation therapy and concurrent and adjuvant temozolomide in patients with newly diagnosed glioblastoma multiforme. Autophagy 10(8):1359–1368. doi:10.4161/auto.28984

119. Wolpin BM, Rubinson DA, Wang X, Chan JA, Cleary JM, Enzinger PC, Fuchs CS, McCleary NJ, Meyerhardt JA, Ng K, Schrag D, Sikora AL, Spicer BA, Killion L, Mamon H, Kimmelman AC (2014) Phase II and pharmacodynamic study of autophagy inhibition using hydroxychloroquine in patients with metastatic pancreatic adenocarcinoma. Oncologist 19(6):637–638. doi:10.1634/theoncologist.2014-0086

120. Amaravadi RK, Yu D, Lum JJ, Bui T, Christophorou MA, Evan GI, Thomas-Tikhonenko A, Thompson CB (2007) Autophagy inhibition enhances therapy-induced apoptosis in a Myc-induced model of lymphoma. J Clin Invest 117(2):326–336. doi:10.1172/JCI28833

121. Bray K, Mathew R, Lau A, Kamphorst JJ, Fan J, Chen J, Chen HY, Ghavami A, Stein M, DiPaola RS, Zhang D, Rabinowitz JD, White E (2012) Autophagy suppresses RIP kinase-dependent necrosis enabling survival to mTOR inhibition. PLoS One 7(7):e41831. doi:10.1371/journal.pone.0041831

122. Xie X, White EP, Mehnert JM (2013) Coordinate autophagy and mTOR pathway inhibition enhances cell death in melanoma. PLoS One 8(1):e55096. doi:10.1371/journal.pone.0055096

123. Carew JS, Nawrocki ST, Kahue CN, Zhang H, Yang C, Chung L, Houghton JA, Huang P, Giles FJ, Cleveland JL (2007) Targeting autophagy augments the anticancer activity of the histone deacetylase inhibitor SAHA to overcome Bcr-Abl–mediated drug resistance. Blood 110(1):313–322

124. Qiu L, Yao M, Gao M, Zhao Q (2012) Doxorubicin and chloroquine coencapsulated liposomes: preparation and improved cytotoxicity on human breast cancer cells. J Liposome Res 22(3):245–253

125. Amaravadi RK, Lippincott-Schwartz J, Yin XM, Weiss WA, Takebe N, Timmer W, DiPaola RS, Lotze MT, White E (2011) Principles and current strategies for targeting autophagy for cancer treatment. Clin Cancer Res 17(4):654–666. doi:10.1158/1078-0432.CCR-10-2634

126. McAfee Q, Zhang Z, Samanta A, Levi SM, Ma XH, Piao S, Lynch JP, Uehara T, Sepulveda AR, Davis LE, Winkler JD, Amaravadi RK (2012) Autophagy inhibitor Lys05 has single-agent antitumor activity and reproduces the phenotype of a genetic autophagy deficiency. Proc Natl Acad Sci U S A 109(21):8253–8258

127. Amaravadi RKWJ (2012) Lys05: a new lysosomal autophagy inhibitor. Autophagy 8(9):1383–1384

Index

Karin Öllinger and Hanna Appelqvist (eds.), *Lysosomes: Methods and Protocols*, Methods in Molecular Biology, vol. 1594, DOI 10.1007/978-1-4939-6934-0, © Springer Science+Business Media LLC 2017

I

J

L

M

MIX
Papier aus verantwortungsvollen Quellen
Paper from responsible sources
FSC® C105338

If you have any concerns about our products,
you can contact us on
ProductSafety@springernature.com

In case Publisher is established outside the EU,
the EU authorized representative is:
Springer Nature Customer Service Center GmbH
Europaplatz 3, 69115 Heidelberg, Germany

Printed by Libri Plureos GmbH
in Hamburg, Germany